Sumit Pramanik, Sandipan Roy, J. Paulo Davim (Eds.)
Advanced Biomedical Composites

Advanced Composites

Edited by
J. Paulo Davim

Volume 22

Advanced Biomedical Composites

Materials, Design, Manufacturing

Edited by
Sumit Pramanik, Sandipan Roy and J. Paulo Davim

DE GRUYTER

Editors
Dr. Sumit Pramanik
Functional and Biomaterials Engineering Lab
Department of Mechanical Engineering
SRM Institute of Science and Technology
Kattankulathur, Chennai 603203
Tamil Nadu
India
sumitprs@srmist.edu.in

Dr. Sandipan Roy
Department of Mechanical Engineering
SRM Institute of Science and Technology
Kattankulathur, Chennai 603203
Tamil Nadu
India
sandipag@srmist.edu.in

Prof. Dr. J. Paulo Davim
Department of Mechanical Engineering
University of Aveiro
Campus Santiago
3810-193 Aveiro
Portugal
pdavim@ua.pt

ISBN 978-3-11-138594-5
e-ISBN (PDF) 978-3-11-138602-7
e-ISBN (EPUB) 978-3-11-138615-7
ISSN 2192-8983

Library of Congress Control Number: 2025931491

Bibliographic information published by the Deutsche Nationalbibliothek
The Deutsche Nationalbibliothek lists this publication in the Deutsche Nationalbibliografie; detailed bibliographic data are available on the internet at http://dnb.dnb.de.

Cover image: gettyimages/thinkstockphotos, Abalone Shell
Typesetting: Integra Software Services Pvt. Ltd.
Printing and binding: CPI books GmbH, Leck

www.degruyter.com
Questions about General Product Safety Regulation:
productsafety@degruyterbrill.com

MIX
Papier | Fördert gute Waldnutzung
FSC www.fsc.org
FSC® C083411

Preface

Advanced biomaterials deal with materials that can be used in modern biological and medical issues, which are related to advanced medical necessity, fundamental research, cutting-edge technical advancement, ethical considerations, industrial participation, and federal regulation. It is intellectually stimulating since engineering covers a remarkably wide range of concepts, designs, sciences, and technologies. Additionally, it meets significant humanitarian needs. An overview of various biomaterials, designs, devices, and technologies is what this book aims to give.

Every chapter has some considerable updates of new biomaterials and technology. Numerous cutting-edge biomaterials and design aspects have been added to the chapters. The curriculum for teaching the core disciplines of manufacturing and computer-aided design of various implants and tissue engineering methods – all of which are essential to the current biomaterials research effort – has been systematically organized.

In advanced biomaterials, three aspects have been included in this book: (1) Materials, (2) Design, and (3) Manufacturing. Both the overall content and the book's size have been intellectually maintained. There are high-quality graphical designs on the three topics in the book. This book is holding significantly different compared to the previous books related to this field, and it has the great potential to discuss cutting-edge biomaterials, designs, and manufacturing in the twenty-first century.

Part 1 of this book is related to various advanced **Materials**. The first chapter introduces various advancements of biomaterials made of ceramic and polymer matrix composites. The second chapter covers several advanced biomedical applications of hydrogels. The third chapter includes recent developments of nanocomposite-based biosensor applications. The fourth chapter covers the use of metallic dental implant materials. Then, the fifth chapter discusses the wear behavior of different dental implant materials.

Part 2 of the book is related to the **Design** of composite biomaterials. In this part, the first chapter covers the patient-specific implant design. The second chapter discusses various modeling techniques of scaffolds for bone tissue engineering. The third to fifth chapters include biomechanics in the musculoskeletal system, bones, muscles, soft tissues, ligaments, articular cartilage, and cell mechanics using computational modeling. The sixth and seventh chapters discuss the osteoporotic fracture implant failure and modeling from experimental and computational analyses.

Part 3 of the book is related to **Manufacturing** of composite biomaterials. In this part, the first chapter discusses the patient-specific implant using additive manufacturing. The second and third chapters cover the fabrication of artificial skin and nerve tissues using composite materials. The final chapter covers the manufacturing of advanced prosthetic limbs using composite materials.

A special thanks to everyone who spent their time and effort for this book. The authors who provided their knowledge and viewpoints are unquestionably the foun-

https://doi.org/10.1515/9783111386027-202

dation of this work, and they deserve greatest recognition and appreciation. Particularly, heartiest thanks to all editors and assistants who spent time and energy in compiling the content, that is, this book.

Opportunities, stimulation, compassion, and philosophical concepts have always abounded in the biomaterials sector. We, the editors and our authors, hope that the overview you are currently approaching will inspire in you the same level of joy and fulfillment that it has in us.

Editors
Sumit Pramanik
Sandipan Roy
J. Paulo Davim

Contents

1 Materials

C. Ayyanar, J. Paulo Davim, and Sumit Pramanik*

Chapter 1.1
Introduction to biomaterials: advances in ceramic and polymer matrix composites

Abstract: The development of advanced biomaterials with unique properties, such as metallic, polymeric, and ceramic biomaterials, is gaining popularity for orthopaedic applications. Researchers in biomedical materials are the primary audience for this chapter, which covers biomaterials and nanocomposite materials knowledge from multiple disciplines. Key features of next-generation biomaterials include biodegradability, bioactivity, and biocompatibility. Composite materials offer advantages over traditional bone grafting, including customization and reduced disease transmission risk. This chapter provides an update on composite ceramic polymers for bone regeneration. Especially, it covers the types of biomaterials, advances in ceramic matrix and polymer matrix composites for various biomedical applications, characterizations of biomaterials, and global market of biomaterials.

Keywords: Biomaterials, Ceramic matrix composite, Polymer matrix composite, Hydroxyapatite, Tissue engineering

1.1.1 Introduction

Advances in human cultivation and biomaterials incorporate an assortment of materials with a large number of size ranges, from nano to micro to macro, with an obvious objective of enhancing human lives and their quality of life. Silver, in various forms, was utilized as an antibacterial substance in avoiding diseases over a millennium ago. In the beginning stages of humanity, many ways of surgery were also performed. Nonetheless, the most significant innovations in the application of nanomaterials might have occurred from 1901 and 2000 [1–3]. Over the last 60 years, resorbable sutures have transformed surgery less difficult, sophisticated cardiovascu-

***Corresponding author: Sumit Pramanik**, Functional and Biomaterials Engineering Lab, Department of Mechanical Engineering, College of Engineering and Technology, SRM Institute of Science and Technology, Kattankulathur, Kancheepuram, Chennai 603 203, Tamil Nadu, India, e-mail: sumitprs@srmist.edu.in, prsumit@gmail.com, ORCID: https://orcid.org/0000-0003-0455-1582
C. Ayyanar, Functional and Biomaterials Engineering Lab, Department of Mechanical Engineering, College of Engineering and Technology, SRM Institute of Science and Technology, Kattankulathur, Kancheepuram, Chennai 603 203, Tamil Nadu, India, e-mail: ayyanar94@gmail.com
J. Paulo Davim, Department of Mechanical Engineering, Campus of Santiago, University of Aveiro, 3810-193 Aveiro, Portugal, e-mail: pdavim@ua.pt

https://doi.org/10.1515/9783111386027-001

lar devices have preserved millions of lives, and artificial joints have enhanced their standards of life for thousands of citizens. Nowadays, intellectual frontiers have been challenged with the advancement of biological tissue engineering and organ regeneration, which will contribute to 2001–2100 [4] more fascinating in the biomaterials sector [5]. However, excellent design of equipment, suitable materials, and device character development are also essential for experiencing all the rewards, in addition to the biomaterials development. Moreover, for a biomedical device to see the commercialization light, for getting authorization from regulators, the rule is also essential to execute testing in accordance with the appropriate rules and regulations. Overall, benefits of investigation in biomaterials might be esteemed. These resources are categorized perfect at the level of materials and device following regulatory guidelines. Considering the multidisciplinary nature of the field, it is also not easy to carry out large variety of experiments using different techniques. Realizing this problem in biomaterials characterization, we have developed this book to offer an insight on various characterization tools focusing on biomaterials and biomedical devices.

1.1.2 Types of biomaterials

An element or combination of materials of either synthetic or natural extraction that communicates with living things can be referred to as a biomaterial. In an attempt to appropriately restore function and occasionally attractive components without exposing human life in danger, biomaterials are typically employed to promote the healing or rehabilitation of an organ in the human body that has been destroyed with a health condition or an unfortunate incident [6]. As shown in Figure 1.1.1, biomaterials can be characterized as metallic, polymeric, ceramic, or composite based on their

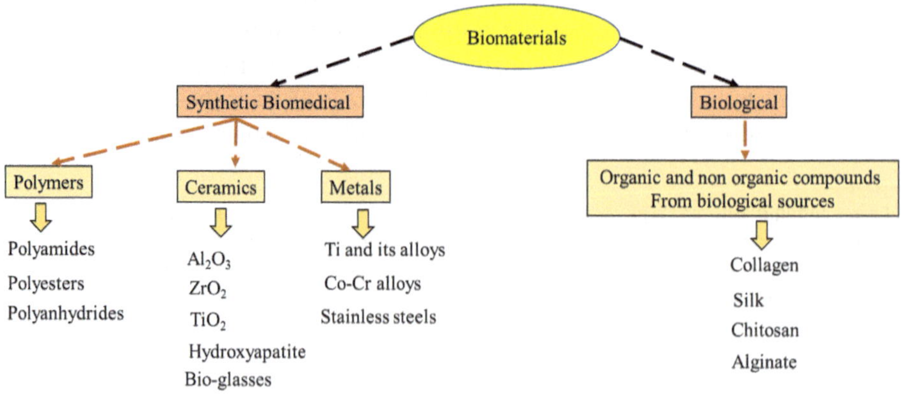

Figure 1.1.1: Types of biomaterials (adapted from [10] – Open access under CC BY 4.0 license).

chemical structure. When categorizing a biomaterial as biocompatible, consideration is essential due to the reality, biocompatibility is an application-specific characteristic: a biomaterial that is adequate or biocompatible for one application could be incompatible for a different one [7]. Through the incorporation of novel materials, coatings, concepts for design, and manufacturing technologies (e.g., tissue engineering, nanotechnologies, biomimetic and functional materials, computational methods for simulation, and additive manufacturing), science and technology have revealed exciting opportunities in the field of implantable biomaterials that are going to change the biomaterials industry in the coming years [8, 9].

(1) ***Metal-based biomaterials:*** Although stainless steel first appeared in 1920, it had been believed to have considerably more resistance to corrosion than any other metals accessible at the time. Therefore, this is recommended for widespread implementation. Metals have been utilized effectively as biomaterials since their discovery. The development in neurological, orthopedic, dental, and cardiovascular disorders that involve implants and surgeries using metals at every stage, contributes to an increase in demand for major biocompatible metals such as titanium, stainless steel, chromium, cobalt, nitinol, gold, and silver. Stainless steel, cobalt and chromium alloys, economically pure titanium and titanium alloys, gold, silver, magnesium, and nitinol are part of the metals. All biomaterials, including those used in dentistry, orthopedics, ophthalmology, and cardiovascular disease, incorporate minerals. From fundamental wires, rods, pins, and screws to fracture fixation plates and total joint prostheses (artificial joints) for the hips, knees, shoulders, and ankles, as well as cardiovascular surgery and dental materials, metals are widely used in all professions, including maxillofacial surgery [11, 12].

(2) ***Polymer-based biomaterials:*** As a result of their low cost, chemical stability, ease of extraction and refurbishment, and excellent corrosion resistance, polymer-based biomaterials are replacing traditional materials, including metals, alloys, and ceramics. The food, personal care products, biotechnology, and pharmaceutical sectors have all made significant use of polymer-based biomaterials. Vascular grafts, implants, wound dressings, sutures, catheters, meshes, stents, ligament and tendon repair, and valves used in heart surgery are only a handful of the medical applications for polymeric biomaterials [13, 14]. Polymeric materials employed in biomedical applications are frequently defined as natural, conventional, or a mix of each of them. Plant and animal sources are the main components of natural polymers, which include cellulose, silk, wool, ribonucleic acid (RNA), deoxyribonucleic acid (DNA) [15], and proteins.

(3) ***Ceramic-based biomaterials:*** The bioceramics are extremely biocompatible due to their stiffness, hardness, chemical stability, and resistance to wear, among other attributes. Biocompatibility and bioactivity, reduced stress shielding, no infec-

tious disease transfer, and easy availability are a few benefits of bioceramics. The material's brittleness is its sole issue, which is problematic for load-bearing applications. The composition of ceramic oxides, which are inert in the body, and resorbable materials, which are eventually replaced by the body after they have aided repair, is different from the range of biocompatibility of bioceramics. Bioceramics are hard materials used in surgical implants and other medical devices. While some bioceramics are extraordinarily durable metal oxides, others are flexible because they are associated with the body's natural components. Bioceramics are often employed in joint replacements, surgical crowns, and dental and bone implants. Bioceramic coatings may at times assist in mitigating reactions of inflammation and wear. Although they are primarily employed as implants, they are also found in respirators, extracorporeal circulation systems, pacemakers, renal dialysis machines, and developed bioreactors [16, 17].

(4) ***Natural biomaterials:*** The two main categories of biomaterials are synthetic and natural biomaterials. Synthetic biomaterials include metals, ceramics, nonbiodegradable polymers, and biodegradable polymers. Natural biomaterials outperform synthetic biomaterials in areas like structure and compositional differences, biocompatibility, and their insufficient ability to induce remodeling in tissues. As a result, research on naturally accomplished biomaterials is being conducted internationally. The following subcategories are applicable to naturally occurring biomaterials: cellulose, silk, hyaluronic acid, chitin, chitosan (CS), gelatin [18], and fibrin [19]. As pharmaceutical delivery systems and medical devices like surgical sutures, they usually serve the purpose for repairing or restoring the structure and function of damaged structures or organs.

(5) ***Inorganic glass-based biomaterials:*** The final product of the super-cooling approach is an amorphous, unambiguous, and hard liquid referred to as inorganic glass. This particular group of biomaterials includes bioactive glasses, phosphates, and silicates [20]. Typically, an assortment of several metallic silicates is mixed together to create inorganic glass. Because glass is a mixture of substances that are inorganic, including sodium, potassium, calcium, and lead silicates, it has been referred to as inorganic glass. Furthermore, no glass is made from organic molecules, which have been referred to as inorganic glass. Inorganic glass may prove to be a very adaptable and effective biological substance. Among additional substances, silicates, phosphates, gallates, rare earth, fluorides, and oxides can be used for the production of glasses. In biotechnology and medical equipment, such as those that analyze infrared radiation or provide high-power mid-infrared lasers, certain glasses are utilized. In dental repair, bone tissue engineering, and bone tissue replacement, inorganic bioactive biomaterials are utilized for the replacement of hard tissues.

(6) ***Regenerative biomaterials:*** The worldwide demand for regenerative biomaterials is being influenced by tissue engineering, when they have discovered their

most important application. Tissue engineering products are made from cells and biomaterials that have been modified in some way. Regenerative biomaterials were previously limited to usage in prosthetic devices. Biomaterials are still utilized in the fabrication of biomaterial scaffolds, wound healing, tissue-engineered blood vessels, bone, liver, cornea, and cardiac tissue engineering [21]. The main application of regenerative biomaterials is in tissue engineering for the restoration of bones and joints. Improvements in technology (nanotechnology-based tissue engineering) have contributed to market growth in this sector.

(7) ***Hybrid biomaterial combinations:*** Combinations of synthetic and naturally accomplished materials have been identified as hybrid biomaterial combinations. Because they create and structure tissues and offer mechanical stability and strength, biomaterials are commonly employed in tissue engineering applications, including scaffolds for regenerative medicine, nanomaterials for biosensing, and the customization of inorganic nanoparticles. Additionally, nanoparticles offer chances to supply inductive chemicals for cell migration and transplanting. For enhanced biological interactions, hybrid biomaterials are appropriate blends of synthetic and natural polymers containing different substances. These materials' biocompatibility promotes robust integrations into the host tissue, which results in appropriate material characteristics and the decomposition kinetics. The specific regions of vascular, neuronal, and bone forms are the main sectors encouraging hybrid materials [22]. In addition, investigations are being carried out on the use of hybrid materials in regenerative medicine and tissue engineering.

1.1.3 Advances in ceramic matrix composites

1.1.3.1 Bioceramics for dentistry applications

The adaptability of ceramic products with tissue that is alive is their primary characteristic for applications in dentistry. Bioceramics such as hydroxyapatite ($Ca_{10}(PO_4)_6(OH)_2$), zirconia, alumina, and bioglasses have appeared in the last few years [23] and have been investigated for the application of dentistry (Figure 1.1.2). Single crystal sapphire, zirconium oxide, and porcelain have previously been extensively used for solving orthodontic complications. The primary difficulties associated with modern bioceramic osteoplastic devices are their brittle nature and inadequate resistance to bending or stretching stresses. With the possible exception of bioactive glass, they cannot be osteoconductive, and bioabsorption can be unanticipated. Interestingly, bioactive glass absorbs moisture quickly, but conventional HA and TCP are not bioresorbable in the ancient period.

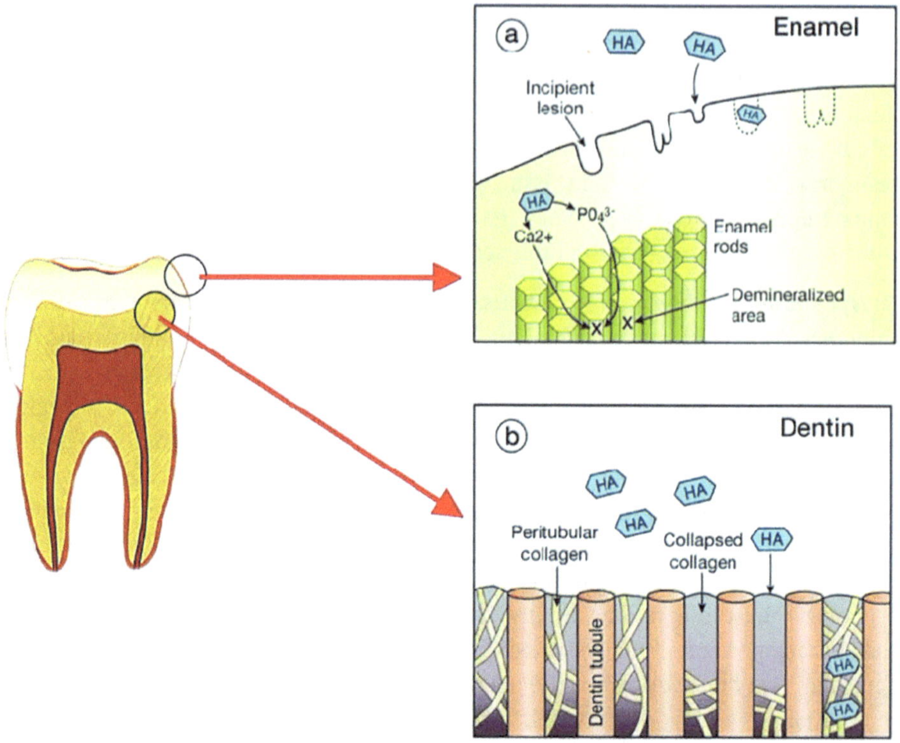

Figure 1.1.2: Hydroxyapatite in dental product (adapted from [24] – Open access under CC BY 4.0 license).

Alumina, zirconia, magnesium spinel ($MgAl_2O_4$), along with additional chemicals in the shape of powders or crystals are normally employed to crystallize a glass mixture in numerous dental applications (Figure 1.1.3). The ultimate result can be accomplished through an appropriate heat treatment, frequently referred to as devitrification. Composite materials called interpenetrating phase composites (IPCs) are employed when crystals are incorporated [25]. They are composed of two interspersed phases (glass and crystals) that continuously expand inside adjacent ones without establishing a chemical bond. There are two procedures associated with developing these IPCs. The ceramic is first sintered to develop a porous core comprised of crystals of alumina or magnesium-spinel ($MgAl_2O_4$) or alumina and zirconia in a 70:30 proportion. After passing through a porous mesh filter, the molten glass occupies all of the slots and voids in a perfect shape. This develops a durable framework that may be employed to deposit and fire distinctive dental porcelain, occasionally referred to as an aesthetic coating. A specific type of ceramic material known as glass ceramic gets formed when oxides are introduced as powder. Crystals of lithium disilicate are particularly often utilized as reinforcing particles.

A combination material system, including clay/kaolin, quartz (silica), and natural feldspar (a mixture composed of potassium and sodium aluminosilicate), is the fundamental structure of glass matrix ceramics. Depending on the amount, potassium feldspar ($K_2A1_2Si_6O_{16}$) can boost a restoration's inherent strength by developing leucite crystals (crystalline phase). In addition to providing an aesthetically appealing monolithic tooth covering material, these bioceramic materials have applications in metal alloys and ceramic substrates. The fine-grained crystalline structure of the polycrystalline ceramic group provides strength and fracture resistance, but in addition it tends to have restricted transparency. Moreover, polycrystalline ceramics are challenging to abrade with hydrofluoric acid due to the absence of a glass phase, demanding longer times at elevated temperatures.

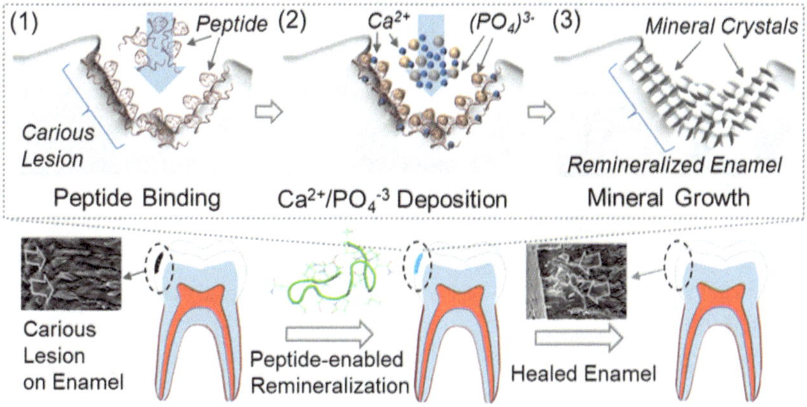

Figure 1.1.3: Biomimetic tooth repair technique (adapted from [26] – Open access under CC BY 4.0 license).

Ceramics are frequently used in applications involving humans because, in addition to their customized physical characteristics, they are easily able to take on the precise form and color, as shown in Figure 1.1.4. Crystalline mineral salts are scattered across a vitreous silicate matrix that generates dental porcelain. Reduced levels of metal oxides have been used as dyes for imitating the color of real teeth as well as to enhance the coefficient of thermal expansion and decrease the melting temperature. Dental porcelain has been used in veneering, the development of prosthetic teeth, indirect aesthetic restorations like facades and inlays/overlays, and in the creation of immovable frames such as metal ceramic rims and bridges. The advanced technology for all ceramic systems has developed in recent years, which has eliminated the requirement for fabricating fixed prosthetic devices (circles and bridges) with a metal frame; for example, using just ceramic biomaterials such as zirconia, alumina, and numerous others [27].

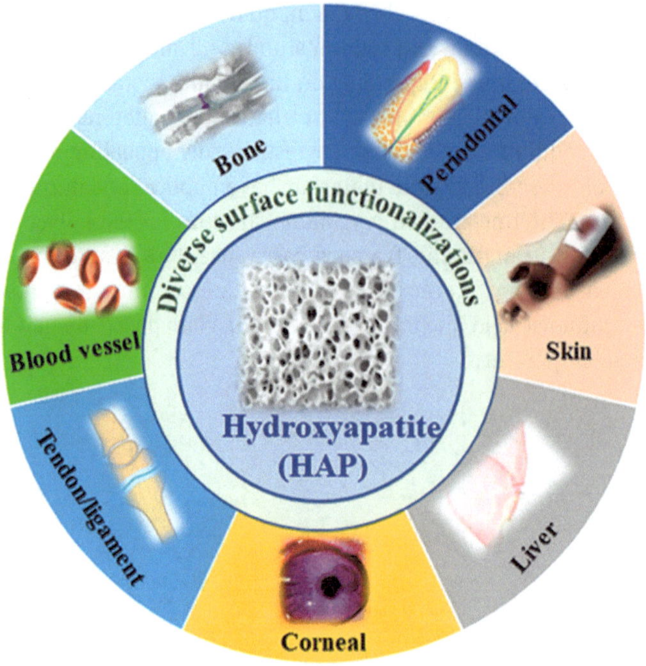

Figure 1.1.4: Hydroxyapatite scaffolds for tissue engineering applications (adapted from [28] – Open access under CC BY 4.0 license).

1.1.3.2 Bioceramics for bone tissue engineering

The ultimate objective of the multidisciplinary field of bone tissue engineering is to create volumetric and porous structures, or "biomimetic scaffolds," through integrating mechanical design principles to biomedical motives. In circumstances where there are significant defects in the bones that cannot be healed on their own, these components are implanted into patients' bodies in order to promote and direct bone tissue regeneration. It is extremely important to note from the physiological standpoint that cells can only migrate randomly to generate two-dimensional layers. They are unable to control the morphology to reassemble and heal a damaged bone region. Although materials are designed for preserving colonies of undifferentiated stem cells and to carefully stimulate their differentiation and proliferation, opportunely established porous scaffolds perform as a biocompatible extracellular matrix.

The fundamental features essential for establishing the best biomimetic scaffold for the regeneration of bone tissue are: (i) sufficient stiffness and mechanical durability for supporting newly formed cells even as they bear stresses during the healing processes; (ii) sufficient characteristics of the surface that encourage osteointegration, differentiation, and attachment of cells; (iii) improved pore architecture and interconnectivity to assure waste material evacuation, vascularization of the structure, and cell migration; (iv) the capacity to avert toxic or inflammatory responses at the implantation sites is recognized as biocompatibility; (v) biodegradability is a characteristic of the implant's physiological environment to dissolve it down and absorb it over an extended period of time; and (vi) convenience of the production in an assortment of size and forms.

Structural ceramics, calcium phosphates, and bioactive glasses are the main categories of ceramic materials that can be employed for substituting the bone tissue [29]. The outstanding hardness and excellent wear resistance of zirconia (ZrO_2) and alumina (Al_2O_3) in the preliminary group may be hazardous if the implant eventually experiences the stress shielding phenomenon. According to Wolff's rule, this occurrence develops when a stiff scaffold material fails to match the mechanical abilities of the tissue to regenerate and, on the other hand, endures a significant portion of the imposed stress. This prohibits the bone tissue in the implantation site from regenerating naturally and self-stiffening. The optimum situation scaffolds should have structural characteristics that are equivalent to or slightly higher than those of the hosting bone. Tricalcium phosphate (TCP, $[Ca_{10}(PO_4)_6]$) and hydroxyapatite (HA, $[Ca_{10}(PO_4)_6(OH)_2]$), the primary part of human bone, are included in the subsequent sector. When embedded in vivo, the latter is more biodegradable than its predecessor and demonstrates higher osteoinductivity and osteoconductivity, which are the ability to trigger osteogenesis during a bone-healing process like a fracture and an inclination of a material for encouraging bone growth on its surface, respectively. This is an ordinary process that occurs following a bone implant. When incorporating carbonates and divalent ions including strontium and magnesium to the bulk material, synthetic HA's bioactivity and biodegradability can be strengthened. The calcium carbonates that can be gathered from fish industry contaminants, especially mussel shells, represent a novel and renewable source of raw biomaterials for highly biocompatible scaffolds, which are intended for the chemical synthesis of HA through the use of phosphorous chemical compounds. Numerous active substances, especially calcium, sodium, and magnesium oxides (CaO, Na_2O, and MgO, respectively), present in silicon dioxide (SiO_2) bulk, are included in the next sector. Because they can dissolve in physiological environments and even form chemical associations with bone substrates, these materials are incredibly biocompatible. However, the comparatively low degree of hardness of the glass bulk is an important disadvantage of glass bioceramics.

1.1.4 Advances in polymer matrix composites

The establishment of medically appropriate polymethacrylate tooth replica implants in 1969 represented the starting point of additional advancements in dental implant polymer development [30]. As shown in Figure 1.1.5, this was implemented in situations like exchanging a natural tooth with its replica, which was found to be beneficial for restoring basic demands (both functional and aesthetic). This emphasized a number of benefits associated with polymers such as the adaptation of physical structures due to modifications to the composition (softer or porous).

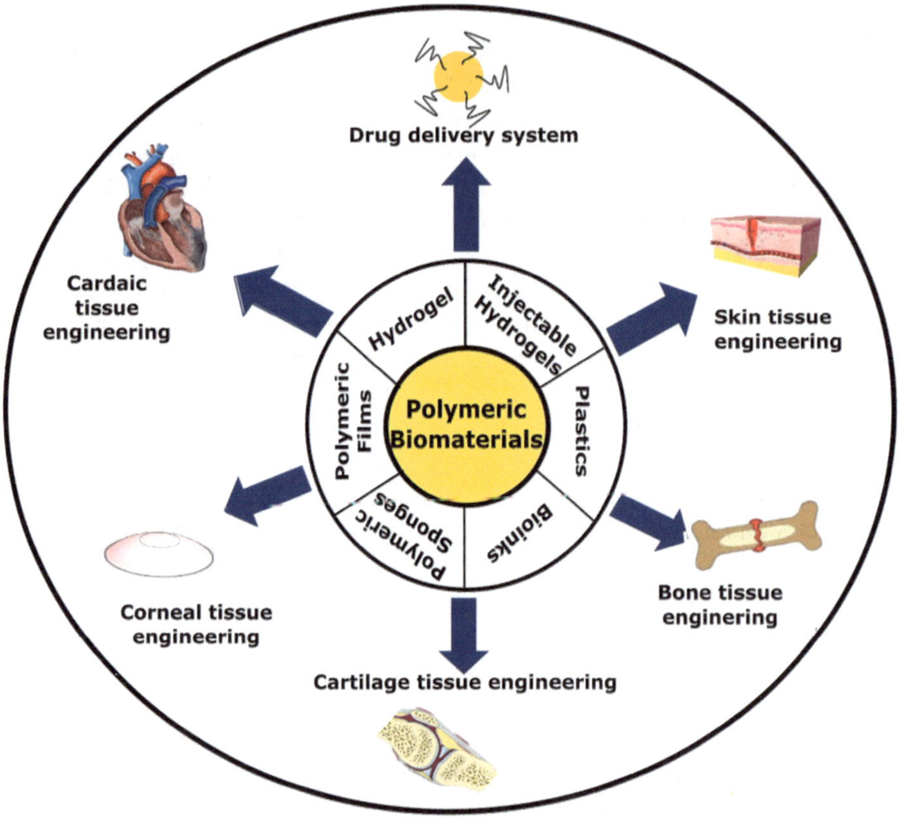

Figure 1.1.5: Different polymer-based biomaterials (adapted from [31] – Open access under CC BY 4.0 license).

1.1.4.1 Natural polymers/biopolymers

Basic biological systems, including living things, plants, and cultured microbes, can be chemically synthesized to generate biopolymers. Compared to synthetic polymers, they offer a number of rewards, including a more complex and well-defined structure, as well as degradability and renewability. When incorporated into other materials, primarily hydroxyapatite and ceramic composites, biopolymers are mostly employed as organic coatings to change the surface appearance of dental implants [32].

As a result of the intrinsic characteristics of these materials, this aids in enhancing their biodegradability and biocompatibility.

1.1.4.2 Polylactic acid (PLA)

Perhaps, the most easily accessible polymer is polylactic acid (PLA), which is produced from biodegradable materials such as potatoes, corn, rice, and starch. PLA has the potential for offering dental implants with the benefits of easy curing and biological absorbability. PLA's biomechanical properties are comparable to those of synthetic polymers such as poly(lactic-co-glycolic acid), but have the advantage of being more readily accessible and less expensive. In view of its adequate capability to contain recombinant bone morphogenetic protein 2, a porous PLA composite material has been examined to promote bone growth in 2 weeks. Because of the strong intermolecular relationship among the composite fillers and polymers, PLA/octadecylamine-functionalized nanodiamond nanocomposites have been reported to demonstrate increased mechanical properties (hardness and Young's modulus of elasticity), noncytotoxicity, and biocompatibility. In addition to their compatibility and inertness to toxicity, implants formed from PLA-PGA (polyglycolic acid) copolymers and compression-molded copolymers are apparent for bone development applications due to their flexibility in changing the rate of deprivation [33, 34].

1.1.4.3 Chitosan

CS is a cationic polymer that is produced through alkalizing treatments at high temperatures. It has the flexibility to be shaped into porous methods and is biocompatible and biodegradable. For this, calcium phosphate-based composites are being used in partnership with CS. It is a 3D macroporous bioceramic structure embedded with porous CS polymers. The oval-shaped CS particles strengthened the biomechanical properties of the ceramic phase (via matrix reinforcement) without compromising the structure of the osteoblast tissue. During 3 months of in vivo testing, hydroxyapatite-chitin material (of osteoconductive type) demonstrated fast deprivation and neovascularization. The ability to tolerate compressive forces, which is influenced by the temperature at the freezing point and the volume of composite suspension, was examined in a study of a macroporous CS-gel/b-TCP composite material, which is primarily employed in the industry of bone tissue engineering. The intramolecular hydrogen bonds of CS, which offer strong resistance to heat generated through macromolecule collision, are among the significant characteristics acquired through its chemical structure. Lower exothermic curing temperatures were noted in CS composites with polymethyl methacrylate, highlighting the advantageous properties of this chemical characteristic and reducing input power [35]. Additionally, it had high interconnected porosity, which facilitated establishing to the surrounding bone and enhanced osteoconduction.

1.1.4.4 Synthetic polymer

In order to develop synthetic polymers with predictable and tunable qualities that may be controlled by changing chemical structures according to the particular applications, it is important to properly design the polymer functional groups. Reduced bioactivity, osteoconductivity, and the presence of cell recognition regions in dental applications are among the outcomes of a comparison study between natural and synthetic polymers. Additionally, as Figure 1.1.6 illustrates, their dilapidation rate can fluctuate in response to changes in crystallinity, chemical composition, and molecular weight.

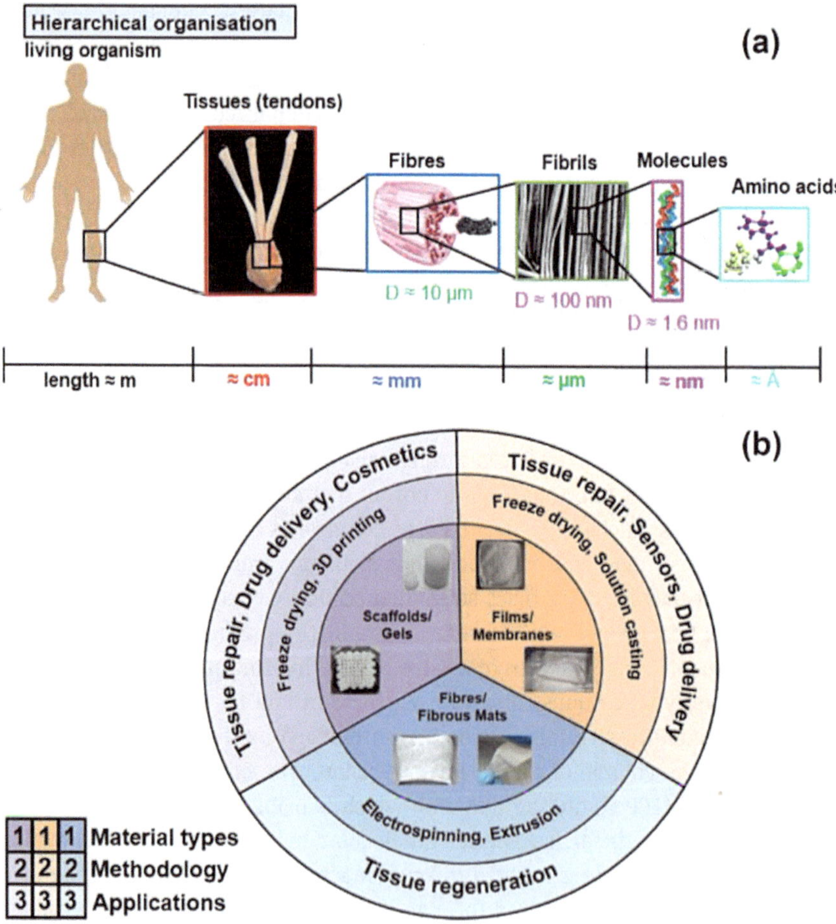

Figure 1.1.6: Advanced biomedical applications of polymer materials (adapted from [4] – Open access under CC BY 4.0 license).

1.1.4.5 Polyetheretherketone (PEEK)

The semicrystalline thermoplastic polymer, or PEEK (polyetheretherketone), combines qualities such as high strength, exceptional stiffness, and the capacity to withstand the chemical and thermal changes brought on by acids and oils. PEEK is a reasonable alternative for the production of dental devices due to its colorlessness and low elastic modulus value. Because of its established modulus of elasticity (3–4 GPa), it is less likely to contribute to osteoporosis and bone absorption, which is primarily brought on by stress shielding effects. However, PEEK's characteristic bioinertness prohibits it from being conductive to the attachment of cells. One of the main justifications for using PEEK and its composites is its inherent radiolucency, chemical resistance, and biocompatibility. By using nanoproduction methodologies, titanium has been embedded to the surface of PEEK implants, improving their biocompatibility. In addition to boosting the bone-in-contact ratio and implant strength, titanium depositions on the PEEK substrate doubled after another assessment of parameters pertaining to initial attachment of cells, osteoblastic differentiation, and proliferation for in vitro cell response. Subsequently, it is asserted that chemical treatment with H_2SO_4 at a suitable concentration was the cause of the porous sulfonated layers formed on the surface of PEEK. This treatment favorably promoted surface bioactivity and osseointegration in both in vitro and in vivo examinations. In addition to the attributes listed above, it offers the right amount of surface roughness to promote cell growth, which increases the ingrowth of essential tissues into the implants and ensures their long-term stability. In vivo studies conducted on animals indicated that the incorporation of graphene oxide in materials promoted the development of new bone more effectively, and the compounds revealed no systemic toxicity or inherent cytotoxicity. Surface-modified carbon fiber/ PEEK (CF/PEEK) composites indicate the characteristics of increased osteogenesis.

1.1.4.6 Bisphenol-a-diglycidyl ether methacrylate/triethylene glycol dimethacrylate (bis-GMA/TEGDMA) composite

Visible light can cure the constant polymer matrix designated as bis-GMA/TEDGMA (bisphenol-a-diglycidyl ether methacrylate/triethylene glycol dimethacrylate). A dispersion medium of barium, silicon, quartz, and zirconia reinforces the polymeric matrix of bisphenol-a-glycidyl methacrylate diluted with triethylene glycol dimethacrylate. Because of their outstanding characteristics, sufficient strength, and significant reduction in expenses over ceramics, resin composites are widely used in dentistry for their aesthetic materials. Additionally, they possess an excellent capacity to adhere to the structure of teeth. When paired with E-glass fibers, these thermoset materials are especially suitable for load-bearing scenarios in dental implants. Because of the 3D polymerized structure, bis-GMA naturally produces a tight physical binding feature. Furthermore, chemical binding between the filler and the base resin is the outcome of the extra silane treatment applied to the filler surface. A less volatile liquid, bis-GMA resin, has been discovered to easily pass through 3D printed objects and solidify into a robust solid after a polymerization reaction. Therefore, it is an excellent option to be exploited as a matrix in composites that are used for resin impregnation and dental implant manufacturing. Using a scanning electron microscope, a study was conducted to ascertain the impact of HA bis-GMA on osteoblast cells. For 5–6 days, these osteoblast cells were cultivated on composites based on HA bis-GMA-g and HA bis-GMA-s. Osteoblast cells adhered to the specimen surface and spread efficiently, based on observations. Additionally, these living cells appear to occupy on a typical shape and adhere to the surfaces of both composites, indicating strong bioactivity. Since bis-GMA-derived composites are one of the more recent materials for dental implants, there is a dearth of study evidence that briefly analyzes their significance.

1.1.5 Characterizations of biomaterials

The interdisciplinary aspect of the biomaterials area is prominent. Scientists from biology, engineering, physical, and medical fields are frequently working together in developing biomaterials that might improve human health. Thus, it has become essential that everyone understands the core idea of this book, which can be used to characterize various biomaterials. It is also important to realize that, in many cases, characterization of the biomaterials alone might not be sufficient; additional systems-level testing could be required for verifying the device's actual functioning. This book has been composed with an emphasis on characterizing the physical, mechanical, surface, in vitro, and in vivo biological properties of biomaterials and biomedical devices in light of all those considerations. A special concentration has been placed in characterizing orthopedic and cardiovascular implants at the instrument level. Researchers from the physical and scientific fields of engineering develop chapters that address the mechanical, surface, and physical characteristics of biomaterials. In order to understand surface-level modifications and their influence on material properties, these chapters emphasize on methods that can disclose fundamental characteristics like atomic structures, bonding, chemical interactions, phase designation and transformation, strength and toughness measurements, and additional information. Figure 1.1.7 illustrates the majority of these approaches implemented for biomaterials researchers in the initial stages of materials development for particular use cases.

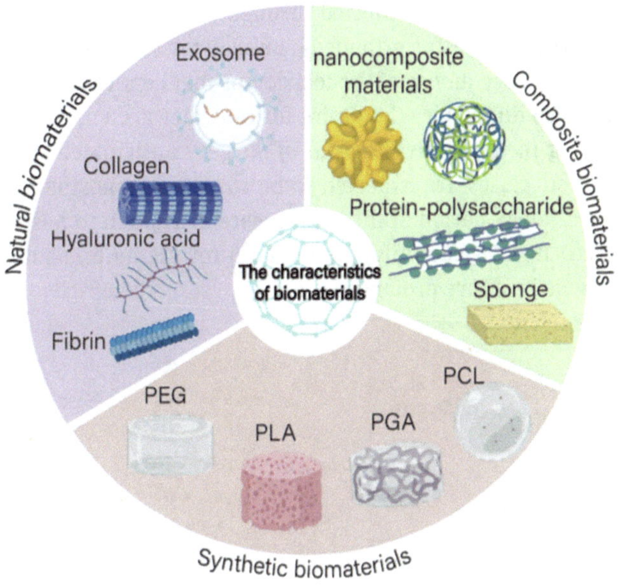

Figure 1.1.7: Biomaterials with different applications (adapted from [36] – Open access under CC BY 4.0 license).

1.1.6 Biomaterials world market

As shown in Figure 1.1.8, the size of the biomaterials market across the globe was projected at $64,874 million in 2020 and is anticipated to continue expanding at a compound annual growth rate of 12.7% from 2021 to 2030, reaching $212,405.2 million. Biomaterials are synthetic or natural materials that are artificially generated and employed in medical technology to improve the function of damaged tissues, bones, and organs. These biological materials are employed in medical devices and implants that supplement or replace natural body functions. Considering the body refuses to embrace any foreign material to fully supplement its biologic functions, implantable biomaterials have the purpose to mitigate an adverse immune response while maintaining optimal functionality. These categories of materials communicate with the biological system for medical objectives, including the treatment of cardiovascular, dental, orthopedic, and neurological diseases.

Globally, the COVID-19 pandemic has triggered difficulties with workflow in the healthcare industry. A number of sectors, including several healthcare-related subdomains were compelled to shut down temporarily as a result of the sickness. The year 2020 observed a fall in worldwide biomaterials market revenue as a consequence of the COVID-19-driven global economic downturn. In addition to having a positive impact on the global economy, the COVID-19 pandemic had an essential impact on how general hospital care for patients who were unaffected through the virus functioned in hospitals all over the world. The market for orthopedic implants is projected to be suppressed in the near future as a consequence of the lockdowns that kept many clinics and hospitals closed, which also diminishes the desire for biomaterials. For example, a May 2020 study appearing in the *British Journal of Surgery* anticipated that around 28.4 million scheduled surgeries are expected to be canceled or postponed worldwide in 2020, based on a 12-week period of the most severe disruption to hospital services attributed to COVID-19. Consequently, the demand for biomaterials declines as these nonemergent procedures are prolonged.

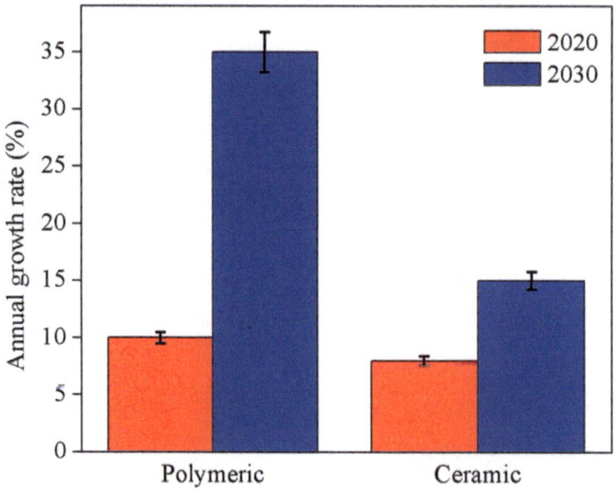

Figure 1.1.8: Biomaterials in the world market.

1.1.7 Biomaterial applications

Cardiovascular, dental, orthopedic, wound healing, plastic surgery, ophthalmology, tissue engineering, neurological disorders, and drug delivery systems depend on application segments where the biomaterials sector is segmented. Due to the expanding use of biomaterials in orthopedic activities and continuing developments for the launch of cutting-edge orthopedic implants by industrialists, the orthopedic sector held the highest share of the market in 2020 and is predicted to continue to do so throughout the duration of the forecast.

The market has been separated into the following segments: natural, ceramic, polymeric, and metallic. As shown in Figure 1.1.9, the market can be categorized into several groups, including cardiovascular, dental, orthopedic, wound healing, plastic surgery, ophthalmology, tissue engineering, neurological disorders, and drug delivery systems.

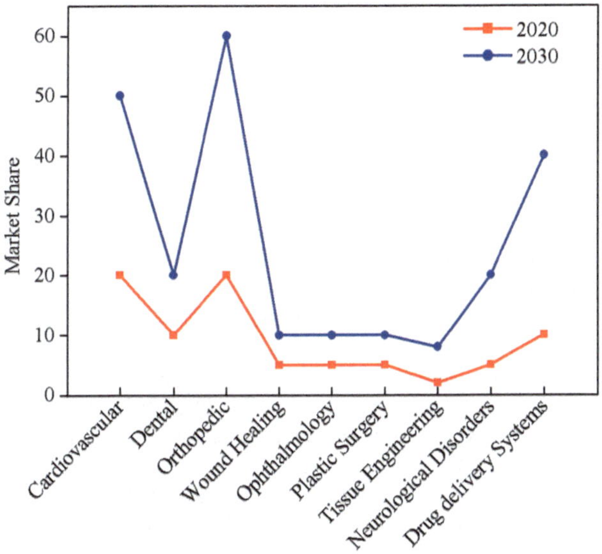

Figure 1.1.9: Applications of biomaterials in various medical fields.

1.1.8 Summary

Materials, biomaterials, and the importance of appreciating different approaches for describing biomaterials are all addressed in Section 1.1. Recently researched biomaterials tend to concentrate on the monitored and controlled release of active chemicals, specifically in tissue regeneration. In the engineering dominion of tissue, there is also increased interest in multilayered hydrogels or scaffolds that have the ability to release active molecules in response to the host reaction or healing stages. To create polymer-based biomaterials with the appropriate structure and activity, a deeper comprehension of the illness is required. When it comes to the future generation of biomaterials, polymer-based, stimuli-responsive smart biomaterials constitute a quickly changing matter. The viewer may observe the manner in which the remaining contents in various segments have been structured to fully understand this naturally diverse focus from the current section. In addition to minimizing labor when examining various biomaterials, the implementation of appropriate characterization technologies can increase the safety of commercial biomedical devices. Safer biomedical equipment can only lessen the suffering of humans over time, which is an intention embraced in biomedical investigations.

References

[1] Hu K, Li Y, Ke Z, Yang H, Lu C, Li Y, Guo Y, Wang W. History, progress and future challenges of artificial blood vessels: A narrative review. Biomaterials Translational. 2022; 3(1): 81.

[2] Moritz WR, Raman S, Pessin S, Martin C, Li X, Westman A, Sacks JM. The history and innovations of blood vessel anastomosis. Bioengineering. 2022; 9(2): 75.

[3] Russo M, Taramasso M, Guidotti A, Pozzoli A, Nietilspach F, Von Segesser L, Maisano F. The evolution of surgical valves. Cardiovascular Medicine. 2017; 20(12): 285–292.

[4] Chelu M, Musuc AM. Advanced biomedical applications of multifunctional natural and synthetic biomaterials. Processes. 2023; 11(9): 2696.

[5] Park J, Lakes RS. Biomaterials: An Introduction, Springer New York, NY, USA: Springer Science & Business Media; 2007.

[6] Patel NR, Gohil PP. A review on biomaterials: Scope, applications & human anatomy significance. International Journal of Emerging Technology and Advanced Engineering. 2012; 2(4): 91–101.

[7] Marin E, Boschetto F, Pezzotti G. Biomaterials and biocompatibility: An historical overview. Journal of Biomedical Materials Research, Part A. 2020; 108(8): 1617–1633.

[8] Taraballi F, Sushnitha M, Tsao C, Bauza G, Liverani C, Shi A, Tasciotti E. Biomimetic tissue engineering: Tuning the immune and inflammatory response to implantable biomaterials. Advanced Healthcare Materials. 2018; 7(17): 1800490.

[9] Dolcimascolo A, Calabrese G, Conoci S, Parenti R. Innovative Biomaterials for Tissue Engineering. In: Mike Barbeck, Ole Jung, Ralf Smeets and Tadas Koržinskas, editors. Biomaterial-Supported Tissue Reconstruction or Regeneration, London, UK: IntechOpen Limited; 2019, pp. 1 18.

[10] Henao J, Poblano-Salas C, Monsalve M, Corona-Castuera J, Barceinas-Sanchez O. Bioactive glass coatings manufactured by thermal spray: A status report. Journal of Materials Research Technology. 2019; 8(5): 4965–4984.

[11] Ijaz M, Héraud L, Castany P, Thibon I, Gloriant T. Superelastic behavior of biomedical metallic alloys. Metallurgical Materials Transactions A. 2020; 51: 3733–3741.

[12] Davis R, Singh A, Jackson MJ, Coelho RT, Prakash D, Charalambous CP, Ahmed W, Da Silva LRR, Lawrence AA. A comprehensive review on metallic implant biomaterials and their subtractive manufacturing. The International Journal of Advanced Manufacturing Technology. 2022; 120(3): 1473–1530.

[13] Teo AJ, Mishra A, Park I, Kim Y-J, Park W-T, Yoon Y-J. Polymeric biomaterials for medical implants and devices. ACS Biomaterials Science Engineering. 2016; 2(4): 454–472.

[14] Trombino S, Curcio F, Cassano R, Curcio M, Cirillo G, Iemma F. Polymeric biomaterials for the treatment of cardiac post-infarction injuries. Pharmaceutics. 2021; 13(7): 1038.

[15] Momin M, Mishra V, Gharat S, Omri A. Recent advancements in cellulose-based biomaterials for management of infected wounds. Expert Opinion on Drug Delivery. 2021; 18(11): 1741–1760.

[16] Montazerian M, Hosseinzadeh F, Migneco C, Fook MV, Baino F. Bioceramic coatings on metallic implants: An overview. Ceramics International. 2022; 48(7): 8987–9005.

[17] Bahati D, Bricha M, El Mabrouk K. Synthesis, characterization, and in vitro apatite formation of strontium-doped sol-gel–derived bioactive glass nanoparticles for bone regeneration applications. Ceramics International. 2023; 49(14): 23020–23034.

[18] Naomi R, Bahari H, Ridzuan PM, Othman F. Natural-based biomaterial for skin wound healing (Gelatin vs. collagen): Expert review. Polymers. 2021; 13(14): 2319.

[19] Roberts IV, Bukhary D, Valdivieso CYL, Tirelli N. Fibrin matrices as (injectable) biomaterials: Formation, clinical use, and molecular engineering. Macromolecular Bioscience. 2020; 20(1): 1900283.

[20] Zhou Y, Shi M, Jones JR, Chen Z, Chang J, Wu C, Xiao Y. Strategies to direct vascularisation using mesoporous bioactive glass-based biomaterials for bone regeneration. International Materials Reviews. 2017; 62(7): 392–414.
[21] Zhu L, Luo D, Liu Y. Effect of the nano/microscale structure of biomaterial scaffolds on bone regeneration. International Journal of Oral Science. 2020; 12(1): 6.
[22] Boni R, Ali A, Shavandi A, Clarkson AN. Current and novel polymeric biomaterials for neural tissue engineering. Journal of Biomedical Science. 2018; 25: 1–21.
[23] Ratnayake JT, Mucalo M, Dias GJ. Substituted hydroxyapatites for bone regeneration: A review of current trends. Journal of Biomedical Materials Research, Part B: Applied Biomaterials. 2017; 105(5): 1285–1299.
[24] Chen L, Al-Bayatee S, Khurshid Z, Shavandi A, Brunton P, Ratnayake J. Hydroxyapatite in oral care products – A review. Materials. 2021; 14(17): 4865.
[25] Yu S, Chi H, Li P, Guo B, Yu Z, Xu Z, Liang P, Zhang Z, Guo Y, Ren L. Interpenetrating phases composites Ti6Al4V/Zn as partially degradable biomaterials to improve bone-implant properties. Additive Manufacturing. 2024; 93: 104411.
[26] Dogan S, Fong H, Yucesoy DT, Cousin T, Gresswell C, Dag S, Huang G, Sarikaya M. Biomimetic tooth repair: Amelogenin-derived peptide enables in vitro remineralization of human enamel. ACS Biomaterials Science Engineering. 2018; 4(5): 1788–1796.
[27] Bhasin SS, Perwez E, Sachdeva S, Mallick R. Trends in prosthetic biomaterials in implant dentistry. Journal of the International Clinical Dental Research Organization. 2015; 7(1): S148–S159.
[28] Bhat S, Uthappa U, Altalhi T, Jung H-Y, Kurkuri MD. Functionalized porous hydroxyapatite scaffolds for tissue engineering applications: A focused review. ACS Biomaterials Science Engineering. 2021; 8(10): 4039–4076.
[29] Hajiali F, Tajbakhsh S, Shojaei A. Fabrication and properties of polycaprolactone composites containing calcium phosphate-based ceramics and bioactive glasses in bone tissue engineering: A review. Polymer Reviews. 2018; 58(1): 164–207.
[30] Kadambi P, Luniya P, Dhatrak P. Current advancements in polymer/polymer matrix composites for dental implants: A systematic review. Materials Today: Proceedings. 2021; 46: 740–745.
[31] Kalirajan C, Dukle A, Nathanael AJ, Oh T-H, Manivasagam G. A critical review on polymeric biomaterials for biomedical applications. Polymers. 2021; 13(17): 3015.
[32] Nathanael AJ, Oh TH. Biopolymer coatings for biomedical applications. Polymers. 2020; 12(12): 3061.
[33] Athanasiou KA, Agrawal CM, Barber FA, Burkhart SS. Orthopaedic applications for PLA–PGA biodegradable polymers, Arthroscopy. The Journal of Arthroscopic Related Surgery. 1998; 14(7): 726–737.
[34] Agrawal CM, Niederauer GG, Athanasiou KA. Fabrication and characterization of PLA-PGA orthopedic implants. Tissue Engineering. 1995; 1(3): 241–252.
[35] Radhakumary C, Nair PD, Mathew S, Nair CR. Biopolymer composite of chitosan and methyl methacrylate for medical applications. Trends in Biomaterials & Artificial Organs. 2005; 18(2): 117–124.
[36] Wu M, Guo Y, Wei S, Xue L, Tang W, Chen D, Xiong J, Huang Y, Fu F, Wu C. Biomaterials and advanced technologies for the evaluation and treatment of ovarian aging. Journal of Nanobiotechnology. 2022; 20(1): 374.

Sudip Kumar Sinha*

Chapter 1.2 Advanced hydrogels for biomedical applications

Abstract: Hydrogels represent a very promising class of material constituted with networks of cross-linked polymers swelling in water that can be created to exhibit attributes like soft biological tissues. Their capability to mimic tissues' native extracellular matrix makes them useful for a range of biomedical uses, including drug delivery, cell culture, contact lenses, and even the replacement of native tissue. Numerous groundbreaking studies have been conducted in the last few decades to investigate the interactions between cells and hydrogel matrixes and determine the underlying mechanisms, opening the door for the translation of hydrogel-based therapeutics from the lab to the clinic.

Even though this field's study is expanding, there are still several obstacles preventing sophisticated applications.

In this chapter, all kinds of biomedical-grade polymeric hydrogels are methodically examined, with a focus on sophisticated production methods from the standpoint of tissue engineering. Hydrogel synthesis techniques are thoroughly reviewed here, and a comprehensive overview of biopolymer-based hydrogels for biomedical applications, particularly in bone regeneration, wound healing, drug administration, bioimaging, and treatment is provided. Lastly, a summary of the challenges encountered in the sector are offered, along with various schemes for new developments and investigations.

Keywords: Hydrogel, tissue, biopolymer, biomedical, cross-link

1.2.1 Introduction

Research on innovative approaches to improve healthcare for the ill and aged continues globally.

Hydrogels are three-dimensional networks of cross-linked polymers that have a strong attraction for water and cannot dissolve, mostly because of their cross-linked structure either chemically or physically. Hydrogels have a high degree of flexibility and elasticity, which helps with cell adhesion, protein adsorption, and tissue friction reduction.

The ability of hydrogels to swell may result from several factors [1] including (i) functional groups that is hydrophilic attached to the polymer bonding, (ii) a high af-

***Corresponding author: Sudip Kumar Sinha,** Department of Metallurgical and Materials Engineering, NIT Raipur, Raipur 492010, Chhattisgarh, India, e-mail: sksinha.mme@nitrr.ac.in

https://doi.org/10.1515/9783111386027-002

finity for thermodynamics, and (iii) an extreme water content, porosity, and uniform uniformity. This characteristic enables them to imitate live tissues quite accurately.

Because of their structural resemblance to the extracellular matrix (ECM) hydrogels found a wide range of applications in the field of healthcare [2]. Moreover, scientists can modify and customize hydrogels' mechanical, chemical, and physical properties to fit a variety of uses.

As a result, hydrogels are emerging as intriguing materials in several technical and biological fields. Among the examples are drug delivery platforms [3–5], stretched electronic devices [6–8], scaffolds for tissue engineering [9–11], antifogging films [12, 13], wound dressings [14, 15], self-healing applications [16, 17], soft robotics [18, 19], and (bio)sensor [20–22].

However, the main drawbacks of hydrogels are their low elongation and poor mechanical properties [23–25]. Therefore, harm could result in the loss of their lifespan, functionality, and fruitful biological applications. Even with all the research being done, contact lenses continue to dominate the biomedical hydrogel market, accounting for around half of all hydrogels. The question that emerges is: Why is there a lag between the invention of hydrogel technologies and their application in clinical settings?

Wichterle and Lim developed the first hydrogel with possible applications in biomedical field in 1960 [26]. The first synthetic hydrogel used in contact lenses was created by these two scientists.

Hydrogels are typically created when the polymeric network's hydrophilic segments or groups are hydrated in aqueous solutions [27]. Cross-linking between the polymer chains is required to stop the hydrophilic group from dissolving in water or bodily fluid. The physical characteristics of the produced hydrogel are mostly determined by the degree of cross-linking. When dissolved in solution, a non-cross-linked hydrophilic polymer will have a low viscosity; limited cross-linking will produce an elastomer; and highly cross-linked polymers will have a substantially higher rigidity [28, 29].

Depending on the type of bond that exists between the polymeric molecules that make up the framework, hydrogels can generally be divided into two categories: physically cross-linked hydrogels and chemically cross-linked hydrogels [30]. Covalent bonding, physical cross-linking, which includes molecular level mix-ups, and/or secondary forces, such as ionic, hydrogen, or hydrophobic forces, can all be used to create hydrogels. Relevant research has reported [31, 32] the use of polysaccharides including chitosan, dextran, and carboxy methyl curdle to generate physically cross-linked hydrogels through hydrophobic alteration.

Cross-linking by chemical means involves the process of joining two or more polymer chains together by grafting monomers onto the polymer support or by applying a cross-linking species. Natural and artificial polymers can be cross-linked by interacting active functional groups (e.g., OH, COOH, and NH_2) with acrylamide, an aldehyde, or methylene cross-linker.

On the other hand, physical hydrogels have the primary benefit of not requiring a sensitive functional group for cross-linking and the formation of a 3D network.

Thus, cross-reactions in vivo are prevented, and the deposition of constituents is simple (no additional alteration with reactive species is needed). However, because of the weak contacts that occur during the creation process, physical hydrogels frequently have poor mechanical properties.

As might be expected, synthetic hydrogels eventually replace naturally occurring hydrogels to produce hydrogels with higher gel strengths, higher water absorption capacities, and longer service lives [33, 34]. Thankfully, hydrogels with specified network architectures, preferred chemical compositions, and adjustable mechanical strength can be created using a variety of sophisticated synthetic techniques.

Hydrogels have exceptional durability in harsh environments such as high temperatures or highly acidic or basic conditions and can be made entirely of synthetic materials. Additionally, the hydrogel's characteristics can be altered by stimuli such as heat, light, magnetic fields, chemical agents, and pH by adding stimuli-responsive functional groups to the polymer chains [35–39].

Over the past few years, excellent review papers and book chapters are available that go into great detail into the synthesis and processes of hydrogels for healthcare applications. Nevertheless, a comprehensive discussion in form of a chapter for this fascinating material that addresses the topic of biomedical application is missing. Therefore, it can be said that researchers and medical professionals who work with or are interested in this fascinating class of biomaterials will probably find this chapter to be a useful resource.

Figure 1.2.1(a) shows the history of hydrogel use in the field healthcare, initiating with the definition of the name "hydrogel" [29] and ending with the development of 4D-printed hydrogel for healthcare and drug delivery applications.

We conducted a search in the "SCOPUS" database using the keywords "hydrogel + biomedical". As shown in Figure 1.2.1(b), we discovered that astonishing and fast advancement has happened during the past 15 years. Figure 1.2.1(c) depicts a pie chart showing an enormous number of research documents published throughout this time frame with research articles dominating amongst others.

1.2.2 Structure of hydrogels

Typically, the solid-state form of a hydrogel is a network structure made of cross-linked polymer chains [40]. Because of the hydrogel's three-dimensional network structure, its molecular weight goes to infinity [41].

The mesh size and the molecular weight of the polymer chain between the cross-links are the most crucial molecular characteristics that shape the hydrogel structure as shown in Figure 1.2.2.

Hydrogels possess both strength and complexity: their ion-rich polymer cross-linking provides structural stability, while the hydrophilic (loving water) aspect of the

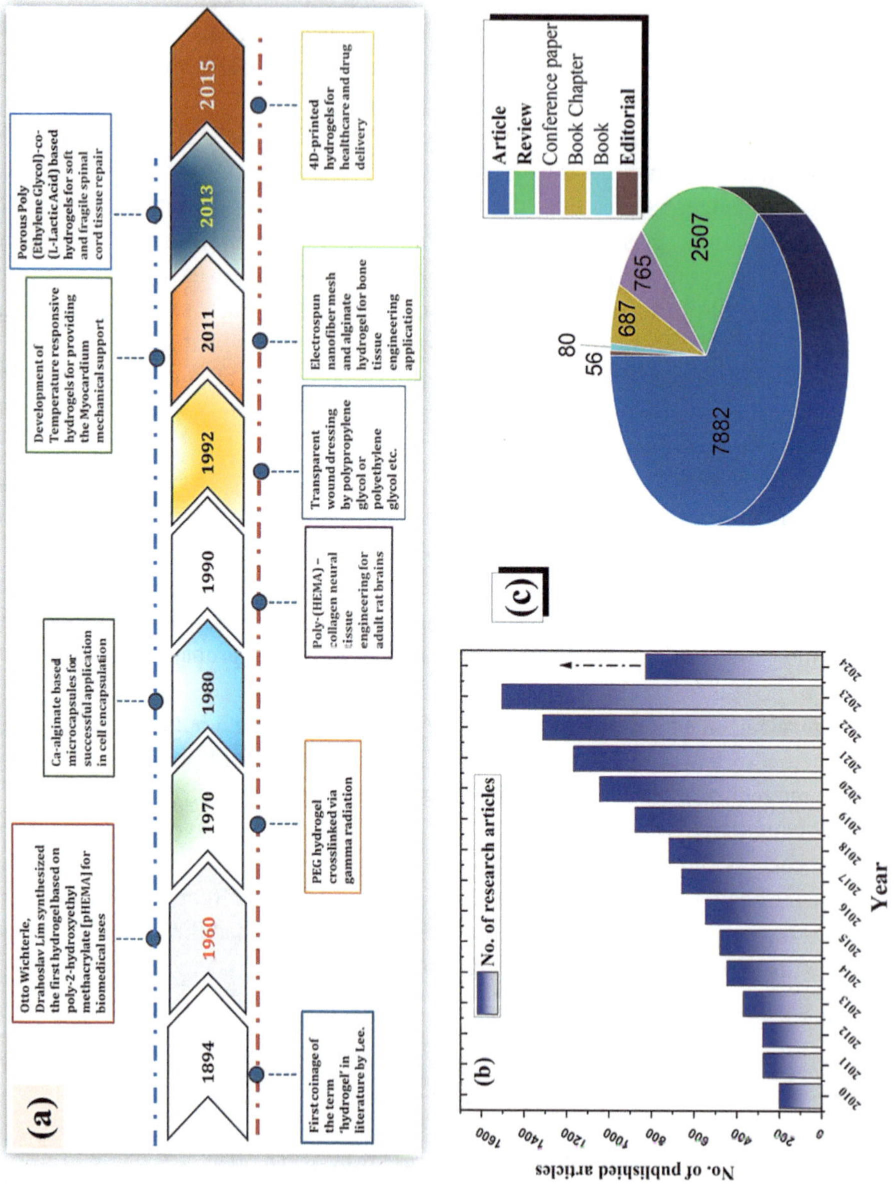

Figure 1.2.1: (a) Historical timeline of hydrogel in biomedical applications; (b) bar diagram depicting the temporal trend of research publications on "hydrogels" that have been published between 2010 and 2024 (mid-year); and (c) 3D pie chart showing the classification of various research documents based on hydrogel research.

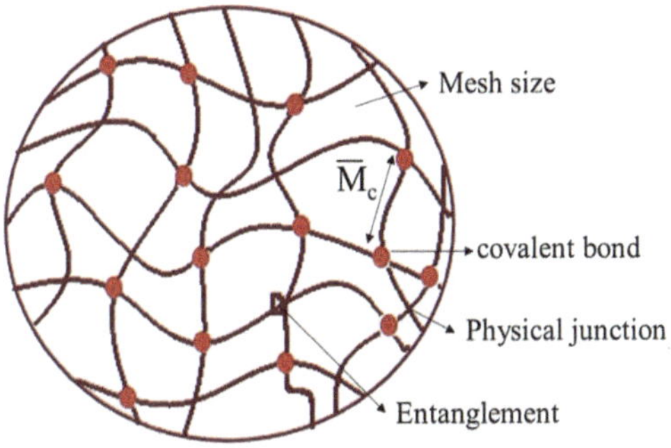

Figure 1.2.2: Molecular level structure of hydrogel [41].

polymers makes them elastic. Hydrogels are also stimuli-responsive, changing their structure completely or remaining the same in response to variations in the pH and temperature of their surroundings. A hydrogel is a remarkable substance that is nearly as promising as Frankenstein when the proper thermodynamic parameters, reaction-made molecular interactions, coupled with production techniques are combined to create a mixture of different macromolecules.

1.2.3 Swelling behavior of hydrogels

Hydrogels are primarily characterized by swelling. The term "swelling" describes the volume change that occurs when a gel absorbs a suitable solvent, in the case of hydrogels, water. The solid component of the hydrogel, which is usually a network of polymers, is both elastic and hydrophilic, allowing water to be soak up to a sufficient degree to expand the volume while generating elastic strain amid the networked polymer chains to avoid complete breakdown. This property makes swelling feasible.

Hydrophilic groups such as hydroxyl (–OH)) groups, carboxyl (COOH) groups, amide (–C(=O)–(N)) bonds, ether (R–O–R) bonds, and charged groups abound in the hydrogel system [6]. The three-dimensional network is completely expanded after swelling, and its volume is extended several times in comparison to the dry gel. These substances have the ability to absorb water 100s or even 1,000s of times their own weight. Figure 1.2.3 shows a diagrammatic representation of hydrogel swelling [42].

The swelling of a hydrogel is regulated by osmotic forces, ionic interactions, and polymer-water interactions. To stop hydrogels from dissolving, the swelling forces are counterbalanced by the elastic forces resulting from network cross-linking.

Researchers looked at polymers that responded to pH changes by swelling or contracting, such as poly(acrylic acid) (PAA) and its derivatives. This created opportunities for controlled release systems and focused medication distribution.

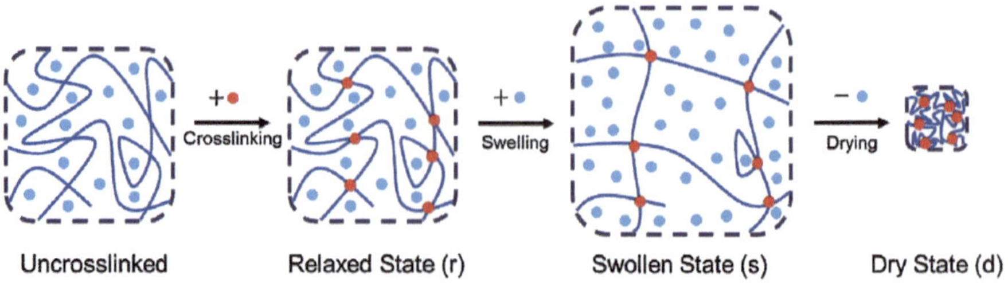

Figure 1.2.3: Illustration of the pertinent state changes in hydrogel swelling [42].

One easy, affordable, and secure method to learn more about the hydrogels is to swell them in a solution liquid. The degree of cross-linking, mechanical and viscoelastic characteristics, refractive index, and degradation rate can all be ascertained from the swelling data [42] Measuring the weight of the hydrogel before and after swelling is typically the simplest method to determine the swelling ratio. Given the variety of uses, such as in implants or contact lenses, there may be a higher demand for the volume of swelling hydrogels than for their mass expansion.

Even though biological tissues are submerged in bodily fluids and are fundamentally hydrogel materials, they can improve their mechanical strength when necessary. For instance, muscles inflate with blood and become stronger (Figure 1.2.4(a)). Figure 1.2.4(b,c) presents a collection of swelling-strengthening hydrogels (SSHs) made possible via a method inspired by biological membrane barriers [43].

Gel kinetics of swelling and shrinking are investigated experimentally and theoretically.

Tanaka and Filmore [44] developed a model based on kinetic theory of absorbent gel swelling. Following this investigation, Li and Tanaka [45] proposed new ways to gel swelling and shrinking, which were later used by several researchers [46, 47].

Hydrogels show swelling performance in an aqueous condition as one of their common characteristics. The exceptional swelling characteristics of hydrogels as a drug delivery technology enable the regulated release of the drug. However, it is not forever beneficial. Hydrogels' swelling actions can degrade mechanical characteristics when employed for underwater motion devices and compress adjacent organs or soft tissue in tissue engineering. Three hydrogels – PAM, polyvinyl alcohol (PVA), and poly (ethylene glycol) (PEG) – had their swelling behavior simulated by Dehkordi et al. [48]. Figure 1.2.5 displays the simulation graphic and swelling volume of these hydrogels [48] over time. The PAM hydrogel shows the biggest swelling volume with time, suggesting that it produced greater voids.

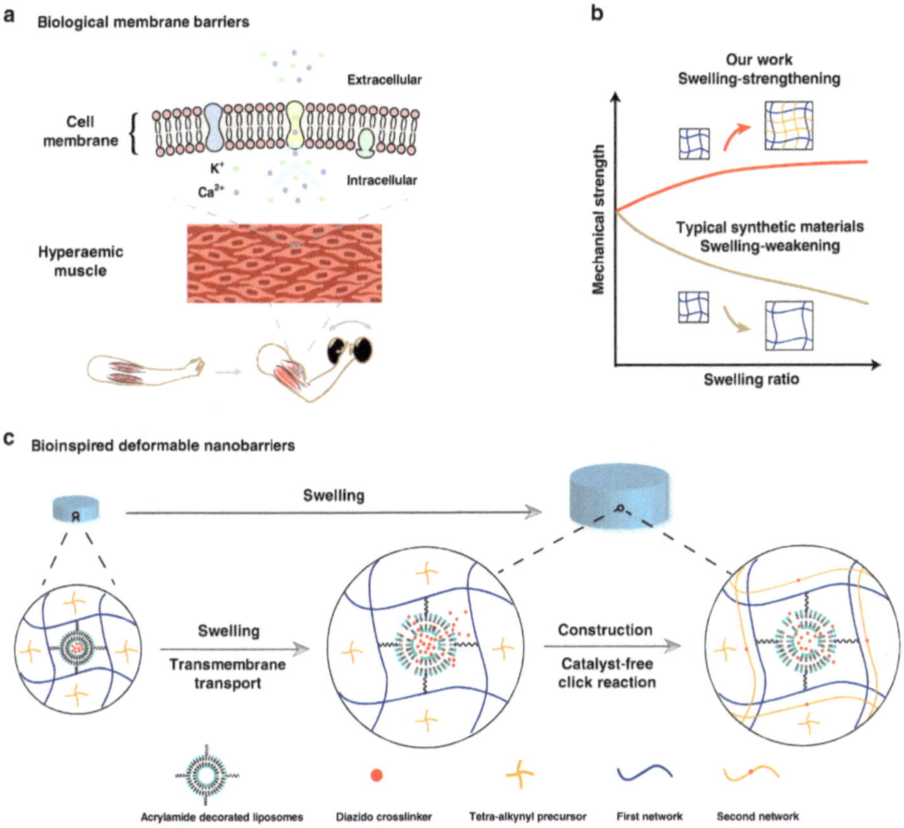

Figure 1.2.4: (a) Living tissues, including muscle, have biological membrane barriers. The hyperemia-mediated increase in muscle stiffness is caused by transmembrane transport, which is dependent on diffusion, ionic channels, transporters, and so on. (b) Difference in mechanical strength following swelling between the swelling-strengthening hydrogels (SSHs) reported in this work and conventional synthetic materials. (c) SSH development with an approach modeled by the biological membrane barrier. Diffusion is primarily responsible for transport over the synthetic liposomal membranes [43].

1.2.4 Swelling analysis of hydrogels

1.2.4.1 Swelling ratio

The swelling ratio of the hydrogel samples can be estimated using eq. (1):

$$S = (W_t - W_o)/W_o \times 100\% = k.t^n \tag{1}$$

where W_o is the initial weight of the xerogel and W_t is the weight of the enlarged gel in a given time interval.

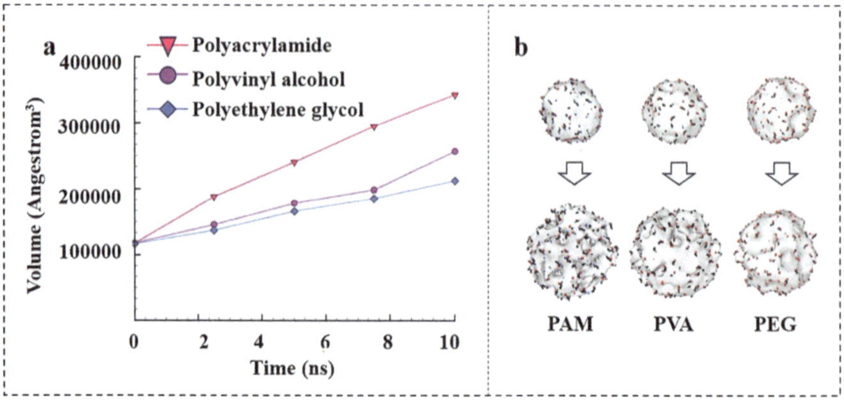

Figure 1.2.5: (a) Hydrogels' swelling volume over time and (b) images of the swelling process [48].

The water transport mechanism is shown by the swelling constant, *k*, and the swelling exponent, *n* [49].

1.2.4.2 Diffusion in hydrogel swelling

The diffusion type was investigated using the following empirical model [49]:

$$F = S/S_e = k.t^n \tag{2}$$

where *F* is the rate of diffusion per area, *n* is the solvent's diffusional exponent, *k* is swelling constant that relies on the polymer network structure, *S* is the swelling ratio at time *t*, and S_e is the equilibrium swelling ratio. A parameter, *n*, needs to be supplied in order to identify the kind of diffusion. The slope of normal logarithm modified equation (2) was used to calculate parameter *n* for the domain where swelling has not yet reached the plateau, which is the case when this equation is valid. The mechanism of diffusion is identified based on the *n* values. Fickian diffusion is the mechanism of choice when $n \le 0.5$; non-Fickian or abnormal diffusion is indicated when $0.5 < n < 1.0$. Diffusion and relaxation in abnormal diffusion are considered to be isochronally effective. A diffusion process is referred to as Case II diffusion if the *n* value is precisely equal to 1. A super Case II diffusion process, which is extremely uncommon [50], is identified when the *n* value is greater than 1 ($n > 1$).

1.2.4.3 Kinetics of hydrogel swelling

The mechanical reliability of hydrogels is controlled by the dynamical swelling and stability in hydrogel solutions. As previously mentioned, the majority of water molecules in hydrogel are bonded by hydrophilic or hydrophobic ensembles. The majority of solutes are limited to diffusing into and through hydrogel from the freed portions of the voids and macropores. Chaotropic solutes can permeate across hydrogel by interfering with the bound water layer contact surrounding the polymer attachment. The kinetics of these changes has been delayed, even though a slight change in an environmental component can cause the swelling ratio to vary significantly. Based on their dimensions and mass, these hydrogels react to the environment in the range of few hours to several days.

The progress in hydrogel research relies heavily on the swelling feature, and a number of methods have been proposed in the published work [51, 52] to investigate the kinetics of this swelling.

Among them, a second-order equation makes use of a basic kinetic analysis equation:

$$ds/dt = k_s\left(S_{\mathrm{eq}} - S\right)^2 \tag{3}$$

where the terms rehydration ratio, swelling rate constant, equilibrium swelling, and swelling ratio at t are denoted by the symbols ds/dt, k_s, S_{eq}, and S, respectively.

Equation (3) becomes the following definite integration by applying the initial condition, such as $S = 0$ at $t = 0$ and $S = S_{\mathrm{eq}}$:

$$t/S = A + Bt \tag{4}$$

Here, $A = 1/k_s S_{\mathrm{eq}}$ [2], and it represents the hydrogel composite's initial swelling rate, the swelling rate constant is k_s, and the magnitude of extreme or equilibrium swelling is represented by B ($=1/S_{\mathrm{eq}}$).

1.2.5 Hydrogels classifications

Hydrogels are classified (Figure 1.2.6) based on a variety of factors including their physical characteristics, mechanism of swelling, manufacturing process, responsiveness, ionic charge, biodegradation rate, and cross-linking characteristics. Well-known hydrogels are categorized according to how they cross-link, including source, preparation, ionic charge, responsiveness, cross-linking, and physical qualities.

1.2.5.1 Classification based on sources

1.2.5.1.1 Natural polymer-based hydrogels

Natural hydrogels are a special class of hydrogels derived from naturally obtained polymers.

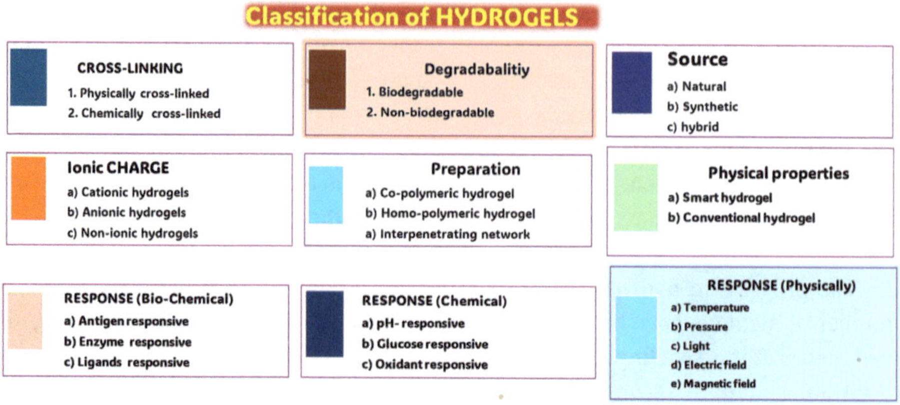

Figure 1.2.6: Schematic description of hydrogel classifications.

Natural polymers are categorized as cationic, anionic, and neutral and can be derived from a variety of natural sources. This class of hydrogels has the advantages of superior biological qualities in addition to being readily available, cheap, plentiful, safe, and biodegradable.

However, their application is limited by their weak mechanical behavior such as stretchability and stiffness.

Natural polymer hydrogels have a wide range of biological uses including tissue engineering, wound healing [53, 54], and the controlled and measured release of medications. Commonly used natural hydrogels are dominated by collagen, gelatin, cellulose, alginate, hyaluronic acid (HA), chitosan, and so on.

1.2.5.1.2 Synthetic hydrogels

Polymeric materials known as synthetic hydrogels closely mimic watery flexible biological tissues. Their soft, porous, and biocompatible nature allows for a physical connection between native biological tissues and innovative biomedical tools. Due to their mechanical plasticity and essential absorptivity, they are considered promising materials for the development of flexible electronics and biomedical devices.

3D swelling grids of covalent or ionically bonded hydrophilic homo- or copolymer hydrogels are known as synthetic polymeric hydrogels. Despite having poor biological response and inadequate biocompatibility, synthetic hydrogels (such as PEG derivatives, polycaprolactone, poly(2-hydroxyethyl methacrylate) (PHEMA), PAA, and PVA) with chemical cross-linking and comparatively superior mechanical properties can endure higher mechanical loads.

1.2.5.1.3 Hybrid hydrogels

In hydrogel research, it has been found that substantial heterogeneity and network defects (for instance, chain culminations, entanglements, and phase-isolated regions) are present in conventional synthetic hydrogels, which have a substantial impact on mechanical characteristics and change the diffusion kinetics and biological action of functioning molecules. To improve on current formulations and broaden their range of applications, hybrid hydrogels are used. Because of their adaptable qualities, hybrid hydrogels find their potential uses that include regenerative medicine, delivery of cancer therapeutics, tissue engineering, medication/gene delivery, and drug discovery. They also find wide use in other applications including optoelectronics and magnetic materials [55].

Natural and artificial polymer hydrogels and other functional moieties are merged to create hybrid hydrogels. Hydrogels that are semisynthetic or hybrid are typically utilized because they combine the benefits of natural and synthetic polymers. Usually heterogeneous, hybrid hydrogels are crucial for ensuring proper cell adhesion, cell-to-cell interactions, and organization in medical settings.

For improved wound healing, Song et al. [56] created bioactive chitosan-built metal-organic polyhedrons/enzyme hybrid hydrogels. The created hydrogels efficiently promote wound healing while having no negative impacts on cells.

1.2.5.2 Classification based on polymeric composition/ preparation

1.2.5.2.1 Homopolymeric hydrogel

These are polymer networks that are created from a lone group of monomers, which is the fundamental structural component of all polymer networks. The cross-linked skeleton of homopolymers can be attributed to the kind of monomer and the method used during polymerization [57].

Single monomer constructed hydrogel could not be able to fully meet the needs of the recommended uses in a variety of biomedical applications. For instance, very high-swelling hydrogels can provide superior controlled release features and good swelling characteristics, but when they are extremely swollen, their mechanical qualities are frequently subpar. Multicomponent hydrogels and various cross-linking techniques can be applied in these circumstances. Common examples of homopolymers, consider PHEMA, poly(3-hydroxypropyl methacrylate) (PHPMA), and poly(hydroxyalkyl methacrylate). Uncross-linked homopolymeric hydrogels include PEG and poly (*N*-vinyl-2-pyrrolidinone) (PNVP) [58].

One homopolymeric hydrogel made from natural polymers is cellulose hydrogel, which is created by dissolving cellulose in a urea-NaOH solution and then proceeding with a single-step polymerization process.

1.2.5.2.2 Copolymeric hydrogel

Two different monomers, at least one of which is hydrophilic, make up copolymeric hydrogels. The component that is hydrophilic in nature is required for the hydrogel to swell to its ideal extent. By polymerizing or cross-linking the two monomers together with an appropriate solvent and initiator, the hydrogel is created.

Copolymer hydrogels' hydrophilic groups can be arranged randomly, alternatingly, or in blocks within the network chain. Among the hydrogels that are available are poly (HEMA-*co*-AA), poly (NVP-*co*-HEMA), and poly (NVP-*co*-HEMA).

The majority of applications for copolymeric hydrogel are in the biomedical field including drug delivery and the creation of hydrogel dressing materials. Copolymeric hydrogels, namely carboxymethyl cellulose, PEG-PEGMA, methacrylic acid, and polyvinylpyrrolidone, are widely utilized in biomedical applications through the utilization of bioprinting technology.

1.2.5.2.3 Interpenetrating network hydrogel (IPN)

An interpenetrating polymer network (IPN) is a polymer made up of two or more networks that are partially integrated at the molecular level, are not covalently bound to one another, and are not possible to be disconnected unless chemical links are broken [59].

The majority of IPNs are diverse systems. Their mechanical and electrochemical qualities are well combined. They consist of a conductive polymer to move anions and a hydrophobic polymer with satisfactory mechanical, chemical, and thermal characteristics. Compared to typical hydrogels, the key benefits of IPNs include their ability to build fairly dense hydrogel bases with mechanical qualities that are stiffer and tougher, controlled physical properties, and more effective drug loading.

An IPN that has one cross-linked polymer and one non-cross-linked polymer component is called a semi-IPN [60]. This type of hydrogel's advantages in drug delivery applications (e.g., superior drug encapsulation capacity, controlled size distributions, adjusting swelling behavior, and gradual drug releasing) and its utility in tissue engineering and other biological applications are well-documented [61].

1.2.5.3 Classification based on behavior and property

1.2.5.3.1 Conventional hydrogel

Conventional hydrogels differ from smart hydrogels in trivial adjustment in swelling with peripheral environmental factors, whereas smart hydrogels can swell or shrink dramatically with minimal temperature or pH changes. These limitations degrade its mechanical strength.

Various physical and chemical cross-linking techniques are used to create hydrogels to biotransform the conventional hydrogel into smart (or intelligent) hydrogel [62].

In biomedical and tissue engineering applications, conventional or traditional hydrogels show the ability to mimic the structure of the ECM and act as carriers for functional moieties (such as drugs or proteins).

Nevertheless, because of their low-to-intermediate strength, conventional hydrogels are not as useful as load-carrying biomaterials. As a possibility, hydrogels enhanced with nanoparticles might have significantly better mechanical characteristics [63].

Additionally, the ineffectiveness of recycling or reprocessing conventional hydrogels makes them a source of environmental degradation. Long-term use also causes hydrogels to break down. The constraints are a result of the structure of the materials according to the researchers. Chemical linkages are what give conventional hydrogels their hardness and solvent-absorbing capacity.

1.2.5.3.2 Smart hydrogel

During the last few decades, understanding organic chemistry has grown, and numerous hydrogels have been created which are chemically cross-linked. Smart hydrogels (often referred to as third-generation hydrogel) are those materials that instantly modify their physical characteristics (e.g., mechanical superiority, swelling ability, and stimuli-sensitivity) in response to even the slightest changes in the outer environmental settings such as temperature, light, pH, composition of solvent, biomolecules (e.g., glucose, DNA, RNA, and proteins), and electric fields.

Other materials are added to the hydrogel to improve the desired characteristic and give it new, intriguing qualities. For example, graphene enhances the performance of supercapacitor hydrogel electrodes, and silver nanoparticles give wound treatment hydrogels' antimicrobial qualities.

In biomedical field, for example, in drug delivery applications, the stimuli-responsive characteristics of polymeric parts in hydrogels have led to enhanced interest in smart hydrogels [64–66].

It is possible to produce smart hydrogels using synthetic or natural polymers. Natural hydrogels have a flaw in that consistency can be challenging to maintain due to their mechanical characteristics.

1.2.5.4 Classification based on electrical charge network

Hydrogel structures can be classified as neutral, amphoteric (comprising both basic and acidic assemblies), zwitterionic (including both anionic and cationic factions), or ionic (i.e., anionic/cationic) depending on their external/electrical charge.

1.2.5.4.1 Ionic (anionic/cationic) hydrogel

Hydrogels that react to changes in pH in the environment can take in or release protons because their polymeric backbones contain ionic pendant groups.

Hydrogels that respond to pH come in two varieties: cationic and anionic hydrogels. Deprotonation happens when the ambient pH is higher than the pK_a in anionic hydrogels containing pendent groups, such as carboxylic or sulfonic acid, which causes the pendent groups to ionize. The hydrogel swells more as a result of this. Conversely, ionization occurs below the pK_b in cationic hydrogels with pendent groups, like amine groups, and this causes an enhancement in electrostatic repulsions, which in turn causes swelling [67].

1.2.5.4.2 Neutral or nonionic hydrogel

There is no charge on the side groups or backbone of a neutral or nonionic hydrogel. The only reason these types of hydrogels swell in an aqueous solution is because of the thermodynamic interactions between the polymer and water, which, in conjunction with elastic polymer and its solution for biomedical use, adds to the total free energy [68].

Polyacrylamide (PAAm), PHEMA, PVA, and PEG are a few instances of nonionic hydrogel.

1.2.6 Application of hydrogels in biomedical engineering

In terms of their mechanics, biological reactions, dynamics of interfaces, and porosities, hydrogels have several distinct qualities that make them extremely relevant to a wide range of possible applications.

In particular, the impact of different tactics to hydrogel formation and subsequently on hydrogel performance in biomedical industry for a variety of purposes is worth mentioning including tissue regeneration, illness therapy, and investigations of physiological and pathological mechanisms. Figure 1.2.7 depicts creation of hydrogels for use in medicine and their relationships to biological reactions and associated signaling pathways [69].

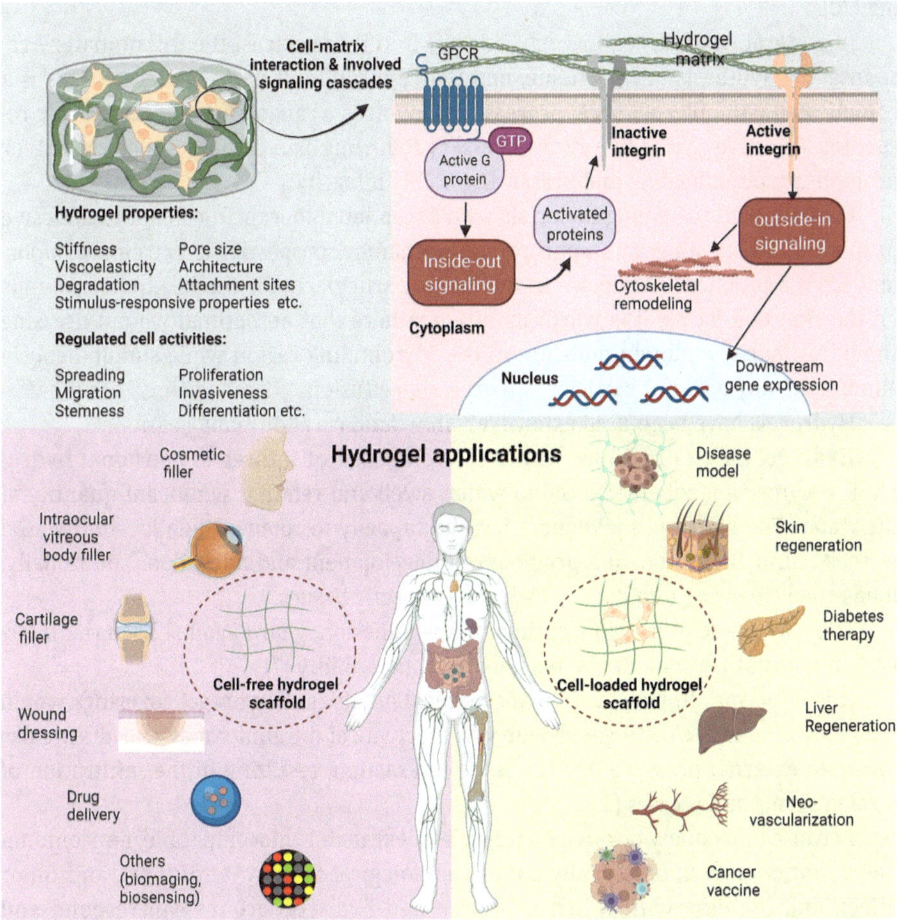

Figure 1.2.7: Diagrams showing how cells interact with hydrogel matrix, how hydrogel physicochemical features affect cell biology by initiating corresponding signaling cascades (e.g., inside-out and outside-in signaling), and demonstrating biomedical application of cell-free and cell-loaded hydrogels [69].

1.2.6.1 Hydrogels for wound dressings

The largest organ in the human body is the skin. It offers a crucial barrier against invading viruses and bacteria. Therefore, serious skin contamination can be harmful or even fatal. When the skin's integrity is damaged, like when wounds appear, it becomes a risk factor for the onset of many diseases.

As a result, it is significant to identify effective techniques to aid wound healing [70].

In general, wound healing may be split into three stages: the inflammatory response, the synthesis of new tissue, and the restoration of damaged tissue. This is a complicated pathophysiological process that involves a range of biological cascade responses. Effective care is therefore necessary following skin damage and is crucial for promoting wound healing and preserving bodily integrity.

Creating wound healing materials with exceptional therapeutic benefits, adhesive properties, self-healing, and appropriate mechanical properties is extremely important from a practical standpoint in medicine, particularly for the healing of wounds on the skin and joints. It is worth mentioning here that an optimal wound dressing should primarily maintain moisture in the surrounding region while simultaneously eliminating any harmful particles and excessive effusion.

Hydrogels have been used extensively in wound care for many years.

Hydrogel-based dressings, which are composed of a three-dimensional hydrophilic polymer net, when exposed to water, swell and retain a significant quantity of the water. The hydrogel's jelly mass has the capacity to retain a high level of wound wetness, absorb effusion, and promote cell development and migration. Additionally, it has good cleansing qualities by hydrating necrotic tissues.

They are made to promote optimal tissue adhesion, guard against bacterial infection, and meet the fundamental need for biocompatibility [71–74].

It is important to note that the mechanical qualities of hydrogel coverings would not be overlooked. Wound gauze with low mechanical durability may tear during use owing to external pressure or skin-tissue interaction, resulting in the infiltration of foreign germs and viruses [75].

Because of its many positive qualities, biodegradable, biocompatible, noncontaminated, antimicrobial, organically adhesive, biological activity, antifungal, antitumor effect, and the ease with which its amino and hydroxyl groups can respond and change in molecular level, chitosan is regarded as the perfect material for hydrogels.

Chitosan increases tissue development and enhances hemostasis, and this way it is advantageous for wound healing [76, 77].

Figure 1.2.8 shows the first three stages of wound healing and the methods by which chitosan-based hydrogels work [78].

By improving the activities of inflammatory cells including polymorphonuclear leukocytes (PMNs), macrophages, fibroblasts, and osteolasts, chitosan and its derivatives may hasten the healing of wounds. The parameters of molecular weight, deace-

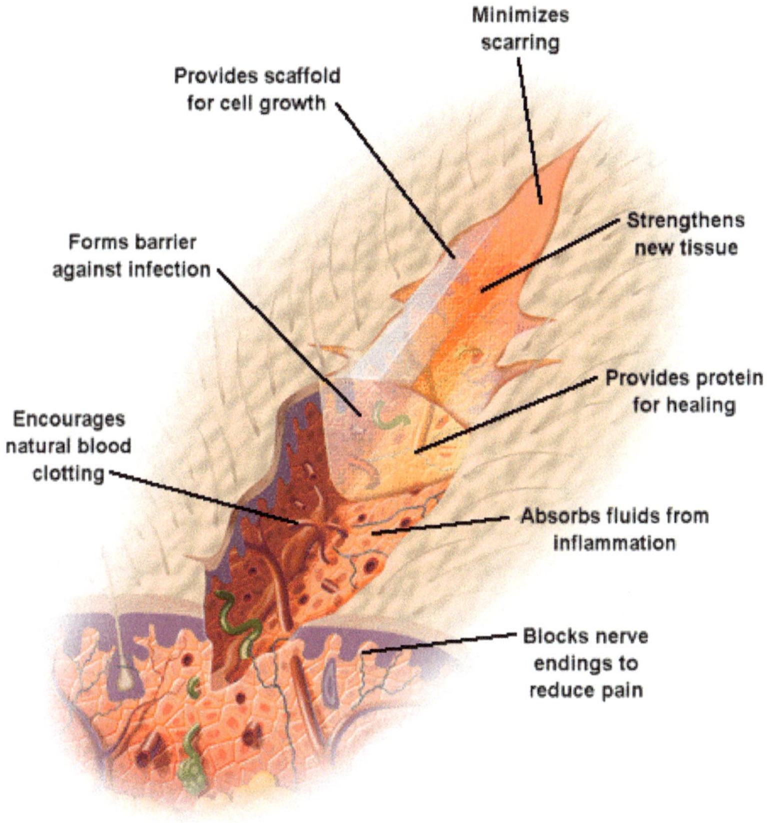

Figure 1.2.8: Illustration of the necessary characteristics of the substance used to treat wounds [78].

tylation degree, and chitosan state may have an impact on the wound-healing properties of chitosan [79].

Researchers have recently shown that hydrogels like carboxymethyl chitosan/gelatin, chitosan/PEG, chitosan/PEG/poly(vinyl pyrrolidone)-coated cotton fibers, chitosan/PVA, chitosan/PVA/poly(ethylene oxide) hydrogels, and chitosan-lactic acid are effective wound dressings that promote healing in view of the length of healing, the extent of tissue granulation, the inflammatory phase, and the synthesis of collagen fibers [80, 81].

For example, Lu and coworkers [82] create a double-network (DN) hydrogels using PAAm that is based on bioadhesive catechol-chitosan hydrogels. They have demonstrated that cell adhesion is promoted and the inflammatory response at the wound site (rat model) is regulated by the addition of catechol groups and (acetalized cyclodextrin-nanoparticle) ACD-NPs loaded with Resolvin E1 (RvE1). Therefore, it is possible that this type of hydrogel composite will be utilized to precisely treat and control the inflammatory milieu of chronic wounds (Figure 1.2.9).

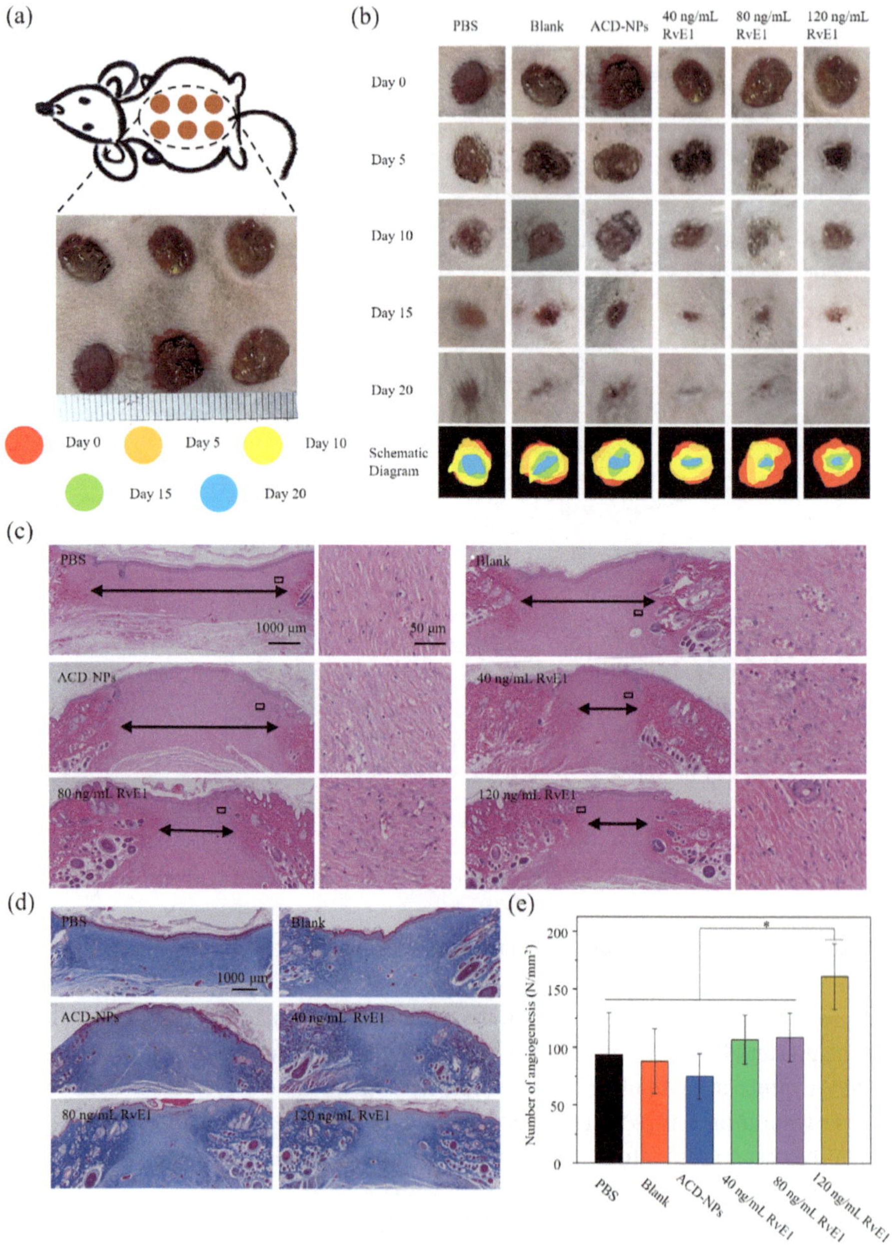

Figure 1.2.9: (a) Schematic illustration of the hydrogel bandage used to treat SD rat back cutaneous injuries; (b) images showing how DN hydrogel dressings helped SD rats' back wounds heal at 0, 5, 10, 15, and 20 days; (c) twenty-day wounds healed by HE staining; (d) Masson staining of wounds that have healed after 20 days; (e) at 20 days, the quantity of newly formed blood vessels in the SD rats' backs (Pic = 6, $n = 5$, $p < 0.05$) [82].

1.2.6.2 Hydrogels for targeted drug delivery applications

Systemic toxicity and recurrent dosage are two problems that are often addressed by conventional drug delivery systems. Hydrogels have been an important drug delivery carrier over the past few decades, helping to reduce the problems associated with traditional drug delivery. Because of their physicochemical characteristics, constructed interaction with living environments, and biocompatibility, they have been regarded as one of the most dependable classes of biomaterials. Hydrogels' unique ability to hold water inside their structural arrangement is another feature that makes them an intelligent medication delivery vehicle.

Exact control over the amount and interval of drug release can be achieved by adjusting the network structure and chemical characteristics of hydrogels, which can improve therapeutic efficacy and minimize side effects. Due to the discovery of responsive hydrogels, drug release behavior can now be adjusted in response to changes in interior conditions, for example, pH, temperature, or certain biomarkers.

For several drug delivery functions, such as oral, transdermal, ophthalmic, and administered delivery systems, hydrogels have been thoroughly studied.

1.2.6.2.1 Oral drug delivery

Hydrogels have been investigated as drug delivery vehicles orally, especially for the regulated release of medicinal substances in the gastrointestinal (GI) system. By shielding the encapsulated medications from the severe environment of the stomach and facilitating their constrained release in the intestine, hydrogels can increase the bioaccessibility and therapeutic ability of the medications. Several excipients, including a medication and a swellable polymer, must be mixed and compacted in precisely the proper amounts throughout their production.

Hydrogels taken orally ought to offer bioavailability contingent on the specifics of the medium such as changes in pH throughout the GI tract [83]. For example, when evaluated for sensitive oral toxicity in BALB/c mice, pH-sensitive hydrogels containing MPEG, caprolactone, and itaconic acid did not exhibit any harmful effects [84]. The metabolic influence that monomers exhibit on the organism is a major consideration when it comes to oral consumption. Healthy metabolism has been demonstrated to be impacted by the hydrogel breakdown of poly-glycolic and poly-lactic-*co*-glycolic acid [85], which has led to some restrictions on the use of hydrogel therapy including digestion-related enzymes may cause denaturation; low circulatory permeability across the epithelial membrane; potential treatment targets may include the finer and mediocre digestive tracts [86].

1.2.6.2.2 Transdermal drug delivery

For more than 30 years, a painless and noninvasive technique for delivering medications via dermal and transdermal administration routes has been growing since it lowers the possibility of drug overdoses that can be connected to oral or injectable administrations. Drugs can be administered via the skin using transdermal drug transfer systems, an alternative to oral, intravascular, subcutaneous, and transmucosal methods.

The main advantage of transdermal drug delivery is its ability to accomplish regulated drug transport through the skin, circumventing the liver and GI tract metabolism. As a result, compared to oral systems, comparatively less is needed for drug administration. Transdermal drugs can maintain stable blood levels for 1–7 days, encouraging better patient compliance [87, 88]. They have triumphed over other polymeric resources owing to their additional benefits such as higher swelling, flexibility of application, improved skin compliance, skin hydration, enhanced drug efficiency, suitability, pliability, protection, and ease of patch production. Hydrogels have been utilized to create covers, gels, or carriers for the transdermal administration of molecules, employing equally inactive and active delivery methods.

Patch dermal therapy is very significant and well-liked when longer administration is necessary to prevent adverse effects. Several synthetic biocompatible polymers, such as PVA, and its derivatives, such as chitosan and carboxymethyl guar gum, as well as natural polymers, have been employed to encapsulate medications as free-standing membranes. To investigate the medication delivery of diclofenac, Giri et al. [89] created nanosilica/acrylic acid-grafted guar gum membranes for transdermal therapy.

1.2.6.2.3 Ocular drug delivery

Because of the eye's distinct anatomy and physiology, developing a drug delivery system to aim at a specific tissue has become a significant problem for today's pharmacologists and formulation scientists. Nowadays, eye drops or ointments are the most commonly used dosage practiced in clinical practice for treating ocular diseases. These formulations present a number of difficulties including reduced bioavailability outlines, quick clearance from the base location and decreased therapeutic efficacy. Figure 1.2.10 shows the obstacles to the transport of drugs into the eyes [90]. In general, ocular medication delivery barriers are classified into anterior and posterior segments according to the eye's anatomical framework.

Ocular drug delivery has been studied using hydrogels as transporters, allowing for continuous release of therapeutic substances and enhanced bioavailability within the eye. Hydrogel-based inserts and contact lenses for the eye can be employed to administer medications to both the front and back parts of the eye, addressing the drawbacks of traditional eye drops, such as prompt removal and limited drug absorption.

Sedlacek [91] was the first to employ soft contact lenses for drug administration in 1965, demonstrating that the application of mesocaine-filled contact lenses produced corneal analgesia more successfully than the application of eye drops.

Numerous medications have been studied for ocular methods through silicone-based and PHEMA-based hydrogel-based contact lenses, which are the two most popular forms of soft contact lenses [92]. The water content of pHEMA-based contacts lenses, which are created by polymerizing HEMA with EGDMA as a cross-linker, is 38%. To improve oxygen absorptivity and enhance the water content of HEMA-based lenses, different hydrophilic monomers can be combined [93].

Nevertheless, ocular medication delivery presents numerous obstacles. These are the outcomes of the several tissue barriers and defensive systems found in the eye including the nasolacrimal drainage arrangement, cornea, conjunctiva, and blinking reflex. This results in low bioavailability accounts, quick clearance from the processing site, and inefficient drug delivery for many traditional ophthalmic preparations including eye drops. Hydrogels have been researched as a delivery method that can get around some of these problems. Hydrogels can be designed to respond to external stimuli by changing their consistency from a liquid to a gel; these biotechnology stages are referred to as “smart” or stimuli-responsive biogels.

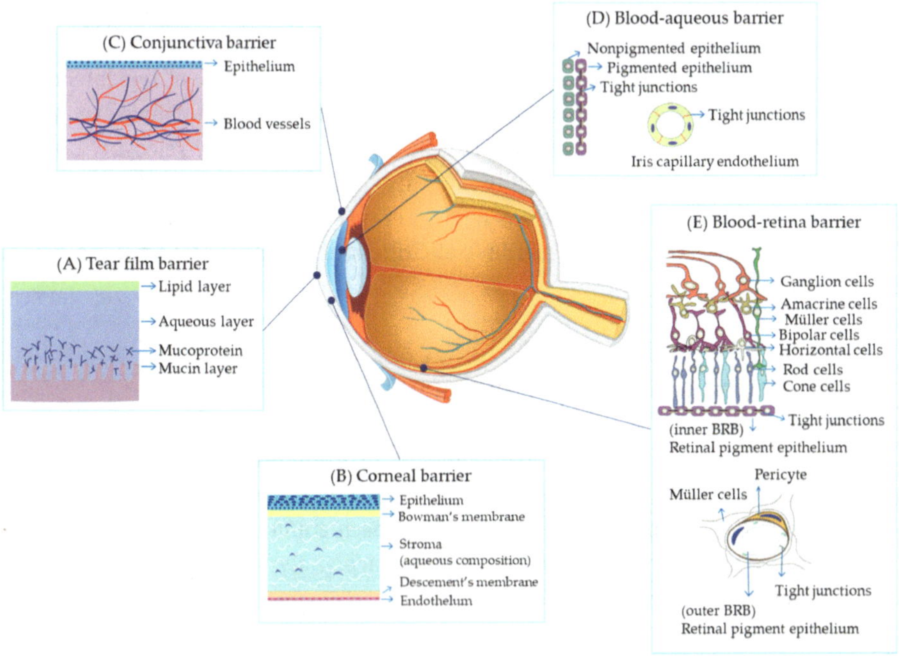

Figure 1.2.10: Drug transport barriers found in the structures of the eyes [90].

1.2.6.2.4 Injectable drug delivery

3D hydrophilic polymeric networks having a high affinity for bodily fluids are known as injectable hydrogels, and they can be directly injected into the body using a syringe or through a catheter [94]. In the field of biomedical engineering, this special type of hydrogel has been recommended as a platform for drug delivery and tissue engineering [95–97].

Typically, in situ chemical polymerization or a rapid sol-gel phase transformation is used to create injectable hydrogels. Among them, "sol-gel" transitioning hydrogels have excellent potential for the creation of injectable drug delivery systems.

They have the benefits of simple cell integration and excellent alignment with the uneven shapes of cells and tissue. Liquid injectable hydrogels that are between 23 and 25 °C can be inserted into the body. They exhibit in situ gelation at temperatures higher than 32 °C after injection.

In recent years, breast cancer has been one of the most frequent types of cancer among women after skin with serious health risks. Chemotherapy, also referred to as systemic therapy, is a treatment for breast cancer that targets cancer cells with strong medications.

Injectable hydrogel-based drug delivery systems have attracted a lot of interest lately because they provide promising options for regulated, targeted, and localized drug distribution to the tumor site. These technologies hold significant promise for enhancing medication efficacy and mitigating the adverse effects associated with prolonged chemotherapy exposure.

Unfortunately, chemotherapeutics is also linked to a number of typical side effects, including neurotoxicity and bone marrow (BM) suppression, which frequently reduce the efficacy of chemotherapy medications. Remarkably, by releasing medication locally, hydrogels can successfully minimize adverse effects [98]. Injectable hydrogel-based administration schemes have been used to locally codeliver immunotherapy and chemotherapy medications, eliciting long-lasting, potent antimalignant immune reactions.

1.2.6.3 Hydrogels for tissue engineering scaffolds

A large number of fatalities or impairments occur each year globally as a result of diseases or traumas that cause tissues and organs to be lost or damaged [99]. Because of this, the interdisciplinary field of tissue engineering and regenerative medicine is growing quickly, encompassing cell biology, surface analysis, biomaterials research, and interactions between materials and cells.

Their goals include using cells [100, 101], stimulating factors [102], genes [103], and scaffolds [104–106] to activate and improve the healing system.

Several contemporary approaches to tissue engineering rely on the use of a material scaffold. In order to arrange cells into a 3D arrangement and to present cues that guide the expansion and production of a designated tissue, these scaffolds function as an artificial ECM.

Hydrogels are a popular scaffold material because they may be supplied, often handled using minimally invasive methods, and physically resemble the ECM of numerous tissues. As a result, hydrogels have been applied widely including scaffold materials for various purposes, medicine administration, and tissue replacement engineering.

A scaffold is a substance that can replicate the ECM found in living organisms to sustain cellular architecture and offer an appropriate environment for the cells to grow in. Scaffolds come in two varieties: natural and synthetic.

1.2.6.3.1 Naturally derived polymer hydrogels

1.2.6.3.1.1 Collagen

The primary component of ECM is collagen, which makes it a prime choice to promote cell proliferation in tissue regeneration (type I collagen). Collagen gels' biofunctional properties have made them useful for tissue engineering a range of organs including the liver, skin, and blood vessels. However, according to Lau et al. [107], collagen gels are frequently soft, similar to adipose tissue (i.e., modulus of elasticity <5 kPa). Consequently, either collagen's density needs to be improved or some amount of chemical modification is essential in order to upgrade its mechanical characteristics and delay the decomposition of hydrogel.

In general, the processes of mature tissue reorganization, and wound recovery depend on the exchanges between collagen and cells, which also sustain cell phenotypic and induce segregation in collagen [108]. Because of its trivial toxicity and low immunogenicity, collagen also possesses another crucial biological characteristic: biocompatibility. Furthermore, collagen has a low antigenicity. The existence of aromatic amino acids, particularly tyrosine, may be the cause of low antigenicity.

Collagen is abundant in blood arteries, but it is incredibly high in capillaries, which only have one level of structure. Consequently, there are several benefits to creating two bioprints from collagen and employing it as bioink for tissue regeneration, followed by vascular network rebuilding. Using collagen as a bioink, Koch et al. [109] created scaffolds with functionalities particular to the skin by encasing fibroblasts and keratinocytes. Collagen was used by Lee et al. [110] as a bioink to encapsulate stem cells.

Tissue engineering has greatly advanced collagen hydrogels in the last few years, laying the foundation for further advancements. These hydrogels still don't have the original ECM's necessary characteristics, though. Therefore, collagen hydrogels must be evaluated and improved in terms of handling, shelf life, and scalability before surgeons will choose to use them as a material of solution.

1.2.6.3.1.2 Gelatin

Collagen, a naturally occurring protein present in the ECM, can be denatured to produce gelatin [111]. As a result, the substance is both bio-friendly and biodegradable for in vitro cell development. Typically, the majority of commercial gelatins are derived from the skin, scale, bones, ligaments, and tendon of cattle or pigs [112]. Gelatin-based hydrogels, which include gelatin alone and gelatin combined with other natural and synthetic polymers, have special qualities that make them useful in medical applications. These characteristics include excellent biocompatibilities, quick biodegradability, and nonimmunogenicity. Gelatin/alginate, gelatin/chitosan, gelatin/hyaluronan, gelatin/fibrinogen, and gelatin/sericin [113] are common examples.

This material is easy to cross-link and offers a wide spectrum of functional groups, gelatin has gained popularity as a hydrogel production alternative for long-term biomedical treatments.

It has been approved by the Food and Drug Administration (FDA) [114] owing to its biocompatible, biodegradable, and low-antigenic nature. The modification of gelatin with photo-cross-linkable methacrylamide groups (GelMA) preserves the distinctive characteristics of gelatin while also enabling the material to permanently solidify from a liquid state through the chemical reaction of the meth acrylamide groups [115].

Hydrogels made from gelatin, a product of collagen, serve as an additional class of scaffolds for the regeneration and repair of several tissues and organs. It i significanttonotethat,incomparisontosyntheticpolymers,collagenandgelatin,whicharecomponents of natural ECM, more closely resemble the native tissue micro [116, 117] environment.

1.2.6.3.1.3 Fibrin

The monomer of fibrin, known as fibrinogen, is made up of two sets of three polypeptide chains called Aα, Bβ, and γ that are connected by six disulfide bridges [118]. Fibrin is produced when fibrinopeptides A and B are broken down by thrombin to release fibrin from their respective Aα and Bβ chains. Due of its high propensity to self-associate and create insoluble fibrin, this produces the fibrin monomer [119]. Fibrin is one of the hydrogels that is most frequently utilized in TE and has the following noteworthy benefits: Fibrillary and porous pattern, low cost, good cell-biomaterial interactions, a regulated rate of disintegration that is compatible with tissue regeneration, nonhazardous degradation products, and outstanding biocompatibility and ease of handling. In contrast, the rapid disintegration rate of collagen-based hydrogel might cause instability in its mechanical properties prior to the completion of tissue restoration or wound healing [120].

Fibrin-based biomaterials are currently being actively developed in several preclinical and clinical methods for the regeneration of injured tissues and wound healing. Specifically, fibrin can be tailored in terms of dimensions (2D or 3D scaffolds) or its phasic properties (injectable or implantable matrices) to facilitate biological interactions in the best possible tissue regeneration for wound healing and functional repairs.

1.2.6.3.1.4 Hyaluronic acid

An extremely important polysaccharide that is cross-linked to hydrogels by enzymes is HA. Largely present in cartilage, HA is a naturally occurring biocompatible and biodegradable linear polysaccharide made up of repeating disaccharide units made of glucuronic acid and *N*-acetylglucosamine [121]. The ECM in many tissues contains HA, which is essential, hence HA hydrogels can mimic the composition and activities of the ECM. Due to its ability to create the perfect environment for cellular growth and maturation, such mimicry is essential for tissue engineering uses. In addition, HA is the most desirable hydrogel precursor for bioinks because of its exceptional chemical and biological qualities including biocompatibility, hydrophilicity, non-immunogenicity, and eco-friendly nature. HA or hyaluronan, is primarily located in cartilage and serves as the primary component of the cartilage ECM, with a "-ve" surface charge [122]. It is anticipated that HA is essential for cartilage tissue regeneration and development because of its appropriate interface with stem cells and cartilage via surface receptors [123].

1.2.6.3.1.5 Chitosan

Chitosan (CS) is a type of natural polysaccharide-based heteropolymer consisting of residues of glucosamine and *N*-acetyl glucosamine. Usually, the shells of shrimp or crabs are used to extract this organically occurring linear cationic heteropolymer. This natural hydrogel offers excellent biocompatibility, strong biodegradability, and a low immunological reaction to human tissues and cells [124].

At pH values below 6.5 (pK_a value ~6.3), the protonated amino groups of D-glucosamine residues allow CS to form a non-Newtonian, shear-thinning fluid in generally diluted acidic solutions [125]. It also helps CS form metal ions added complexes, polymers, lipids, proteins, DNA, and other materials.

Cell regeneration, bone, cartilage, intervertebral disk, blood vessel, corneal, skin, periodontal, and tissue fixation are among the tissue engineering applications of chitosan that are significant [126]. Although chitosan hydrogels have a lot going for them, one major drawback is still their lack of superior mechanical property. Scientists have investigated several strategies to get over this restriction, such as changing the scaffold's fabrication and composition to increase its strength [127]. Furthermore, they exhibit limited cell adhesion, a mismatching disintegration rate, and no protherapeutic benefits in vascular regeneration.

1.2.6.3.1.6 Alginate

Polysaccharides are a significant part of natural ECM and can be found in many green sources in a variety of forms. They also serve a number of physiological purposes. Consequently, they have found widespread application in numerous fields such as drug delivery and tissue engineering. Alginate, chitosan, glucan, carrageenan, starch HA, and cellulose are the most often used among these [128].

Blocks comprising successive or alternating β-D-mannuronic acid (M) and α-L-guluronic (G) units joined by (1 → 4) links produce alginate (ALG), an anionic linear polysaccharide. Different sources, including bacteria and marine brown algae, have different M/G contents, molecular weights, and block lengths, which affect the physicochemical characteristics of alginate. It is a naturally occurring biopolymer that is better suited for use in diverse biomedical fields due to its many exceptional qualities, including low toxicity, biodegradability, biocompatibility, relatively little extraction and handling costs, minor gelation, and superior hydrogel-forming capacity owing to the incorporation of divalent cations like Ca^{2+}.

Many studies in the theme of regenerative medicine have examined sodium alginate hydrogels, which show great promise for use in cell and tissue engineering. The ability of sodium alginate films to treat wounds has been well-recognized [129], with results indicating quick, effective healing and instant wound closure.

It has been demonstrated that using calcium ions to cross-link sodium alginate hydrogels greatly increases osteoblast proliferation and differentiation. In addition to this, alginate based hydrogels are widely used in other areas of tissue engineering for example, cartilage repair [130], bone restructuring and repair [131], and cardiac tissue regeneration [132].

1.2.6.3.2 Synthetic polymer hydrogels

Tissue engineering has become more interested in synthetic hydrogels because of their ability to overcome specific application obstacles, mechanical support for cells, and adjustable properties [133]. Synthetic hydrogels can be enzymatically cross-linked to enable innocuous encapsulation of cells and cytokines, minimally invasive surgery, and defect conformation.

Hydrogel scaffolds that can carry drugs or bioactive substances to enhance cell growth, differentiation, and angiogenesis are frequently made using synthetic polymers including poly(glycolic acid) (PGA), poly(lactic-*co*-glycolic acid) (PLGA), and poly (lactic acid) (PLA). These cellular processes can be regulated by releasing bioactive substances through synthetic polymer matrices.

A widely used synthetic hydrogel that can be mentioned in this aspect is PHEMA hydrogels. Since their introduction in the 1960s for use in contact lenses, PHEMA hydrogels have been extensively investigated in tissue engineering due to their many of the same benefits as previously mentioned [134] PVA and PEG.

Typically, there are several possible uses for synthetic polymers in tissue engineering.

When an alien substance is introduced into a living organism, the material's surface becomes colonized by immune cells. Immune cell adhesion to hydrogels is poor, nevertheless because the majority of synthetic hydrogels lack molecules that encourage cell adherence. Previous research indicates [135] that functional groups on the surface, such as $(CH_3)_2N$–, –OH, –CONH–, $–SO_3H$, and –COOH, are mostly responsible for cell adhesion. Here, the gels with carboxylic acid or acid salt groups exhibit the lowest cell adhesion qualities, whereas the hydrogels with $(CH_3)_2N$– functionality exhibit the strongest cell adhesion.

Heterogeneity and network defects are intrinsic shortcomings of traditional synthetic hydrogels that negatively impact their mechanical characteristics, diffusion rates, and biological activity. To overcome these difficulties, hybrid hydrogels have developed with the goal of improving their potency, therapeutic efficacy, and drug release efficiency.

1.2.6.3.3 3D bioprinting of polymer hydrogel

With the use of computers three-dimensional (3D), bioprinting is an additive manufacturing technique that rapidly prints biofunctional constituents and the components that support them layer by layer onto a substrate or a tissue culture bowl to produce multifaceted active tissues/ components (such as drugs, growth factors, or viable cells) and organs with the required 3D cellular structure and roles [136]. The capacity of 3D bioprinting techniques to create scaffolds containing a variety of biomaterials and cells has unquestionably been shown. To heal lesions of various sizes and thicknesses, the scaffolds employed in regenerative therapies are supposed to fuse with the original tissue already present.

Several developments have been made, including vascularized cell-filled constructions [137], cartilaginous tissues with chondrocytes, PEG/HA, and [138], and scaffolds for bone regeneration that are coated with collagen and composed of calcium phosphate [139]. The assembly and formation of scaffolds as well as the creation and alteration of biomaterials are the main subjects of study.

Representative examples of various artificial and biological hydrogels utilized in 3D bioprinting are shown in Figure 1.2.11 [140].

1.2.6.4 Biopolymer hydrogels for bone regeneration

Bone, which makes up the majority of the body and provides support throughout our lifetimes, is a compact, porous, mineralized connective tissue [141]. A baby has roughly 270 bones [141], which typically fuse together to become 206 in adults [142].

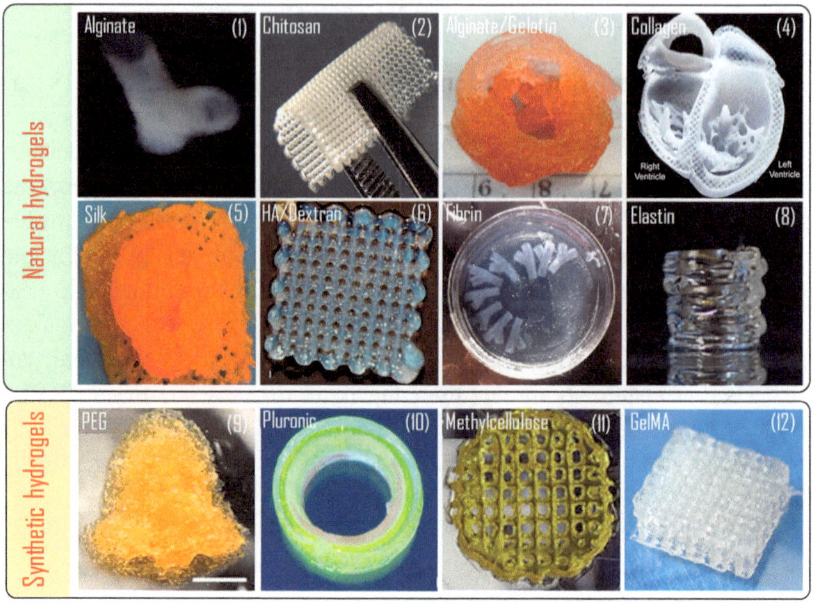

Figure 1.2.11: Examples of various natural and artificial hydrogels used in 3D bioprinting [140].

One of the greatest examples of a natural composite is bone, which is mostly made up of water (10%), solid organic matrix (30%), and inorganic components (60%) [143]. The matrix structure resembling honeycomb is what gives bone its rigidity.

Biomaterials that are frequently utilized as 3D scaffolds can be categorized based on their chemical composition: metals, ceramics and glass-ceramics, natural and artificial polymers, and composite aggregates [144]. Originating biomaterials are typically chosen from physiologically inert materials that are least harmful to humans like ceramics, synthetic polymers, cobalt chromium alloys, zirconia, Ti- or Ti-alloys, stainless steel, and zirconia and other metals [145]. Later, both natural and synthetic biodegradable polymers are frequently employed in biomaterials, such as collagen, Ca-phosphate, and silica, because many natural biomaterials have significant tissue constituents and have strong biocompatibility and biodegradability [146]. Due to their unique advantages – such as their low cost, processability, and biocompatibility – polymeric materials, which are frequently employed in the biomedical industry, have garnered interest thus far [147].

On the other hand, there has been much research on the use of bulk hydrogels in bone diseases as scaffolds for bone tissue creation, drug delivery vehicles, and bone regeneration materials. It is important to highlight one special benefit of the hydrogels: their particular porosity architectures, which resemble the ECM of bone and cartilage, can act as transporters for different growth agents [148, 149]. However, bulk hydrogels have certain drawbacks such as their size, low specific surface area, absence of micropores, and sluggish rate of breakdown. Low cell infiltration and vitality are the results of those drawbacks [150].

1.2.6.4.1 Hydrogels in bone diseases – bone fracture

One of the leading musculoskeletal system injuries is a bone fracture, which affects over 18.3 million individuals annually and costs around $12,000 for each problem in the United States [151].

A strong pressure or shock is typically the cause of the bone fracture. Bone fracture treatment (including avulsion, grind, and squash fractures) depends on bone growth and production.

In the field of fracture therapy, hydrogels with bioactive characteristics have gained significant attention recently [152], mostly due to their ability to serve as carriers for bioactive small molecules that stimulate osteogenesis and angiogenesis, ultimately contributing to the repair and remodeling phase [153]. In addition, hydrogels can absorb large quantities of water or bodily fluids and mimic the ECM of primary bone; they can also facilitate cell replacement and propagation. This property of hydrogels allows for several advantages [154, 155].

A molecular transporter, injectable, and sticky HA@SDF-1α/M2D-Exos hydrogel was devised and created by Chen et al. [156]. It was demonstrated to provide a local antibacterial milieu that promotes fracture recovery. In their study, hydrogel with varying amounts of HA was produced and analyzed at 2%, 4%, and 6%, to investigate the potential differences in antibacterial action and biocompatibility. All in all, the design expedited fracture repair while lowering infection because it was in harmony with the body's natural healing mechanism.

In a different study, a gelatin/CS composite hydrogel incorporating TiO_2 nanoparticles was created by Guo et al. [157] to regulate the speed at which bone fractures heal. Interestingly, the hydrogel was joined in a tubular shape with the addition of TiO_2. In the meantime, the TiO_2 nanoparticles dispersed throughout the matrix, enhancing its heating resistance and offering support as well as numerous active locations for the nucleation of polymer units. The gelatin served as a stabilizing agent. In a similar vein, Han and colleagues [158] produced a gelatin hydrogel containing silver nanoparticles (AgNPs) to enhance bone replacement and fracture healing. In addition to integrating the antibacterial properties of the AgNPs, the results showed that the combined AgNPs/Gel hydrogels were not toxic to osteoblasts and showed increased cell viability.

Scientists discovered [159] that osteoblasts were not harmed by synthetic hydrogels. Additionally, they significantly increased the survival rate and spread of osteoblasts, demonstrating their capacity to control the repair of fractures.

1.2.6.4.2 Cartilage damage

There are several disorders that can lead to injury to cartilage and subchondral bone including trauma, arthritis, and sports-related injuries [160, 161]. Researchers have found [162] that cartilage loss is present in 60% of individuals who undergo knee arthroscopy, and approximately 15% of adults over 60 have some clinical signs of this condition. Surgery can be used to partially restore cartilage function through methods like stem cell stimulation, autologous replacement, and allogeneic transfer. However, these traditional methods have significant drawbacks such as poor host tissue integration, cartilage degradation, and the requirement for additional surgeries [163]. Consequently, new approaches to treating cartilage restoration are being studied such as tissue engineering and synthetic cartilage materials.

Through the application of logical design and control techniques, hydrogels can be made to have desirable mechanical characteristics, biocompatibility, and joined lubrication, in addition to the capacity to promote cell adhesion, proliferation, and migration while offering a physiological scaffold. These properties allow hydrogels to imitate cartilage's ECM or even to create cartilage in a way that is biomimetic [164].

A new hydrogel called GelMA was investigated by research team [165] to improve cartilage regeneration. It was modified with a TGF-β1-affinity peptide (HSNGLPL). Cartilage regeneration is promoted by the hydrogel's peptide because of its strong affinity for TGF-β1.

This work opens the door to a more pragmatic strategy by tackling the drawbacks of exogenous TGF-β1 delivery including the requirement for supraphysiological concentrations and regulatory obstacles. This innovative method offers a bioactive scaffold with superior mechanical qualities and biocompatibility, showing promise for clinical cartilage restoration treatments.

A biocompatible hydrogel scaffold loaded with icariin and based on PVA has been reported [166] for the healing of cartilage defects (Figure 1.2.12). PVA, tannic acid (TA), gelatin (gel), HA, and icariin were the ingredients of the PTGH + icariin hydrogel, which was made using a straightforward physical cross-linking technique. The outcomes demonstrated the porosity microstructures and robust mechanical qualities of the PTGH + icariin hydrogel. In a rat model of cartilage defect, the PTGH + icariin hydrogel showed promise in promoting cartilage regeneration. Specifically, the insertion of PTGH + icariin hydrogel may make it easier for rats to regain their motor function. This work will offer a possible method for treating cartilage abnormalities.

1.2.6.4.3 Osteosarcoma and rheumatoid arthritis (RA)

The bone surrounding a joint is frequently affected by both rheumatoid arthritis (RA) and osteosarcoma (OS), which are complicated disorders that mostly affect the knee region in OS [167] and the wrist joint in RA [2]. Although OS primarily affects children

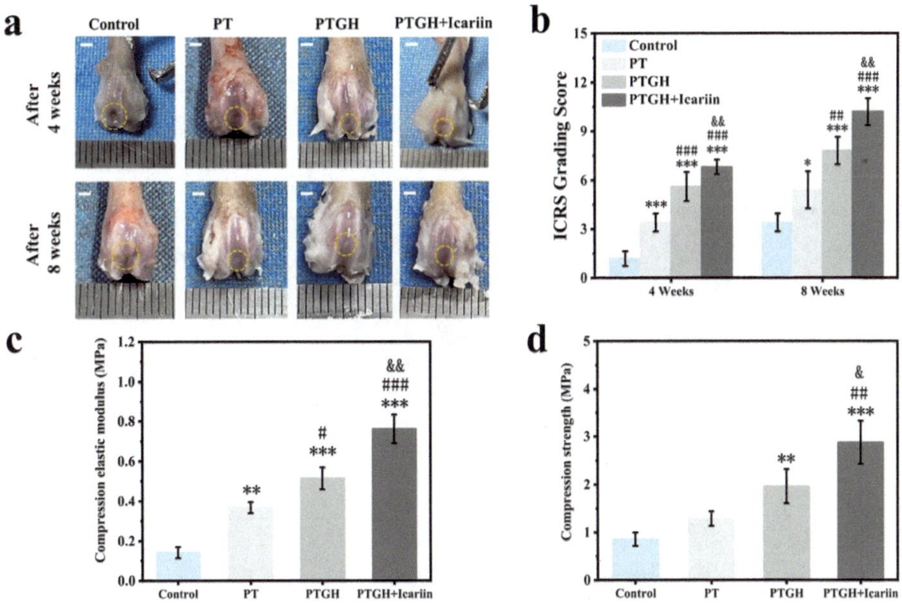

Figure 1.2.12: (a) The integrity of the knee cartilage was grossly observed (2 cm is the scan bar); (b) the cartilage-deficient regions' ICRS grading score. Eight weeks following surgery, the cartilage-deficient regions' (c) compressive strength; and (d) compressive elastic modulus were measured. $^{*}p < 0.05$, $^{**}p < 0.05$, $^{***}p < 0.001$ versus the control group; $^{\#}p < 0.05$, $^{\#\#}p < 0.01$, $^{\#\#\#}p < 0.001$ versus the PT group; $^{\&}p < 0.05$, $^{\&\&}p < 0.01$ versus the PTGH group [166].

and adolescents, RA primarily affects adults. Strong chemotherapy and surgery are often necessary for treating OS, yet these treatments have a great deal of adverse effects and can be harmful to healthy tissues. immunosuppressive drugs, which are disease-modifying anti-rheumatic medications, can have harmful side effects when used to treat RA, an autoimmune condition [168].

Hydrogels are likely to play a significant role in the domains of bone tissue engineering and cartilage regeneration in the future. They enhance the way cartilage defects heal by delivering drugs and changing physicochemical behaviors. Because of their permeable structure, diverse biocompatibility, biodegradability, and the capacity to load growth variables, hydrogels have been shown in numerous studies to be effective in the treatment of cancers. Hydrogels can facilitate the administration of therapeutic medicines or operating cells into the respected OS region [169]. For the treatment of tumor-related bone abnormalities, a hydrogel known as nanohydroxyapatite hybrid reduced graphene oxide (nHA-rGO) was created [101–105]. Using photothermal treatment, the nHA-rGO hydrogel virtually eliminated all MG-63 OS cells. Moreover, in a rat cranial wound prototype, this hydrogel stimulated collagen accumulation and osteoblast mineralization, which led to bone restoration [170].

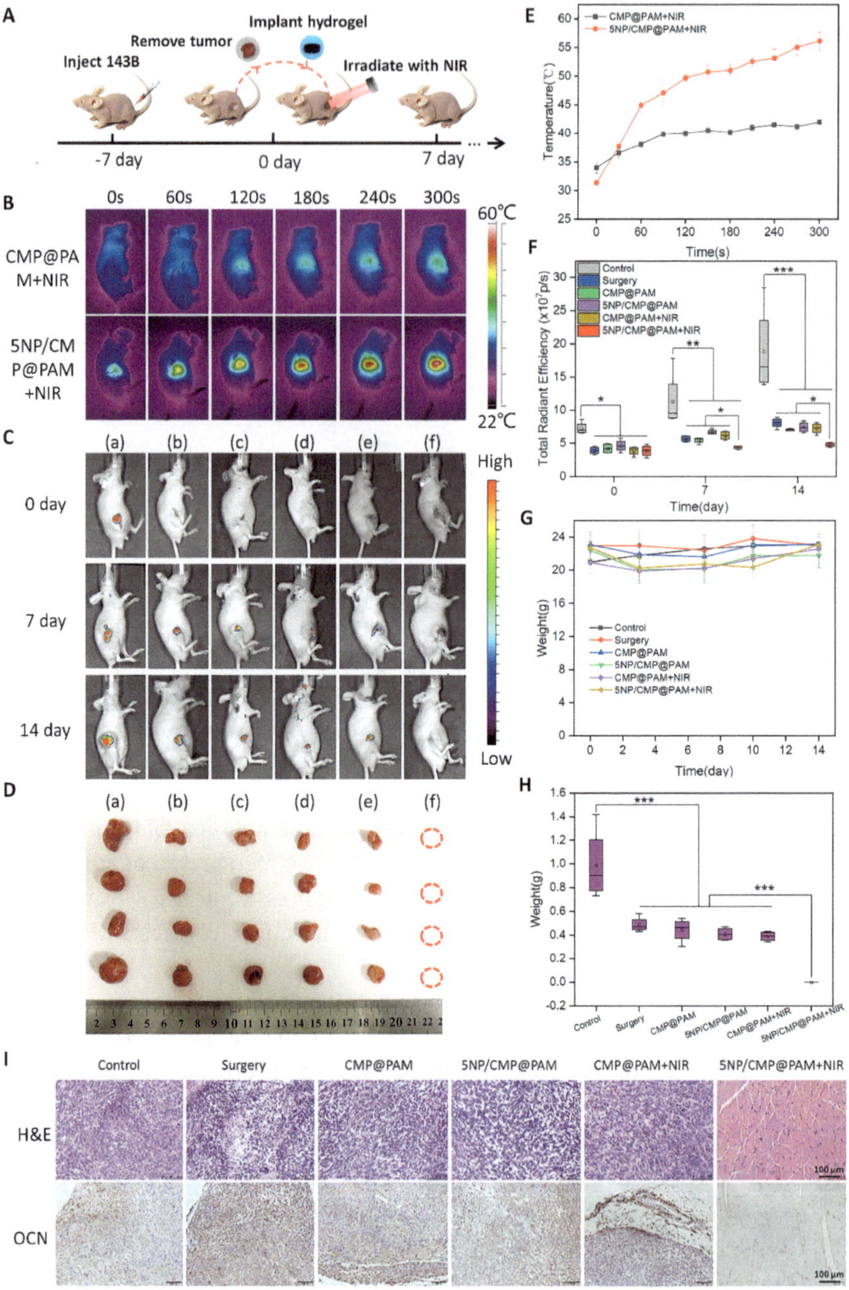

Figure 1.2.13: The effectiveness of 5NP/CMP@PAM as a photothermal therapy in vivo: (A) schematic diagram showing how subcutaneous osteosarcoma in Balb/c nude mice is prevented postoperatively. 143B tumor-bearing mice were implanted with CMP@PAM and 5NP/CMP@PAM hydrogels at different times after being exposed to NIR (808 nm, 1 W/cm^2, 5 min). The heating curve and NIR thermal images

In a different study, Li et al. [171] created a novel multifunctional bioactive hydrogel based on MgO nanoparticles (M-NPs) with a 2-aminoethyl methacrylate (2-AM)-grafted polydopamine (PDAM) shell to form a bioactive nanoparticle known as MDA-NPs. A sequence of MDA-NPs nanocomposite with 0, 0.5, 2.5, 5, and 10 mg/mL were then incorporated and coded as CMP@PAM, 0.5NP/CMP@PAM, 2.5NP/CMP@PAM, 5NP/CMP@PAM, and 10NP/CMP@PAM, respectively. Excellent dual activities of mending bone defects and reducing OS recurrence were exhibited by the 5NP/CMP@PAM hydrogel. Figure 1.2.13 shows the in vivo postoperative treatment of OS for the 5NP/CMP@PAM specimens.

For the purpose of treating RA, hydrogels must possess certain qualities. First and foremost, the fundamental and most significant characteristic of all hydrogel types is high biocompatibility. Second, hydrogels need to possess good mechanical qualities such as toughness, swelling ratio, Young's modulus, rheological qualities, and fatigue properties [172]. Thirdly, tribology characteristics and wear resistance are also significant attributes of hydrogels used in rheumatoid arthritis therapy. The capacity to load and discharge medications or cells, as well as biodegradability, rank fourth. Last but not least, the transdermal hydrogels' capacity to penetrate the skin barrier is critical.

The most widely used methods for treating RA center on injectable hydrogels. Hydrogels loaded with diclofenac sodium were developed intra-articularly by Qi et al. [173] in an attempt to evaluate the hydrogels' potential as drug delivery vehicles for enhancing the anti-inflammatory impact. The findings demonstrated that, after 3 weeks, the anti-inflammatory efficacy of hydrogels outperformed that of chitosan hydrogels and diclofenac solution alone.

Zhao et al. [174] have combined 3DPMS with infliximab-based hydrogels to create an improved scaffold for managing RA. It was recommended to have stiff 3DPMS for extensive RA-caused bone deficiencies. This hybrid system combining inorganic and organic components offers a viable approach for stem cell-based treatments and could change the direction of research on RA.

All things considered, hydrogels offer more precise, secure, and efficient drug delivery mechanisms, which could completely transform the way RA is treated.

Figure 1.2.13 (continued)
are shown in (E) and (B) accordingly. (A) Control group, (B) surgery group, (C) CMP@PAM group, (D) 5NP/CMP@PAM group, (E) CMP@PAM + NIR group, and (F) 5NP/CMP@PAM + NIR group at days 0, 7, and 14 ($n = 4$); (C) whole-body fluorescence imaging for the tumor; (G) the weight of the Balb/c nude mice ($n = 4$) following the therapeutic process was recorded; (I) tumor paraffin section pictures stained with H&E and OCN. $^{*}p < 0.05$, $^{**}p < 0.01$, and $^{***}p < 0.001$ [171].

1.2.6.5 Hydrogels for advanced stem cell therapies

The use of stem cells in therapy shows promise in promoting the regeneration of diseased and damaged tissues, and it is considered as one of the biggest developments in contemporary medicine [175, 176]. Being able to differentiate into specialized cells such as bone, muscle, nerve, and blood cells that subsequently take part in distinct body tasks, stem cells possess the ability to self-renew.

However, because of their high operational costs and limited stem cell survival, stem cell therapies still confront obstacles.

Hydrogels are networks of porous polymers that are heavily saturated with water. Hydrogels with the right polymer composition and gelation techniques can be biocompatible and offer the best mechanical and biochemical characteristics for 3D stem cell cultivation.

As designed habitats, hydrogels can also supply biophysical and biochemical cues to encourage stem cell proliferation, preserve quiescence, and guide differentiation in vitro [177].

Despite significant advancements in the creation of hydrogel-based cell culture methods for the growth of stem cells, it is still difficult to completely replicate the biochemical and mechanical cues of the natural stem cell milieu. Here, we present a list of recently produced hydrogels for pluripotent stem cells (PSCs), mesenchymal stem/stromal cells (MSCs), and hematopoietic stem cells (HSCs) in the following sections. All cell types share certain common characteristics, nevertheless.

1.2.6.5.1 Mesenchymal stem/stromal cells (MSCs)

The human body contains a range of tissue sources, including BM, dental pulp, umbilical cord, and adipose tissue, that are home to a diverse population of multipotent progenitor cells known as MSCs. MSCs' multipotent properties first attracted a lot of attention in the field of biomedical engineering because they can differentiate into families that are osteogenic, chondrogenic, adipogenic, nervous, skeletal, and cardiac, which may allow them to replace diseased or injured tissue within the body with neo-tissue or an entire organ [178, 179]. Hydrogel scaffolds encapsulating MSCs exhibit improved cell viability and committed segregation as well as sustained fundamental and operational reliability. These findings present encouraging opportunities for tissue engineering and the resulting field of regenerative medicine.

For various reasons, it is crucial to take into account the precise technique for adding cells and bioactive/small molecules to the hydrogel.

Two methods of combining hydrogels, MSCs, and bioactive/small molecules are depicted in Figure 1.2.14: (i) by the compound's encapsulation in gel or nano/microparticles and (ii) through the compound's covalent conjugation to the hydrogel [180].

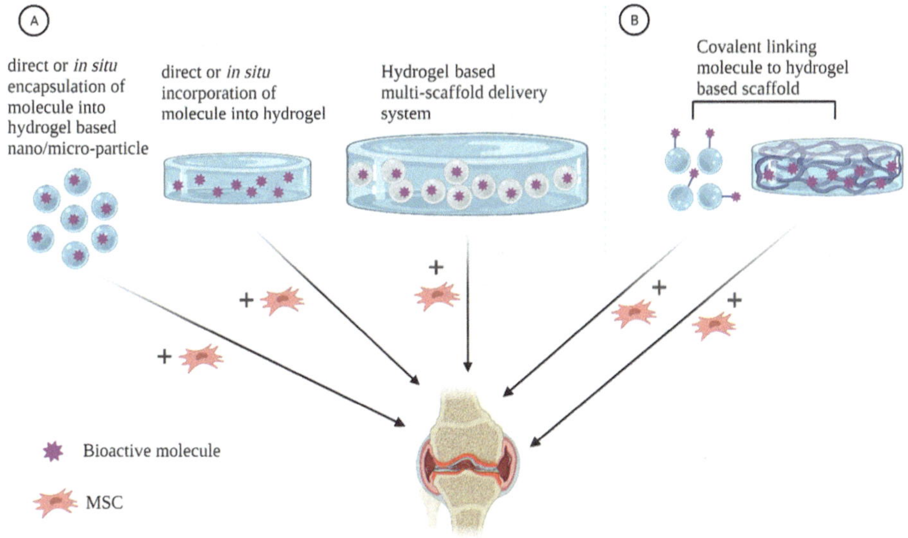

Figure 1.2.14: Techniques for adding bioactive or tiny compounds to a hydrogel system that contains MSCs. Hydrogels can be filled with bioactive or tiny compounds through direct encapsulation or integration in one or more polymers (A).Additionally, the molecules may covalently bind to various hydrogel moieties (B) [180]. With permission from P.G. Fernandez, Copyright© 2022, Elsevier.

The flexibility of the tissue that contains stem cell residents affects most stem cells. Because tissues have varying degrees of elasticity, stem cells preferentially develop into resident tissue cells whose elasticity is similar to that of the original tissue.

The ability of MSCs to differentiate into many pluripotent forms, such as osteogenic or adipogenic difference, which may be directly influenced by substrate elasticity, makes them a promising source of adult stem cells for stem-cell-based therapies. According to Engler et al., an elastically adjustable gel system demonstrated MSC differentiation. Their findings indicate that growing MSC on soft gels (less than 1 kPa) that replicate brain elasticity can stimulate neurogenic variation, transitional gels (8–17 kPa) that replicate striated muscle elasticity can stimulate myocytes, and stiff gels (more than 25 kPa) that stimulate osteoblasts [181].

Many researchers have tried using different approaches to adjust the elasticity of hydrogels because of the impact of matrix viscoelasticity on stem cell fate commitment. Viscoelasticity has been found to be optimized in ECM-derived proteins including HA, chitosan, and alginate [182], while the viscoelastic characteristics of synthetic hydrogels have been used to control elasticity.

For the purpose of cultivating human MSCs (hMSCs) on TCPS, Schmitt et al. created a thin, cross-linked copolymer covering [183]. With the help of this copolymer, hMSC culture can be supported by the integrin-binding cell adhesion peptide Arg-Gly-Asp (RGD). The ability of this coating to undergo fast peptide attachment in aqueous solution at low peptide concentrations, good compatibility with wide area plastic substrates, and long-term stability in cell culture conditions were discovered.

In a different study, Jeon and colleagues [184] synthesized coacervated microdroplet-laden photo-cross-linked hydrogels by merely combining oxidized, methacrylated alginate (OMA) and GelMA. This allowed for the controlled release of the loaded growth factor, which in this case was human bone morphogenetic protein-2 (BMP-2). In the vicinity of low-intensity UV light and a photo-initiator, photo-cross-linking was used to trigger the synthesis of coacervates and coacervate-laden hydrogels, which formed the hydrogel immediately after mixing. Combining the results showed that long-term BMP-2 hosting in the coacervate hydrogel matrix promotes osteogenic differentiation of photoencapsulated stem cells and extracellular mineralization associated with bone formation. For medicinal uses like bone tissue engineering, these gels have the potential to be used as a medium for the localized, sustained administration of bioactive molecules to encapsulated stem cells.

2.6.5.2 Human pluripotent stem cells (HPSCs)

Human pluripotent stem cells (HPSCs), also known as induced and embryonic pluripotent stem cells (hESCs and human-induced pluripotent stem cells (hiPSCs)), are essential for disease modeling, regenerative medicine, toxicological screening, and fundamental developmental biology research.

The generation of hiPSCs from patients' skin fibroblasts was demonstrated in 2007 [185]. Since HPSCs are normal primary cell lines with the capacity for infinite self-renewal and the flexibility to assume any cellular destiny, they represent an excellent option for therapeutic research [186]. Vasculogenesis and angiogenesis are two major uses of PSCs in cardiac engineering [187].

In light of this, hydrogels made of both naturally occurring and synthetic macromolecules – such as gelatin, heparin, and HA – are being created more frequently for PSC culture and differentiation. Examples of these materials include PVA and PEG. In an ideal world, tailored matrices would encourage lineage-specific differentiation and enable PSC populations to grow and self-renew. To assist hPSCs' long-term endurance and self-renewal, for instance, Ekerdt and associates [188] created a thermoreversible HA-poly(*N*-isopropylacrylamide) (HA/PNIPAAm) hydrogel. In a related instance, Nazari et al. [189] demonstrated that, in contrast to a two-dimensional (2D) culture, the application of fibrin hydrogels greatly enhanced the differentiation of hiPSCs into oligodendrocytes. In a different study concerning improvement in retinal tissue development, Hunt et al. [190] have shown that culture in an RGD-alginate (RGD: arginine-glycine-aspartic acid) hydrogel scaffold improves the derivation of retinal pigmented epithelium both independently and in combination with neural retina from hESCs and hiPSCs. It is possible that the results of this study could have an impact on the development of additional pigmented, neuronal, or epithelial tissue folds. Specifically,

RGD-alginate may be preferred over HA-based hydrogels for the in vivo transport of retinal cells. Based on the above-selected observations, it can be interpreted that hydrogels are ideal material that can be tuned for physical, chemical, and biological stimuli provide the perfect medium for PSC development and differentiation.

Spheroid cultures of human PSCs have garnered significant attention. Part of the reason for this interest stems from the growing idea that 3D or spheroid cultures are physiologically superior and favorable in mimicking of in vivo signaling [191].

Fattahi and coworkers [192] have shown the creation of spheroids using microfluidic encapsulation of hPSCs. Four hundred micrometer diameter capsules with an aqueous core and a PEG hydrogel shell were created using a coaxial droplet microfluidic device.Their research emphasizes the usefulness and possible advantages of using microfabricated capsules to create hPSC suspension cultures as shown in Figure 1.2.15.

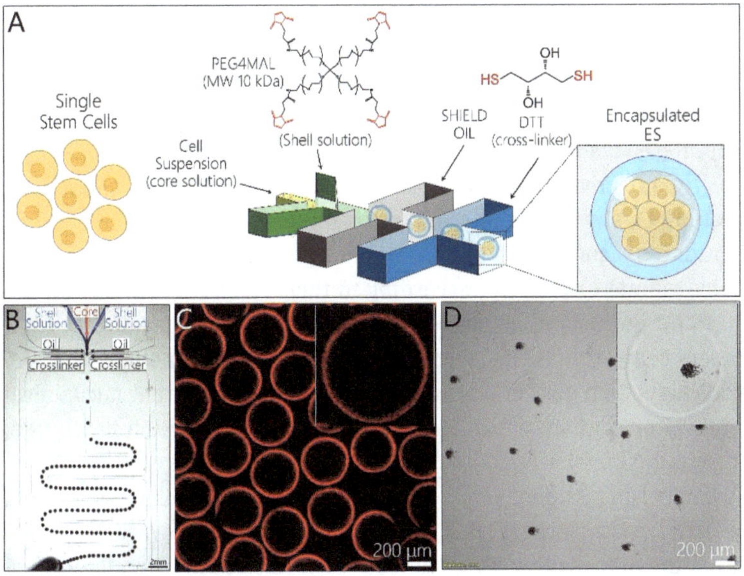

Figure 1.2.15: Design of aqueous core and PEG gel shell core-shell microcapsules: (A) schematic representation of the process of encapsulating cells; (B) top view of the coaxial flow-focusing apparatus producing the aqueous droplet train; (C) images of fluorescence demonstrating how water droplets gelled into uniformly sized microcapsules with a thin (5–10 µm) shell. To aid with visibility, rhodamine-labeled PEG was added to the hydrogel shell. The mean diameter of microcapsules is 392.6 ± 8.2 µm (n = 109); (D) in imaging, microbeads incorporated into the core stream during encapsulation were seen to circulate spontaneously inside capsules and aggregate in the middle of the capsules. This indicates that the microcapsules' aqueous core was present [192].

1.2.6.5.3 Hematopoietic stem cells (HSCs)

PSCs that can renew themselves are known as HSCs. All types of blood cells are produced by differentiation from HSCs. Maintaining the HSC pool throughout life depends on their ability to self-renew. The microenvironment of HSCs in the BM strongly controls both processes through interactions between cells and between cells and a matrix [193, 194]. It continues to be the most successful treatment for either benign or malignant hematologic and lymphoid illnesses. In addition, HSC treatment is frequently used to treat autoimmune diseases, amyloidosis, and aplastic anemia and other related conditions [195].

In view of the demand for a high cell count combined with stem cells' scarcity, it is necessary to prepare or increase the rare populations of stem cells using in vitro culture techniques before using them [196, 197]. Hydrogel as advanced biomaterials offer a versatile approach to introducing biological and biophysical activators as well as good biocompatibility and some of the physical characteristics of the natural matrix. Hydrogel-based culture techniques are often the focus of several ongoing efforts to create a designed BM-like^ niche for the in vitro proliferation of HSCs.

Liu et al. [198] offered a straightforward technique for $CD34^+$ cell in vitro culture, which allowed for the effective growth of $CD34^+$ cells using a special gel-embedding cell culture method. HSCs are unable to retain normal morphology and prematurely lost stemness as traditional suspension culture technologies did not offer a mushy 3D in vitro culture environment. However, according to their study, this flaw might be worked around by the permeable alginate hydrogel method they developed. Non-anchorage-dependent cells, such as HSC, were able to retain their phenotypic and proliferation in porous alginate hydrogel scaffolds because of their inherent mechanical characteristics and water content, which were comparable to living tissues with decent permeability and flexibility.

A 3D scaffold comprising graphene foams (GFs) and gelatin-HA (gel-HA) hydrogels was designed by Hong and coworkers [199] to imitate the bone marrow niche, drawing inspiration from the areas with varying degrees of stiffness. This scaffold was intended to facilitate the in vitro proliferation of HSCs. A gel-HA hydrogel that was photo-cross-linked was formed around the GF to create the composite scaffold. The vascular niche and endosteal niche in the bone marrow are modeled by the "stiff" GF and "soft" gel-HA hydrogel, respectively, in terms of their stiffness and formation. HSCs cultured in the gel-HA/GF scaffold also exhibited good proliferation, pluripotency, and the ability to maintain the $CD34^+$ $CD38^-$ immunophenotype, indicating that the gel-HA/GF composite scaffold supported HSCs' in vitro expansion while preserving their primitive phenotype and capacity to differentiate into functional blood cells.

On the other hand, cell membrane lipids that are hydrophilic and amphiphilic predominate in the 3D hematopoietic niche that exists in vivo. The hydrophobic groups and different biomolecules can be easily modified in hydrogels due to their high hydrophilicity and abundance of functional groups. A habitat that is more similar to the niche may be created by their regulated physical and chemical properties such as flexibility, porosity, rigidity, and the functionalization of important cytokines.

Numerous studies have demonstrated that greater hydrophobic levels preferentially result in higher concentrations of reactive oxygen species (ROS), which are known to cause damage and differentiation in HSCs [200]. For instance, LT-HSC expansion in 3D zwitterionic hydrogels was found to enhance 73-fold by the prevention of extreme ROS [201]. According to the equivalent in vivo data in immuno conceded mice, the enlarged HSCs could reconstruct the hematopoiesis for a minimum of 24 weeks.

1.2.7 Conclusions and future perspectives

For approximately 50 years, hydrogels have been widely used as one of the most exciting and innovative biomaterials in the medical industry. Hydrogels are undoubtedly a diverse biomaterial in terms of their features; their high-water retention ability and superior functional properties make them suitable for diffusive drug delivery, cell culture, molecular imprinting, cancer therapy, and tissue interface applications.

Although hydrogels are widely used in products nowadays, their full potential is yet unknown. Tissue engineering and drug delivery applications have not thoroughly investigated via commercially available hydrogels.

Over the coming decades, there will probably be more progress made in identifying new synthetic resources or upgraded biopolymers for the formation of hydrogels, comprehending the properties of hydrogels and how to process them into anticipated structures, and developing novel or advanced applications to address particular biomedical issues.

Here, the authors attempt to concentrate on hydrogel-based biomaterials and their uses in tissue engineering, wound healing, bone regeneration, stem cell therapies, and drug delivery. In summary, as the development of increasingly sophisticated and intelligent biopolymers continues, hydrogel design in the future will be guided by the challenges of the present and the results of the clinical studies.

References

[1] Kasai RD, Radhika D, Archana S, Shanavaz H, Koutavarapu R, Lee D, Shim J. A review on hydrogels classification and recent developments in biomedical applications. International Journal of Polymeric Materials. 2023; 72(13): 1059–1069.

[2] Cao H, Duan L, Zhang Y, et al. Current hydrogel advances in physicochemical and biological response-driven biomedical application diversity. Signal Transduction and Targeted Therapy. 2021; 6: 426.

[3] Li C, Chen P, Shao Y, Zhou X, Wu Y, Yang Z, Li Z, Weil T, Liu D. A writable polypeptide-DNA hydrogel with rationally designed multi-modification sites. Small. 2015; 11: 1138–1143.
[4] Yang B, Zhao B,Z, Pan Y, Xie J, Zhou B, Li Y, Dong Y, Liu D. Shear-thinning and designable responsive supramolecular DNA hydrogels based on chemically branched DNA. ACS Applied Materials and Interfaces. 2021; 13: 48414–48422.
[5] Kumar N, Ghosh B, Kumar A, Koley R, Dhara S, Chattopadhyay S. Multilayered "SMART" hydrogel systems for on-site drug delivery applications. Journal of Drug Delivery Science and Technology. 2023; 80: 104111.
[6] Tanaka T, J D. Filmore, kinetics of swelling of gels. Journal of Chemical Physics. 1979; 70(3): 1214–1218.
[7] Li Y, Tanaka T. Kinetics of swelling and shrinking of gels. Journal of Chemical Physics. 1990; 92(2): 1365–1371.
[8] Aktas DK, Evingur GA, Pekcan O. Study on swelling of hydrogels (PAAm) at various temperatures by using fluorescence technique. Journal of Materials Science. 2007; 42(20): 8481–8488.
[9] Xing J, Liu D, Zhou G, Li Y, Wang P, Hu K, Gu N, Ji M. Liposomally formulated phospholipid-conjugated novel near-infrared fluorescence probe for particle size effect on cellular uptake and biodistribution in vivo. Colloids and Surfaces B: Biointerfaces. 2018; 161: 588–596.
[10] Amiryaghoubi N, Fathi M, Barar J, Omidi Y. Hydrogel-based scaffolds for bone and cartilage tissue engineering and regeneration. Reactive and Functional Polymers. 2022; 177: 105313.
[11] Zhang W, Zhang K, Yan S, Wu J, Yin J. A tough and self-healing poly (l-glutamic acid)-based composite hydrogel for tissue engineering. Journal of Materials Chemical B. 2018; 6(42): 6865–6876.
[12] Ha N, Lee SJ. Enhanced fog collection of hydrogel-coated wires inspired by water film-trapping characteristics of tillandsia. Chemical Engineering Journal. 2023; 466: 143185.
[13] Zhang Y, Wang F, Yu Y, Wu J, Cai Y, Shi J, Morikawa H, Zhu C. Multi-bioinspired hierarchical integrated hydrogel for passive fog harvesting and solar-driven seawater desalination. Chemical Engineering Journal. 2023; 15(466): 143330.
[14] Qu J, Zhao X, Liang Y, Zhang T, Ma Peter X, Guo B. Antibacterial adhesive injectable hydrogels with rapid self-healing, extensibility and compressibility as wound dressing for joints skin wound healing. Biomaterials. 2018; 183: 185–199.
[15] Liang Y, He J, Guo B. Functional hydrogels as wound dressing to enhance wound healing. ACS Nano. 2021; 15(8): 12687–12722.
[16] Talebian S, Mehrali M, Taebnia N, Pennisi CP, Kadumudi FB, Foroughi J, Hasany M, Nikkhah M, Akbari M, Orive G, Dolatshahi-Pirouz A. Self-healing hydrogels: The next paradigm shift in tissue engineering? Advanced Science. 2019; 6: 1801664.
[17] Uman S, Dhand A, Burdick JA. Recent advances in shear-thinning and self-healing hydrogels for biomedical applications. Journal of Applied Polymer Science. 2020; 137: 48668.
[18] Jiao D, Li Zhu Q, Yu Li C, Zheng Q, Zheng Q, Zi Liang W. Programmable morphing hydrogels for soft actuators and robots: From structure designs to active functions. Accounts of Chemical Research. 2022; 55(11): 1533–1545.
[19] Han D, Farino C, Yang C, Scott T, Browe D, Choi W, Freeman JW, Lee H. Soft robotic manipulation and locomotion with a 3D printed electroactive hydrogel. ACS Applied Materials and Interfaces. 2018; 10(17): 512–17518.
[20] Zhang YZ, Lee KH, Anjum DH, Sougrat R, Jiang Q, Kim H, Alshareef HN. Mxenes stretch hydrogel sensor performance to new limits. Science Advances. 2018; 4: eaat0098.
[21] Lei Z, Wang Q, Sun S, Zhu W, Wu P. A bioinspired mineral hydrogel as a self-healable, mechanically adaptable ionic skin for highly sensitive pressure sensing. Advances in Materials. 2017; 29: 1700321.
[22] Darabi MA, Khosrozadeh A, Mbeleck R, Liu Y, Chang Q, Jiang J, Cai J, Wang Q, Luo G, Xing M. Skin-inspired multifunctional autonomic-intrinsic conductive self-healing hydrogels with pressure sensitivity, stretchability, and 3D printability. Advances in Materials. 2017; 29: 1700533.

[23] Parenteau-Bareil R, Gauvin R, Berthod F. Collagen-based biomaterials for tissue engineering applications. Materials (Basel). 2010; 3: 1863–1887.
[24] GhavamiNejad A, Hashmi S, Vatankhah-Varnoosfaderani M, Stadler FJ. Effect of H_2O and reduced graphene oxide on the structure and rheology of self-healing, stimuli responsive catecholic gels. Rheologica Acta. 2016; 55: 163–176.
[25] Xue K, Wang X, Yong PW, Young DJ, Wu YL, Li ZB, Loh XJ. Hydrogels as emerging materials for translational biomedicine. Advances in Therapy. 2019; 2: 1800088.
[26] Wichterle O, L'ım D. Hydrophilic gels for biological use. Nature. 1960; 185: 117–118.
[27] Hennink WE, Van Nostrum CF. Novel crosslinking methods to design hydrogels. Advanced Drug Delivery Reviews. 2002 Jan 17; 54(1): 13–36.
[28] Zhao X. Multi-scale multi-mechanism design of tough hydrogels: Building dissipation into stretchy networks. Soft Matter. 2014; 5(10): 672–687.
[29] Li CH, Wang C, Keplinger C, Zuo JL, Jin L, Sun Y, Zeng P, Cao Y, Lissel F, Linder C, You XZ, Bao Z. A highly stretchable autonomous self-healing elastomer. Nature Chemistry. 2016; 8: 618–624.
[30] Sikdar P, Uddin MM, Dip TM, Islam S, Hoque MS, Dhar AK, Wu SS. Recent advances in the synthesis of smart hydrogels. Materials Advances. 2021; 2: 4532–4573.
[31] Maitra J, Shukla VK. Crosslinking in hydrogels – A review American. Journal of Polymer Science: Polymer Letters Edition. 2014; 4(2): 25–31.
[32] Hennink WE, Van Nostrum CF. Department of pharmaceutics. Advanced Drug Delivery Reviews. 2002; 54: 13–36.
[33] Guo G, Chen Y, Liu X, Zhu DY, Zhang B, Lin N, Gao L. Tough and durable hydrogels with robust skin layers formed via soaking treatment. Journal of Materials Chemical B. 2018; 6: 8043–8054.
[34] Bashir S, Hina M, Iqbal J, Rajpar A, Mujtaba M, Alghamdi N, Wageh S, Ramesh K, Ramesh S. Fundamental concepts of hydrogels: Synthesis, properties, and their applications. Polymers. 2020; 12: 2702.
[35] Guo W, Lu CH, Orbach R, Wang F, Qi XJ, Cecconello A, Seliktar D, Willner T. pH-stimulated DNA hydrogels exhibiting shape-memory properties. Advances in Materials. 2015; 27: 73–78.
[36] Chen H, Gu Y, Hu Y, Qian Z. Characterization of pH-and temperature-sensitive hydrogel nanoparticles for controlled drug release. PDA Journal of Pharmaceutical Science and Technology. 2007; 61: 303–313.
[37] Baroli B. Photopolymerization of biomaterials: Issues and potentialities in drug delivery, tissue engineering, and cell encapsulation applications. Journal of Chemical Technology and Biotechnology. 2006; 81: 491–499.
[38] Koetting MC, Peters JT, Steichen SD, Peppas NA. Stimulus-responsive hydrogels: Theory, modern advances, and applications. Materials Science and Engineering R: Reports. 2015; 93: 1–49.
[39] Meleties M, Katyal P, Lin P, Britton D, Montclare JK. Self-assembly of stimuli-responsive coiled-coil fibrous hydrogels. Soft Matter. 2021; 17(6): 470–6476.
[40] Ullah F, Bisyrul M, Othman H, Javed F, Ahmad Z, Md. Akil, H. Classification, processing and application of hydrogels: A review. Materials Science and Engineering: C 2015; 57(1): 414–433.
[41] Aswathy SH, Narendrakumar U, Manjubala I. Commercial hydrogels for biomedical applications, Commercial hydrogels for biomedical applications. Heliyon. 2020; e03719.
[42] Richbourg NR, Peppas NA. The swollen polymer network hypothesis: Quantitative models of hydrogel swelling, stiffness, and solute transport. Progress in Polymer Science. 2020; 105: 101243.
[43] Wu F, Pang Y, Liu J. Swelling-strengthening hydrogels by embedding with deformable nanobarriers. Nature Communications. 2020; 11: 4502.
[44] Tanaka T, J D. Filmore, kinetics of swelling of gels. The Journal of Chemical Physics. 1979; 70(3): 1214–1218.
[45] Peters A, Candau SJ. Kinetics of swelling of polyacrylamide gels. Macromolecules. 1986; 19(7):1952–1955.

[46] Aktas DK, Evingur GA, Pekcan O. Study on swelling of hydrogels (PAAm) at various temperatures by using fluorescence technique. Journal of Materials Science. 2007; 42(20): 8481–8488.
[47] Aktas DK, Evingur GA, Pekcan O. O. A fluorescence study on swelling of hydrogels (PAAm) at various cross-linker contents. Advances in Polymer Technology. 2009; 28(4): 215–223.
[48] Dehkordi NK, Shojaei S, Asefnejad A, Hassani K, Benisi SZ. The effect of three types of cross-linked hydrogels and volume fraction of polyacrylamide on the swelling and thermal behavior using molecular dynamics simulation. Jouranal of Materials Research and Technology. 2023; 24: 4627–4638.
[49] Yiamsawas D, Kangwansupamonkon W, Chailapakul O, Kiatkamjornwong S. Synthesis and swelling properties of poly[acrylamide-*co*-(crotonic acid)] superabsorbents. Reactive & Functional Polymers. 2007; 67: 865–882.
[50] Bajpai AK, Shrivastava M. Swelling kinetics of a hydrogel of Poly(ethylene glycol) and Poly (acrylamide-*co*-styrene). Journal of Applied Polymer Science. 2002; 85: 1419–1428.
[51] Wang C, Li Y, Hu Z. Swelling kinetics of polymer gels. Macromolecules. 1997; 30: 4727–4732.
[52] Plunkett KN, Kraft ML, Yu Q, Moore JS. Swelling kinetics of disulfide cross-linked microgels. Macromolecules. 2003; 36: 3960–3966.
[53] Hu Y, Zhang Z, Li Y, Ding X, Li D, Shen C, Xu FJ. Dual-crosslinked amorphous polysaccharide hydrogels based on chitosan/alginate for wound healing applications. Macromol Rapid Commun. 2018; 39: 1800069.
[54] Kesharwani P, Bisht A, Alexander A, Dave V, Sharma S. Biomedical applications of hydrogels in drug delivery system: An update. Journal of Drug Delivery Science and Technology. 2021; 66: 102914.
[55] Palmese LL, Thapa RK, Sullivan MO, Kiick KL. Hybrid hydrogels for biomedical applications. Current Opinion in Chemical Engineering. 2019; 24: 143–157.
[56] Song J, Zhang C, Kong S, Liu F, Hu W, Su F, Li S. Novel chitosan-based metal-organic polyhedrons/ enzyme hybrid hydrogel with antibacterial activity to promote wound healing, carbohydr. Polymer. 2022; 291: 119522.
[57] Ahmed EM. Hydrogel: Preparation, characterization, and applications: A review. Journal of Advanced Research. 2015; 6: 105e121.
[58] Ahmad Z, Salman S, Khan SA, Amin A, Rahman ZU, Al-Ghamdi YO, Akhtar K, Bakhsh EM, Khan SB. Versatility of hydrogels: From synthetic strategies, classification, and properties to biomedical applications. Gels. 2022; 8: 167.
[59] Salehi Dashtebayaz MS, Nourbakhsh MS. Interpenetrating networks hydrogels based on hyaluronic acid for drug delivery and tissue engineering. International Journal of Polymeric Materials and Polymeric Biomaterials. 2019; 68: 442–451.
[60] Sperling LH. Interpenetrating Polymer Networks and Related Materials, Media: Springer Science & Business; Berlin/Heidelberg, Germany. 2012.
[61] Ignat L, Stanciu A. Advanced Polymers: Interpenetrating Polymer Networks. In: Kulashreshtha AK, Vasile C, editor. Handbook of Polymer Blends and Composites, Shrewsbury, UK;: Rapra Technology; 2003, 275e280.
[62] Işıkver Y, Saraydın D. Gels smart hydrogels: Preparation, characterization, and determination of transition points of crosslinked *N*-sopropylcrylamide/acrylamide/carboxylic acids polymers. Gels. 2021; 7(3): 113.
[63] Zhao H, Liu M, Zhang Y, Yin J, Pei R. Nanocomposite hydrogels for tissue engineering applications. Nanoscale. 2020; 12: 14976–14995.
[64] Zhao G, Ren R, Wei X, Jia Z, Chen N, Sun Y, Zhao Z, Lele SM, Zhong HA. Thermoresponsive polymeric dexamethasone prodrug for arthritis pain. Journal of Controlled Release. 2021; 339: 484–497.
[65] Xing R, Mustapha O, Ali T, Rehman M, Zaidi SS, Baseer A, Batool S, Mukhtiar M, Shafique S. Development, characterization, and evaluation of sln-loaded thermoresponsive hydrogel system

of topotecan as biological macromolecule for colorectal delivery. BioMed Research International. 2021; 2021: 9968602.
[66] Witika BA, Stander JC, Smith VJ, Walker RB. Nano co-crystal embedded stimuli-responsive hydrogels: A potential approach to treat HIV/AIDS. Pharmaceutics. 2021; 13: 127.
[67] A. Patel K. Mequanint Hydrogel Biomaterials. In: Prof. Fazel R, editor. Biomedical Engineering – Frontiers and Challenges, Rijeka, Croatia: IntechOpen; 2011, pp. 275. ISBN: 978 -953 -307 -309 -5.
[68] Hennink W, Van Nostrum C. Novel crosslinking methods to design hydrogels. Advances Drug Deliveryl Reviews. 2012; 64: 223–236.
[69] Cao H, Duan L, Zhang Y, Cao J, Zhang K. Current hydrogel advances in physicochemical and biological response-driven biomedical application diversity. Signal Transduction and Targeted Therapy. 2021; 6: 426.
[70] Tao G, Cai R, Wang Y, Zuo H, He H. Fabrication of antibacterial sericin based hydrogel as an injectable and mouldable wound dressing. Materials Science and Engineering, C: Materials for Biological Applications. 2021; 119: 111597.
[71] Chen W, Chu R, Li H, Hua T, Chen H, Li R, Zhou D, Cao S, Ye S, Li H. A novel wound dressing based on a gold nanoparticle self-assembled hydrogel to promote wound healing. Materials Advances. 2023; 4: 2918–2925.
[72] Tavakoli S, Klar AS. Advanced hydrogels as wound dressings. Biomolecules. 2020-08-11; 10(8): 1169.
[73] Qu J, Zhao X, Liang Y, Zhang T, Ma PX, Guo B. Antibacterial adhesive injectable hydrogels with rapid self-healing, extensibility, and compressibility as wound dressing for joints skin wound healing. Biomaterials. 2018; 183: 185–199.
[74] Li Y, Song W, Kong L, He Y, Li H. Injectable and microporous microgel-fiber granular hydrogel loaded with bioglass and siRNA for promoting diabetic wound healing. Small. 2023; 20(19): e2309599.
[75] Afewerki S, Wang X, Ruiz-Esparza GU, Tai CW, Kong X, Zhou S, *et al.* Combined catalysis for engineering bioinspired, lignin-based, long-lasting, adhesive, self-mending, antimicrobial hydrogels. ACS Nano. 2020; 14: 17004–17017.
[76] Bazmandeh AZ, Mirzaei E, Fadaie M, Shirian S, Ghasemi Y. Dual spinneret electrospun nanofibrous/gel structure of chitosan-gelatin/chitosan-hyaluronic acid as a wound dressing: In-vitro and in-vivo studies. International Journal of Biological Macromolecules. 2020; 162: 359–373.
[77] Bilal M, Rasheed T, Zhao Y, Iqbal HMN. Agarose-chitosan hydrogel-immobilized horseradish peroxidase with sustainable bio-catalytic and dye degradation properties. International Journal of Biological Macromolecules. 2019; 124: 742–749.
[78] Jayakumar R, Prabaharan M, Sudheesh Kumar PT, Nair SV, Tamura H. Biomaterials based on chitin and chitosan in wound dressing applications. Biotechnology Advances 2011; 29(3): 322–337.
[79] Dai T, Tanaka M, Huang YY, R M. Hamblin chitosan preparations for wounds and burns: Antimicrobial and wound-healing effects. Expert Review of Anti-infective Therapy. 2011; 9(7): 857–879.
[80] Fan L, Yang J, Wu H, Hu Z, Yi J, Tong J, Zhu X. Preparation, and characterization of quaternary ammonium chitosan hydrogel with significant antibacterial activity. International Journal of Biological Macromolecules. 2015; 79: 830–836.
[81] Baxter RM, Dai T, Kimball J, Wang E, Hamblin MR, Wiesmann WP, McCarthy SJ, Baker SM. Chitosan dressing promotes healing in third degree burns in mice: Gene expression analysis shows biphasic effects for rapid tissue regeneration and decreased fibrotic signaling. Journal of Biomedical Materials Research Part A. 2013; 101: 340–348.
[82] Lu B, Han X, Zou D, Luo X, Liu L, Wang J, Maitz MF, Yang P, Huang N, Zhao A. Catechol-chitosan/polyacrylamide hydrogel wound dressing for regulating local inflammation. Materials Today Bio. 2022; 16: 100392.

[83] Tulain UR, Ahmad M, Rashid A, Malik MZ, Iqbal FM. Fabrication of pH-responsive hydrogel and its in vitro and in vivo evaluation. Advances in Polymer Technology. 2018; 37: 290–304.
[84] Tan L, Xu X, Song J, Luo F, Qian Z. Synthesis, characterization, and acute oral toxicity evaluation of pH-sensitive hydrogel based on MPEG, poly (ε-caprolactone), and itaconic acid. BioMed Research International. 2013; 2013: 239838.
[85] Naahidi S, Jafari M, Logan M, Wang Y, Yuan Y, Bae H, Dixon B, Chen P. Biocompatibility of hydrogel-based scaffolds for tissue engineering applications. Biotechnology Advances. 2017; 35: 530–544.
[86] Sharpe LA, Daily AM, Horava SD, Peppas NA. Therapeutic applications of hydrogels in oral drug delivery. Expert Opinion Drug Delivery. 2014; 11: 901–915.
[87] Nayak AK, Ahmad SA, Beg S, Ara TJ, S M. Hasnain, Drug Delivery: Present, Past, and Future of Medicine. In: Inamuddin, Asiri AM, Mohammad A, editors. Applications of Nanocomposite Materials in Drug Delivery, Eds: Inamuddin, Asiri AM, Mohammad A, Cambridge, UK: Woodhead Publishing; 2018, pp. 255–282.
[88] Jain KK. An overview of drug delivery systems. Methods in Molecular Biology. 2020; 2059: 1–54.
[89] Giri A, Bhunia T, Mishra SR, Goswami L, Panda AB, Pal S, Bandyopadhyay A. Acrylic acid grafted guargum–nanosilica membranes for transdermal diclofenac delivery. Carbohydrate Polymers. 2013; 91: 492–501.
[90] Lee H, Noh H. Advancements in nanogels for enhanced ocular drug delivery: Cutting-edge strategies to overcome eye barriers. Gels. 2023; 9(9): 718.
[91] Torres-Luna C, Fan X, Domszy R, Hu N, Wang NS, Yang A. Hydrogel-based ocular drug delivery systems for hydrophobic drugs. Hydrogel-based ocular drug delivery systems for hydrophobic drugs. European Journal of Pharmaceutical Sciences 2020; 154: Article. 105503.
[92] Pimenta AFR, Ascenso J, Fernandes JCS, Colaço R, Serro AP, Saramago B. Controlled drug release from hydrogels for contact lenses: Drug partitioning and diffusion. International Journal of Pharmaceutics. 2016; 515: 467–475.
[93] Garcia-Millan E, Koprivnik S, Otero-Espinar FJ. Drug loading optimization and extended drug delivery of corticoids from pHEMA based soft contact lenses hydrogels via chemical and microstructural modifications. International Journal of Pharmaceutics. 2015; 487(1–2): 260–269.
[94] Li Y, Rodrigues J, Tomás H. Injectable and biodegradable hydrogels: Gelation, biodegradation and biomedical applications. Chemical Society Reviews. 2012; 41(2): 193–2221.
[95] Norouzi M, Nazari B, Miller DW. Injectable hydrogel-based drug delivery systems for local cancer therapy. Drug Discovery Today. 2016; 21: 1835–1849.
[96] Mathew AP, Uthaman S, Cho KH, Cho CS, Park IK. Injectable hydrogels for delivering biotherapeutic molecules. International Journal of Biological Macromolecules. 2018; 110: 17–29.
[97] Yu S, He C, Chen X. Injectable hydrogels as unique platforms for local chemotherapeutics-based combination antitumor therapy. Macromolecular Bioscience. 2018; 18(12): 1800240.
[98] Fan DY, Tian Y, J Z. Liu injectable hydrogels for localized cancer therapy. Frontiers in Chemistry. 2019; 7: 675.
[99] Chapekar MS. Tissue engineering: Challenges and opportunities. Journal of Biomedical Materials Research. 2000; 53: 617–620.
[100] Kumari S, Mondal P, Chatterjee K. Digital light processing-based 3D bioprinting of κ-carrageenan hydrogels for engineering cell-loaded tissue scaffolds. Carbohydrate Polymers. 2022; 290: 119508.
[101] He Y, Jiang T, Li C, Zhou C, Yang G, Nie J, Wang F, Lu C, Yin D, Yang X, Chen Z. Thiol-ene-mediated degradable POSS-PEG/PEG hybrid hydrogels as potential cell scaffolds in tissue engineering. Polymer Degradation and Stability. 2023; 211: 110316.
[102] El-Husseiny HM, Mady EA, Doghish AS, Zewail MB, Abdelfatah AM, Noshy M, Mohammed OA, El-Dakroury WA. Smart/stimuli-responsive chitosan/gelatin and other polymeric macromolecules natural hydrogels vs. synthetic hydrogels systems for brain tissue engineering: A state-of-the-art review. International Journal of Biological Macromolecules. 2024; 260(1): 129323.

[103] Enayati M, Liu W, Madry H, Neisiany RE, Cucchiarini M. Functionalized hydrogels as smart gene delivery systems to treat musculoskeletal disorders. Advances in Colloid Interface Science. 2024; 331: 103232.
[104] Motasadizadeh H, Tavakoli M, Damoogh S, Mottaghitalab F, Gholami M, Atyabi F, Farokhi M, Dinarvand R. Dual drug delivery system of teicoplanin and phenamil based on pH-sensitive silk fibroin/sodium alginate hydrogel scaffold for treating chronic bone infection. Biomaterials Advances. 2022; 139: Article. 213032.
[105] Li J, Weber E, Guthundel S, Schuleit M, Kuttler A, Halleux C, Accart N, Doelemeyer A, Basler A, Tigani B, Wuersch K, Fornaro M, Kneissel M, Stafford A, Freedman BR, Mooney DJ. Tough composite hydrogels with high loading and local release of biological drugs. Advanced Healthcare Materials 2018; 7: Article. 1701393.
[106] Yao Q, Liu Y, Pan Y, Li Y, Xu L, Zhong Y, Wang W, Zuo J, Yu H, Lv Z, Chen H, Zhang L, Wang B, Yao H, Meng Y. Long-term induction of endogenous BMPS growth factor from antibacterial dual network hydrogels for fast large bone defect repair. Journal of Colloid and Interface Science. 2022; 607: 1500–1515.
[107] Lau YK, Gobin AM, West JL. Overexpression of lysyl oxidase to increase matrix crosslinking and improve tissue strength in dermal wound healing Ann. Biomedical Engineering 2006; 34: 1239–1246.
[108] Sassi ML, Eriksen H, Risteli L, Niemi S, Mansell J, Gowen M, Risteli J. Immunochemical characterization of assay for carboxyterminal telopeptide of human type I collagen: Loss of antigenicity by treatment with cathepsin. Bone. 2000; 26: 367–373.
[109] Koch L, Deiwick A, Schlie S, Michael S, Gruene M, Coger V, Zychlinski D, Schambach A, Reimers K, Vogt PM, Chichkov B. Skin tissue generation by laser cell printing. Biotechnology & Bioengineering. 2012; 109(7): 1855–1863.
[110] Lee YB, Polio S, Lee W, Dai G, Menon L, Carroll RS, Yoo SS. Bio-printing of collagen and VEGF-releasing fibrin gel scaffolds for neural stem cell culture. Experimental Neurology. 2010; 223(2): 645–652.
[111] Bello AB, Kim D, Kim D, Park H, Lee SH. Engineering and functionalization of gelatin biomaterials: From cell culture to medical applications. Tissue Engineering Part B: Reviews. 2020; 26: 164–180.
[112] Okonkwo E, Ezeanyanaso C, Ofoegbu O, Ali E, Iyamu N. Extraction and characterization of gelatin from cow bones: Using substitute demineratizing agent (H2SO4). Global Journal of Pure and Applied Sciences. 2007; 13: 189–191.
[113] Wang X, Ao Q, Tian X, Fan J, Tong H, Hou W, Bai S. Gelatin-based hydrogels for organ 3D bioprinting. Polymers. 2017; 9(9): 401.
[114] Kuijpers JA, Engbers HMG, Krijgsveld J, Sebastian Zaat AJ, Dankert J, Feijen J. Crosslinking and characterization of gelatin matrixes for biomedical applications. Journal of Biomaterials Science, Polymer Edition. 2000; 11(3): 225–243.
[115] Nichol JW, Koshy ST, Bae H, Hwang CM, Yamanlar S, Khademhosseini A. Cell-laden microengineered gelatin methacrylate hydrogels. Biomaterials. 2010; 31: 5536.
[116] Antoine EE, Vlachos PP, Rylander MN. Review of collagen I hydrogels for bioengineered tissue microenvironments: Characterization of mechanics, structure, and transport. Tissue Engineering Part B: Reviews. 2014; 20: 683–696.
[117] Gasperini L, Mano JF, Reis RL. Natural polymers for the microencapsulation of cells. Journal of the Royal Society Interface. 2014; 11: 20140817.
[118] Mosesson MW. Fibrinogen and fibrin structure and functions. Journal of Thrombosis and Haemostasis. 2005; 3: 1894–1904.
[119] Mosesson MW, Siebenlist KR, Meh DA. The structure and biological features of fibrinogen and fibrin. Annals of the New York Academy of Sciences. 2001; 936: 11–30.
[120] Nimni ME, Cheung D, Strates B, Kodama M, Sheikh K. Chemically modified collagen: A natural biomaterial for tissue replacement. Journal of Biomedical Materials Research. 1987; 21(6): 741–771.

[121] Yue K, Trujillo-de Santiago G, Alvarez MM, Tamayol A, Annabi N, Khademhosseini A. Synthesis, properties, and biomedical applications of gelatin methacryloyl (GelMA) hydrogels. Biomaterials. 2015; 73: 254–271.

[122] Bulpitt P, Aeschlimann D. New strategy for chemical modification of hyaluronic acid: Preparation of functionalized derivatives and their use in the formation of novel biocompatible hydrogels. Journal of Biomedical Materials Research 1999; 47: 152–169.

[123] Li L, Duan X, Fan Z, Chen L, Xing F, Xu Z, Chen Q, Xiang Z. Mesenchymal stem cells in combination with hyaluronic acid for articular cartilage defects. Scientific Reports. 2018; 8: 1–11.

[124] Keller L, Regiel-Futyra A, Gimeno M, Eap S, Mendoza G, Andreu V. Chitosan-based nanocomposites for the repair of bone defects. Nanomedicine: Nanotechnology, Biology and Medicine. 2017; 13(7): 2231–2240.

[125] Berillo D, Cundy A. 3D-macroporous chitosan-based scaffolds within situ formed Pd and Pt nanoparticles for nitrophenol reduction. Carbohydrate Polymers. 2018; 192: 166–175.

[126] Islam MS, Haque P, Rashid TU, Khan MN, Mallik AK, Khan MNI, Khan M, Rahman MM. Core-shell drug carrier from folate conjugated chitosan obtained from prawn shell for targeted doxorubicin delivery. Journal of Materials Science Materials in Medicine. 2017; 28(4): 55.

[127] Kuang W, Liu C, Xu H. The application of decellularized nucleus pulposus matrix/chitosan with transforming growth factor beta 3 for nucleus pulposus tissue engineering. Cytotechnology. 2021; 73: 447–456.

[128] Van Vlierberghe S, Dubruel P, Schacht E. Biopolymer-based hydrogels as scaffolds for tissue engineering applications: A review. Biomacromolecules. 2011; 12(5): 1387–1408.

[129] Summa M, Russo D, Penna I, Margaroli N, Bayer IS, Bandiera T. A biocompatible sodium alginate/ povidone iodine film enhances wound healing. European Journal of Pharmaceutics and Biopharmaceutics. 2018; 122: 17–24.

[130] Balakrishnan B, Joshi N, Jayakrishnan A, Banerjee R. Self-crosslinked oxidized alginate/gelatin hydrogel as injectable, adhesive biomimetic scaffolds for cartilage regeneration. Acta Biomaterial. 2014; 10: 3650–3663.

[131] Venkatesan J, Bhatnagar I, Manivasagan P, Kang KH, Kim SK. Alginate composites for bone tissue engineering: A review. International Journal of Biological Macromolecules. 2015; 72: 269–281.

[132] Mostafavi A, Daemi H, Rajabi S, Baharvand H. Highly tough and ultrafast self-healable dual physically crosslinked sulfated alginate-based polyurethane elastomers for vascular tissue engineering. Carbohydrate Polymers. 2021; 257: 117632.

[133] Kethar J. NanoMedicine; An analysis of hydrogels scaffolds for tissue engineering application. Journal of Student Research. 2023; 12(1).

[134] Smetana K Jr. Cell Biology of Hydrogels. Biomaterials. 1993; 14(14): 1046–1050.

[135] Zare M, Bigham A, Zare M, Luo H, Rezvani EG, Ramakrishna S. pHEMA: An overview for biomedical applications. International Journal of Molecular Sciences. 2021; 22: 637.

[136] Murphy SV, Atala A. 3D bioprinting of tissues and organs. Nature Biotechnology. 2014 Aug; 32(8): 773–785.

[137] Lee VK, Lanzi AM, Haygan N, Yoo SS, Vincent P, Dai G. Generation of multi-scale vascular network system within 3D hydrogel using 3D bio-printing technology. Cellular and Molecular Bioengineering. 2014; 7: 460–472.

[138] Sharma B, Fermanian S, Gibson M, Unterman S, Herzka DA, Cascio B, Coburn J, Hui AY, Marcus N, Gold GE, Elisseeff JH. Human cartilage repair with a photoreactive adhesive-hydrogel composite. Science Translational Medicine. 2013; 5: 167ra166.

[139] Inzana JA, Olvera D, Fuller SM, Kelly JP, Graeve OA, Schwarz EM, Kates SL, Awad HA. 3D printing of composite calcium phosphate and collagen scaffolds for bone regeneration. Biomaterials. 2014; 35: 4026–4034.

[140] Farhat W, Chatelain F, Marret A, Faivre L, Arakelian L, Cattan P, Fuchs A. Trends in 3D bioprinting for esophageal tissue repair and reconstruction. Biomaterials. 2021; 267: 120465.
[141] Gong T, Xie J, Liao J, Zhang T, Lin S, Lin Y. Nanomaterials and bone regeneration. Bone Research. 2015; 3: 15029.
[142] Kim TG, Shin H, Dong WL. Biomimetic scaffolds for tissue engineering. Advanced Functional Materials. 2012; 22(2): 446–246.
[143] Feng X. Chemical and biochemical basis of cell-bone matrix interaction in health and disease. Current Chemical Biology. 2009; 3(2): 189–196.
[144] Geetha M, Singh AK, Asokamani R, Gogia AK. Ti based biomaterials, the ultimate choice for orthopaedic implants – A review. Progress in Materials Science. 2009; 54: 397–425.
[145] Hench LL. Biomaterials. Science. 1980; 208: 826–831.
[146] Ulery BD, Nair LS, Laurencin CT. Biomedical applications of biodegradable polymers. Journal of Polymer Sicence: Part B. 2011; 49: 832–864.
[147] Xue X, Hu Y, Deng Y, Su J. Recent advances in design of functional biocompatible hydrogels for bone tissue engineering. Advanced Functional Materials 2021; 19(31) p.2009432.
[148] Ulery BD, Nair LS, Laurencin CT. Biomedical applications of biodegradable polymers. Journal of Polymer Science Part B: Polymer Physics. 2011; 49: 832–864.
[149] Buwalda SJ, Vermonden T, Hennink WF. Hydrogels for therapeutic delivery: Current developments and future directions. Biomacromolecules. 2016; 18: 316–330.
[150] Feng Q, Li D, Li Q, Cao X, Dong H. Microgel assembly: Fabrication, characteristics and application in tissue engineering and regenerative medicine. Bioactive Materials 2022; 9: 105–119.
[151] Nielsen JJ, Low SA, Ramseier NT, Hadap RV, Young NA, Wang M, Low PS. Analysis of the bone fracture targeting properties of osteotropic ligands. Journal of Controlled Release. 2021; 329: 570–584.
[152] Bai X, Gao M, Syed S, Zhuang J, Xu X, Zhang XQ. Bioactive hydrogels for bone regeneration. Bioactive Materials. 2018; 3: 401–417.
[153] Ji X, Yuan X, Ma L, Bi B, Zhu H, Lei Z, Liu W, Pu H, Jiang J, Jiang X, Zhang Y. Mesenchymal stem cell-loaded thermosensitive hydroxypropyl chitin hydrogel combined with a three-dimensional-printed poly (ε-caprolactone)/nano-hydroxyapatite scaffold to repair bone defects via osteogenesis, angiogenesis and immunomodulation. Theranostics. 2020; 10(2):725.
[154] Garg V, Gupta T, Rani S, Bandyopadhyay-Ghosh S, Ghosh SB, Qiao L, Liu G. A hierarchically designed nanocomposite hydrogel with multisensory capabilities towards wearable devices for human-body motion and glucose concentration detection. Composites Science and Technology. 2021; 213: 108894.
[155] Yue S, He H, Li B, Hou T. Hydrogel as a biomaterial for bone tissue engineering: A review. Nanomaterials. 2020; 10: 1511.
[156] Chen L, Yu C, Xiong Y, Chen K, Liu P, Panayi AC, Xiao X, Feng Q, Mi B, Liu G. Multifunctional hydrogel enhances bone regeneration through sustained release of stromal cell-derived factor-1α and exosomes. Bioactive Materials. 2023; 25: 460–471.
[157] Guo N, Zhang L, Wang J, Wang S, Zou Y, Wang X. Novel fabrication of morphology tailored nanostructures with Gelatin/Chitosan Co-polymeric bio-composited hydrogel system to accelerate bone fracture healing and hard tissue nursing care management. Process Biochemistry. 2020; 90: 177–183.
[158] Han X, He J, Wang Z, Bai Z, Qu P, Song Z, Wang W. Fabrication of silver nanoparticles/gelatin hydrogel system for bone regeneration and fracture treatment. Drug Delivery. 2021; 28: 319–324.
[159] Han X, He J, Wang Z, Bai Z, Qu P, Song Z, Wang W. Fabrication of silver nanoparticles/gelatin hydrogel system for bone regeneration and fracture treatment. Drug Delivery. (London). 2021; 28: 319–324.

[160] Söntjens SHM, Nettles DL, Carnahan MA, Carnahan MA, Setton LA, Grinstaff MW. Biodendrimer-based hydrogel scaffolds for cartilage tissue repair. Biomacromolecules. 2006; 7: 310–316.

[161] Ren K, He C, Xiao C, Li G, Chen X. Injectable glycopolypeptide hydrogels as biomimetic scaffolds for cartilage tissue engineering. Biomaterials. 2015; 51: 238–249.

[162] Vilela CA, Correia C, Oliveira JM, Sousa RA, Mendes JE, Reis RL. Cartilage repair using hydrogels: A critical review of in vivo experimental designs. ACS Biomaterials Science & Engineering. 2015; 1: 726–739.

[163] Camarero-Espinosa S, Rutishauser BR, Weder C, Foster EJ. Directed cell growth in multi-zonal scaffolds for cartilage tissue engineering. Biomaterials. 2016; 74: 42–52.

[164] Leppiniemi J, Lahtinen P, Paajanen A. 3D-printable bioactivated nanocellulose-alginate hydrogels. ACS Applied Materials & Interfaces. 2017; 9: 21,959–21970.

[165] Ju X, Liu X, Zhang Y, Chen X, Chen M, Shen H, Feng Y, Liu J, Wang M, Shi Q. A photo-crosslinked proteinogenic hydrogel enabling self-recruitment of endogenous TGF-β1 for cartilage regeneration. Smart Materials in Medicine. 2022; 3: 85–93.

[166] Xiang C, Guo Z, Zhang Q, Wang Z, Li X, Chen W, Wei X, Li P, Xiang C. Physically crosslinked poly(vinyl alcohol)-based hydrogels for cartilage tissue engineering. Materials and Design. 2024; 243: 113048.

[167] Messerschmitt PJ, Garcia RM, Abdul Krim WF, Greenfield EM, Getty PJ. Osteosarcoma. JAAOS – Journal of the American Academy of Orthopaedic Surgeons. 2009; 17(8): 515–527.

[168] Schiff MH, Whelton A. Renal toxicity associated with disease-modifying antirheumatic drugs used for the treatment of rheumatoid arthritis. Seminars in Arthritis & Rheumatism. 2000; 30(3): 196–208.

[169] Yang Z, Liu J, Lu Y. Doxorubicin and CDCUR inclusion complex coloaded in thermosensitive hydrogel PLGAPEGPLGA localized administration for osteosarcoma. International journal of oncology. 2020; 57(2): 433–444.

[170] Li D, Nie W, Chen L, McCoul D, Liu D, Zhang X, Li Y, Yu B, He C. Self-assembled hydroxyapatite-graphene scaffold for photothermal cancer therapy and bone regeneration. Journal of Biomedical Nanotechnology. 2018; 14: 2003–2017.

[171] Li C, Zhang W, Wang R, Du X-F, Jiang D, Liu B, Nie Y, Liao J, Chen Y, Liang X, Qin L, Lai Y. Nanocomposite multifunctional hydrogel for suppressing osteosarcoma recurrence and enhancing bone regeneration. Chemical Engineering Journal. 2022 1; 435(2): 134896.

[172] Diego T, Lorenzo V, Eti T, Madina T, Gilbert N, Saverio A, Al-Haddad H, Lisignoli G, Ricotti L. Graphene oxide-doped gellan gum-PEGDA bilayered hydrogel mimicking the mechanical and lubrication properties of articular cartilage. Advanced Healthcare Materials. 2021; 10(7): 2001434.

[173] Qi X, Qin X, Yang R, Qin J, Li W, Luan K, Wu Z, Song L. Intra-articular administration of chitosan thermosensitive in situ hydrogels combined with diclofenac sodium-loaded alginate microspheres. Journal of Pharmaceutical Sciences. 2016; 105: 122–130.

[174] Zhao Y, Gao C, Liu H, Feng Y, Li Z, Liu H, Wang J, Lin Q, Yang B. Infliximab-based self-healing hydrogel composite scaffold enhances stem cell survival, engraftment, and function in rheumatoid arthritis treatment. Acta Biomateriala. 2021; 121: 653–664.

[175] Chen FM, Wu LA, Zhang M, Zhang R, Sun HH. Homing of endogenous stem/progenitor cells for in situ tissue regeneration: Promises, strategies, and translational perspectives. Biomaterials. 2011; 32(12): 3189–3209.

[176] Kharbikar BN, Mohindra P, Desai TA. Biomaterials to enhance stem cell transplantation. Cell Stem Cell. 2022; 29: 692–721.

[177] Madl CM, Flaig IA, Holbrook CA, Wang YX, Blau HM. Biophysical matrix cues from the regenerating niche direct muscle stem cell fate in engineered microenvironments. Biomaterials. 2021; 275: 120973.

[178] Hass R, Kasper C, Böhm S, Jacobs R. Different populations and sources of human mesenchymal stem cells (MSC): A comparison of adult and neonatal tissue-derived MSC. CCS. 2011; 9: 12.

[179] Pittenger MF, Discher DE, Péault BM, Phinney DG, Hare JM, Caplan AI. Mesenchymal stem cell perspective: Cell biology to clinical progress. NPJ Regenerative Medicine. 2019; 4: 22.
[180] Fernandez PG, Nogales CR, Jordan O, Allémann E. Combination of mesenchymal stem cells and bioactive molecules in hydrogels for osteoarthritis treatment. European Journal of Pharmaceutics and Biopharmaceutics. 2022; 172: 41–52.
[181] Engler AJ, Sen S, Sweeney HL, Discher DE. Matrix elasticity directs stem cell lineage specification. Cell. 2006; 126: 677–689.
[182] Nam S, Lee J, Brownfield DG, Chaudhuri O. Viscoplasticity enables mechanical remodeling of matrix by cells. Biophysical Journal. 2016; 111: 2296–2308.
[183] Schmitt SK, Xie AW, Ghassemi RM, Trebatoski DJ, Murphy WL, Gopalan P. Polyethylene glycol coatings on plastic substrates for chemically defined stem cell culture. Advanced Healthcare Materials. 2015; 4: 1555–1564.
[184] Jeon O, Wolfson DW, Alsberg E. In-situ formation of growthfactor-loaded coacervate microparticle-embedded hydrogels for directing encapsulated stem cell fate. Advances in Materials. 2015; 27: 2216–2223.
[185] Takahashi K, Tanabe K, Ohnuki M, Narita M, Ichisaka T, Tomoda K. Induction of pluripotent stem cells from adult human fibroblasts by defined factors. Cell. 2007; 131: 861–872.
[186] Avior Y, Sagi I, Benvenisty N. Pluripotent stem cells in disease modelling and drug discovery. Nature Reviews Molecular Cell Biology. 2016; 17: 170–182.
[187] Madden LR, Mortisen DJ, Sussman EM, Dupras SK, Fugate JA, Cuy JL, Hauch KD, Laflamme MA, Murry CE, Ratner BD. Proangiogenic scaffolds as functional templates for cardiac tissue engineering. Proceedings of the National Academy of Sciences of the United States of America. 2010; 107: 15211–15216.
[188] Ekerdt BL, Fuentes CM, Lei Y, Adil MM, Ramasubramanian A, Segalman RA, Schaffer DV. Thermoreversible hyaluronic acid-PNIPAAm hydrogel systems for 3D stem cell culture. Advanced Healthcare Materials. 2018; 7: e1800225.
[189] Nazari B, Kazemi M, Kamyab A, Nazari B, Ebrahimi-Barough S, Hadjighassem M, Norouzi-Javidan A, Ai A, Ahmadi A, Ai J. Fibrin hydrogel as a scaffold for differentiation of induced pluripotent stem cells into oligodendrocytes. Journal of Biomedical Materials Research Part B. 2020; 108: 192–200.
[190] Hunt NC, Hallam D, Karimi A, Mellough CB, Chen J, Steel DHW, Lako M. 3D culture of human pluripotent stem cells in RGD-alginate hydrogel improves retinal tissue development. Acta Biomaterialia. 2017; 49: 329–343.
[191] Glorevski N, Sachs N, Manfrin A, Giger S, Bragina ME, Moran PO, Clevers H, Lutolf MP. Designer matrices for intestinal stem cell and organoid culture. Nature. 2016; 539: 560–564.
[192] Fattahi P, Rahimian A, Slama MQ, Gwaon K, Suarez AMG, Wolf J, Baskaran H, Duffy CD, Stybayeva G, Peterson QP, Revzin A. Core–shell hydrogel microcapsules enable formation of human pluripotent stem cell spheroids and their cultivation in a stirred bioreactor. Scientific Reports. 2021; 11: 7177.
[193] Morrison SJ, Spradling AC. Stem cells and niches: Mechanisms that promote stem cell maintenance throughout life. Cell. 2008; 132: 598–611.
[194] Lee-Thedieck C, Schertl P, Klein G. The extracellular matrix of hematopoietic stem cell niches. Advanced Drug Delivery Reviews. 2021; 181: 114069.
[195] Copelan EA. Hematopoietic stem-cell transplantation. The New England Journal of Medicine. 2006; 354(17): 1813–1826.
[196] Nikiforow S, Ritz J. Dramatic expansion of HSCs: New possibilities for HSC transplants?. Cell Stem Cell. 2016; 18(1): 10–12.
[197] Rocha V, Gluckman E. Improving outcomes of cord blood transplantation: HLA matching, cell dose and other graft- and transplantation-related factors. British Journal of Haematology. 2009; 147(2): 262–274.

[198] Liu B, Jin M, Wang DA. In vitro expansion of hematopoietic stem cells in a porous hydrogel-based 3D culture system. Acta Biomaterialia. 2023; 161: 67–79.
[199] Hong J, Zhu Z, Cui L, Wang Z, Hao Y, Tian X, Cheng G. Bone marrow-inspired hydrogel/graphene composite scaffolds to support in vitro expansion of hematopoietic stem cells. Journal of Materials Chemical B. 2024; 12: 2354–2363.
[200] Hong J, Zheng W, Wang X, Hao Y, Cheng G. Biomedical polymer scaffolds mimicking bone marrow niches to advance in vitro expansion of hematopoietic stem cells. Journal of Materials Chemistry. 2022; 10: 9,755–9769.
[201] Bai B,T, Li J, Sinclair A, Imren S, Merriam F, Sun F, O'Kelly MB, Nourigat C, Jain P, Delrow JJ. Expansion of primitive human hematopoietic stem cells by culture in a zwitterionic hydrogel. Nature Medicine. 2019; 25: 1566–1575.

T. Senthamil Selvan and Sumit Pramanik*

Chapter 1.3
Recent developments of nanocomposites and fabrications for biosensor applications

Abstract: A biosensor is a particular kind of sensor that recognizes biomaterials or biological entities. In recent, polymer composites have significantly increased their flexibility, sensitivity, and sensitivity for controlling biomolecules. Thus, plenty of desired qualities, a wide range of industries, including biomedicine and environmental analysis, are interested in creating biosensors using polymer composites. In this chapter, the types of composites, biosensors, and the fabrications of biosensors have discussed extensively. More specifically, the direct growth, coating, direct deposition, and printing technologies have been discussed with several informative illustrations. It has been observed that the deposition of thin films of nanoparticles controls a small number of the challenges. In order to create a variety of devices, including biosensors, energy storage devices, and photonics, the electrode's surface is currently coated or modified using a variety of processes. In this context, proper deposition techniques can solve a number of problems with conventional techniques. Researchers interested in creating cost-effective biosensors will be benefited by this chapter.

Keywords: Biosensors, Electrode, Coating, Printing, Nanomaterial

1.3.1 Introduction

The medical field, environmental protection, the packaging sector, and other sectors all use biosensors as diagnostic equipment. Virus, bacterial, and microbe identification, illness determination, medication development, food safety, monitoring the environment, and quick diagnostics are all possible using biosensors [1, 2]. A biosensor is a type of sensor that uses a biological particle as its recognition element [3]. Previously, a wide range of biosensing devices were developed by incorporating diverse biologically derived chemicals into mechanical, electrochemical, optical, and acoustic detectors. These substances included nucleic acids, enzymes, cells, antigen antibodies,

***Corresponding author: Sumit Pramanik**, Functional and Biomaterials Engineering Lab, Department of Mechanical Engineering, College of Engineering and Technology, SRM Institute of Science and Technology, Kattankulathur, Kancheepuram, Chennai 603 203, Tamil Nadu, India, e-mail: sumitprs@srmist.edu.in, prsumit@gmail.com, ORCID: 0000-0003-0455-1582
T. Senthamil Selvan, Functional and Biomaterials Engineering Lab, Department of Mechanical Engineering, College of Engineering and Technology, SRM Institute of Science and Technology, Kattankulathur, Kancheepuram, Chennai 603 203, Tamil Nadu, India, e-mail: st1544@srmist.edu.in

https://doi.org/10.1515/9783111386027-003

and microbes. The intention was to create devices with superior accuracy and sensitivity [4]. Excellent sensitivity, selectivity, perfect response, stability, reversibility, affordability, and user-friendliness are some of the best qualities of modern sensors. Its rapid response to the appearance of the monitored component enables rapid intervention, which is a major advantage [5].

Natural polymers such as chitosan, cellulose, and starch are commonly utilized in sensor applications. Due to their features, biopolymers have a wide range of uses. For example, they can accumulate analytes on sensor surfaces and can be used in many ways like thin sheets, hydrogels, and sponges. Every year, researchers are interested in developing the biosensor using biopolymer composites in several industries [1]. Biosensors are classified into different types based on various applications of bioreceptors and transducers [6].

Recently, polymers have greatly improved their sensitivity, sensitivity, and flexibility for control of biomolecules. The fluorinated ionomer Nafion was one of the first polymer architectures used in biosensing tools [7]. In the field of analytical sciences, polymers have long been regarded as promising materials for the development of biomolecule setup and immobilization systems. Many industries, from environmental analysis to biomedicine, are interested in developing biosensors using polymers because of their many desirable properties such as superior conductivity, biofunctionalization, endurance, biocompatibility, and easily adjustable chemical functionalities [8].

1.3.2 Types of composites

A combination of two or more materials chemically and physically bonded together is called composite materials. The composites are classified according to the matrix such as metal, polymer, and ceramic matrix composites. The classification of composite and types of polymer matrix composites are shown in Figure 1.3.1. Chemical and phys-

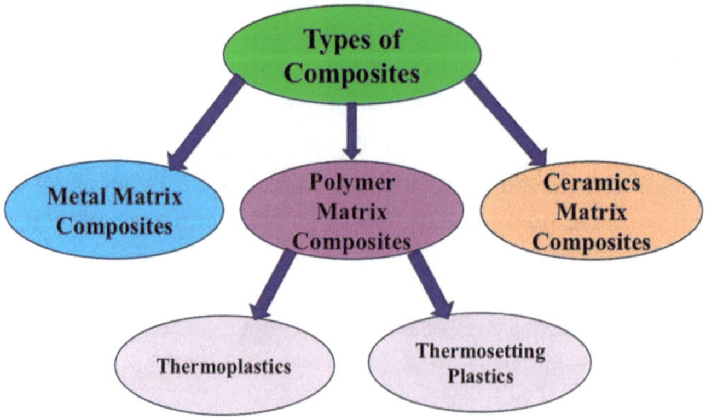

Figure 1.3.1: Types of polymer matrix composites.

ical characteristics of conductive and nonconductive binding materials used to produce the composite biosensors. The carbon-based materials naturally acted as electrically conductive materials with the polymer matrix for example (carbon nanotubes, carbon, graphene, and graphite) [1, 9].

Polymers are essential to modern engineering as well as the biological world. While various naturally occurring polymers, like proteins and nucleic acids, convey crucial biological information, other polymers, like polysaccharides, provide energy for cell activity and function as building blocks for biological structures [5]. The various polymeric composite materials are shown in Table 1.3.1.

Table 1.3.1: The different composites of polymers.

S. no.	Matrix	Reinforcement	Reference
1	Bacterial cellulose	Prussian blue (PB) nanocubes	[10]
2	ABS	Carbon nanofibers	[11]
3	Polylactic acid	Almond skin powder	[12]
4	Starch	Graphene	[13]
5	Starch	Cellulose	[14]
6	Starch	AuNP (silver nanoparticles)	[15]
7	Starch	SWCNTs	[16]
8	Methacrylated chitosan	Graphene oxide	[17]

Polymers are crucial for creating sensor devices using composites implemented in various sectors of the industrial and economic domains. Biopolymers from biological chemical sensing systems exhibit flawless performance, making them ideal candidates for sensitive components in the development of sensors that use biopolymers for chemical sensing in intricate settings. Chemical sensing sensors that use biopolymers or biomaterials have made significant strides, provided useful viewpoints, and shown possible uses [18].

1.3.3 Types of biosensors

The sensor is a device that could be applied to gather the physical activity and convert into a predictable signal. It needs to be able to react in the intended manner with various biological parts. For instance, the biosensor can identify biomolecules like urea, estrogen, glucose, or cholesterol [19]. Biosensors are classified as stated by the signal transduction mode and biological mode as shown in Figure 1.3.2. The biosensor device that could measure analytes constantly and quickly by both one-use and multiuse biosensors [20]. Choosing an appropriate immobilization technique is essential to maintain biorecognition components during sensor readings and to provide a robust sensor for repeated uses [21].

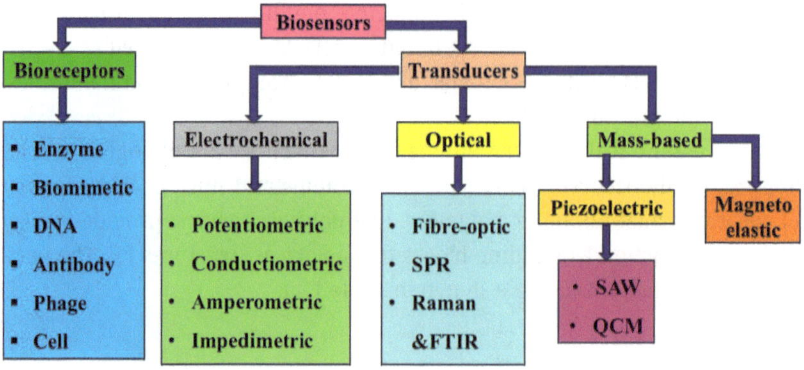

Figure 1.3.2: Classifications of biosensors.

1.3.4 Types of fabrications of biosensors

The fabrication types of biosensors are shown in Figure 1.3.3. The oxygen electrode was used as a glucose biosensor in 1962. Many conductive electrodes have been exploited in the fabrication of biosensors such as silver, carbon, platinum, fluorine-doped tin oxide (FTO), and gold [22]. Nevertheless, biosensor outcomes are limited when using bare electrodes.

Therefore, to enhance sensitivity, electrode alteration using nanomaterials is necessary to increase electrode area on the surface. By using modified electrodes, catalytic processes may also benefit from improved electron transport, less nonspecific binding, and improved immobilization of functional recognition molecules [21]. Biosensors have been fabricated by several methods like direct growth, screen printing, coating, laser scribing, inkjet printing, deposition, and nozzle-jet printing methods. The various fabrication methods of biosensors using composites are illustrated in Table 1.3.2.

1.3.4.1 Direct growth method

The surface area and shape of nanoparticles are crucial for creating biosensors. The direct growth method has improved the surface contact between the materials and the electrode, stability, and electronics. Due to its straightforward manufacturing method, the direct growth of nanomaterials on electrode surfaces is gaining more and more attention. Different methods are used to grow nanomaterials like anodization, hydrothermal, chemical, and thermal decomposition [21]. Figure 1.3.4 shows the fabrication of glucose sensors by direct growth method. The various fabrication methods of biosensors using composites are illustrated in Table 1.3.2.

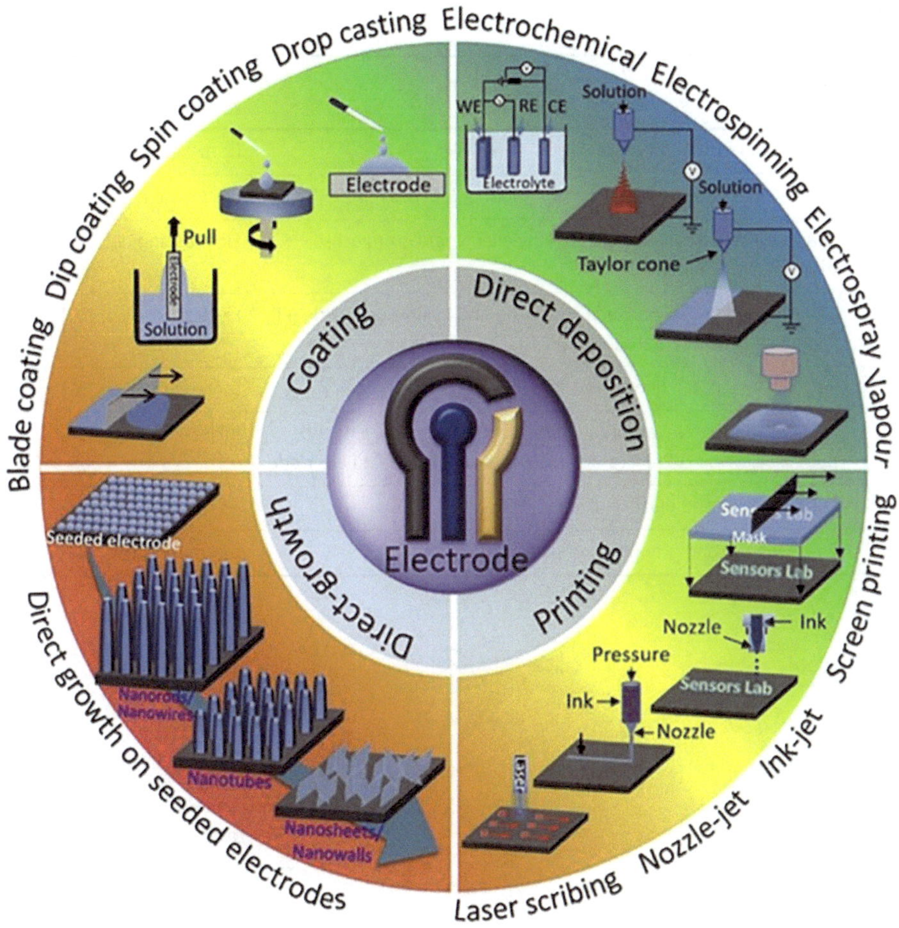

Figure 1.3.3: The various fabrication methods of biosensors using nanomaterials (adapted from ref. [21]. with permission).

Table 1.3.2: Various fabrication methods of biosensors using composites.

S. no.	Materials	Fabrication technique	Reference
1	ZnO/Au/Cr/glass substrate	Direct growth	[23]
2	F-doped SnO_2-conducting glass (Fe_3O_4@SiO2@NH_2/SnO_2:F)	Dip coating	[24]
3	Urease/nano-ZnO/TiO_2/FTO	Blade coating	[25]
4	PDMS/PEDOT:PSS (poly(ethylene dioxythiophene):polystyrene sulfonate)	Spin coating	[26]

Table 1.3.2 (continued)

S. no.	Materials	Fabrication technique	Reference
5	Rectangular ytterbium oxide (Yb_2O_3) nanodisks	Drop casting	[27]
6	Cu/Cu_2O nanocomposite coated on fluorine-doped tin oxide (FTO)	Electrochemical	[28]
7	PVA (poly(vinyl alcohol))/PEI Poly(ethyleneimine)/Gox (glucose oxidase) NFs (nanofibers)	Electrospinning	[29]
8	GO, rGO, and CNF-deposited FTO, silicon and flexible polyimide (PI), and polyethylene terephthalate (PET))	Electrospray	[30]
9	Polyurethane (PU), conductive silver-flake-based ink	Screen printing	[31]
10	AQ-labeled acpcPNA (anthraquinone-labeled pyrrolidinyl peptide nucleic acid) and G-PANI (graphene-polyaniline)-modified electrode	Inkjet	[32]
11	ZnO nanorods and glucose oxidase (GOx)	Nozzle jet	[33]
12	Polyimide (PI) and graphitic carbon	Laser scribing	[34]

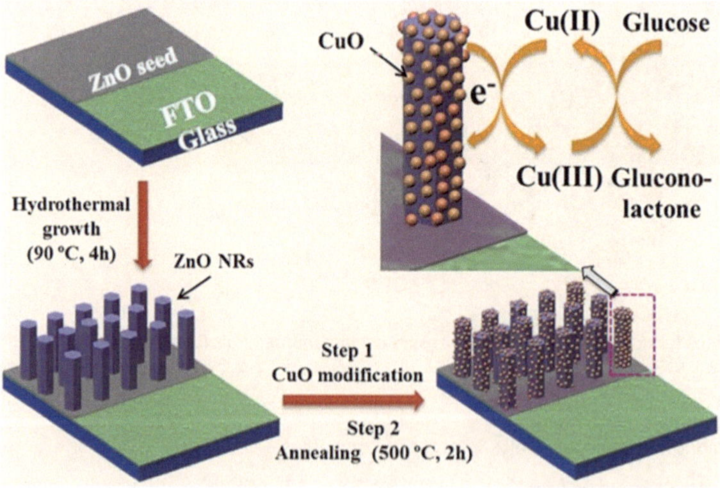

Figure 1.3.4: Fabrication procedure of glucose sensor and detection mechanism (adapted from ref. [35] with permission).

In this method, the hydrothermal process is used to grow the ZnO nanorods on the FTO electrode surface. The ZnO nanorods are vertically grown on the surface of the electrode. To improve the performance of the detection of glucose, ZnO nanorods and the FTO layer were functionalized with copper oxide. Primarily, FTO was coated on the glass plate as a thin layer. Continuously, ZnO was deposited on the coated FTO layer. Then, ZnO nanorods were fixed vertically and grew on the seeded area. Finally,

CuO evenly coated on the nanorods [35]. This method enhances the efficiency of the sensor due to the ease of fabrication of nanorods, and the fabrication cost is less [36].

1.3.4.2 Coating method

Coating techniques are used for electrodes with nanomaterials. The coating technique does not have difficult instruments. So, this is a cheap fabrication technique to make a biosensor. In this technique, solution-based nanomaterials are deposited on the surface of the electrode using various coating processes such as drop casting, dip coating, spin coating, and blade coating. These methods are shown in Figure 1.3.5.

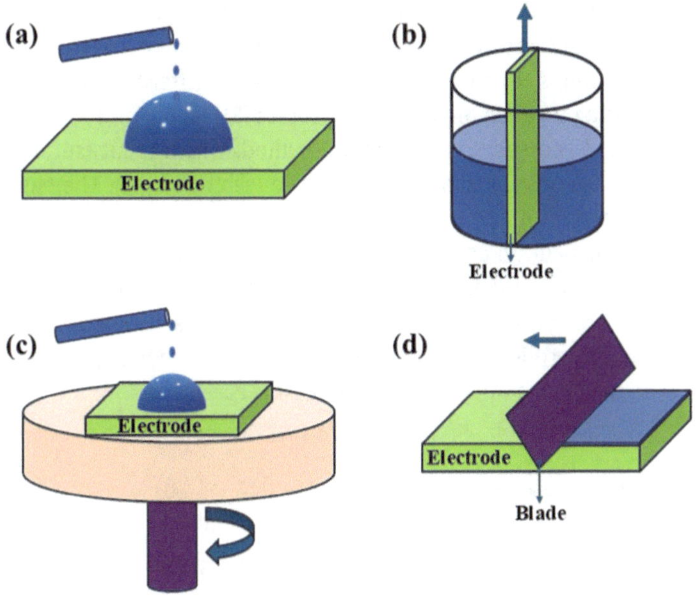

Figure 1.3.5: Systematic diagrams of coating methods of deposition on electrode: (a) drop casting, (b) dip coating, (c) spin coating, and (d) blade coating.

In this drop-casting technique, nanomaterials are uniformly dispersed and dropped on the cleaned electrode's surface. Then the dropped surface will be evaporated and dried to bond the solvents. However, this process has less waste of materials, and many parameters such as drying, materials mixing, and temperature are optimized to get a smooth surface of the coated electrode [37]. The systematic diagram of the drop-casting method is displayed in Figure 1.3.5(a). In this method, the electrode was cleaned, and then the prepared nanomaterial solution was released dropwise by dropper until the solution was evenly spread to the total electrode's surface and dried at an atmosphere temperature [38].

The dip-coating process is one of the coating processes which is another easier process based on the sol-gel method for spreading the coating homogeneously on the huge electrode surfaces [39, 40]. The systematic diagram of the dip-coating method is displayed in Figure 1.3.5(b). Initially, the prepared electrode is immersed in the nanomaterial solution for a particular duration and taken out at a constant speed. The immersion time, electrode return-taking speed, and stability of force at the electrode edge control the thickness of the film and regularity [41]. This method is not appropriate for extreme enzyme immobilization. It is used to fabricate the very tiny layer on the surface of the electrode. It has the main advantage of creating a homogeneous spreading layer on the electrode over the drop-casting technique [21].

Among the coating techniques, spin coating is a vital method for fabricating the polymer film assembly. Figure 1.3.5(c) shows the working principle of the spin-coating method. In the spin-coating method, sol-gel formation is used to coat the solution on the electrode surface. The solution is dropped on the electrode surface when it is spinning. Then, the dropped solution is covered quickly as a thin layer. After that moisture will evaporate from the coating through an annealing process [21].

The blade-coating method is a solution-deposition method, which is measurable, needs less temperature, less fabrication charge, and a facile technique [42]. The basic diagram of the blade-coating technique is shown in Figure 1.3.5(d). This method is used to coat the nanomaterial solution on big surfaces with the same thickness, quick deposition, and spread to uniform thickness without wasting the materials [43]. In this process, nanoparticles are synthesized with polymer, and the solution is transferred to an electrode. Then, the roller or blade is moved to the electrode to get a uniform thickness. After that, the drying and annealing process is carried out.

1.3.4.3 Direct deposition method

In the last 10 years, biosensor electrodes were manufactured by direct deposition techniques. In this method, nanomaterials are straightly pasted onto electrodes. This direct deposition technique is classified into four types and has shown in Figure 1.3.6:

1. Electrochemical
2. Electrospinning
3. Electrospray
4. Sputtering and vapor

Electrochemical deposition is an easy and less expensive process mainly applied to coating on a conductive electrode. This process does not need an extreme vacuum and works at an ambient temperature. Three types of electrodes reference, working, and counter electrodes are used in this process [44]. Electrochemical deposition has established general consideration for depositing nanomaterials and manufacturing various biosensors due to its simple method, less fabrication time, and low production

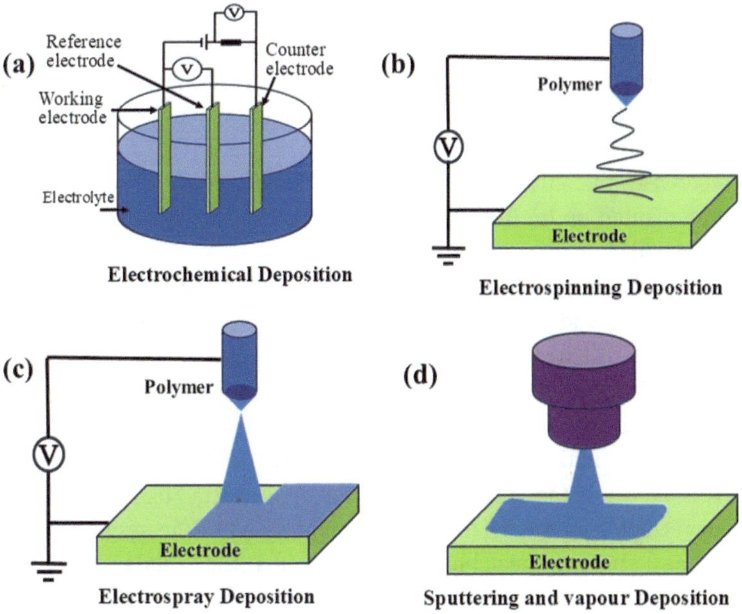

Figure 1.3.6: Systematic diagrams of direct deposition methods on the electrode: (a) electrochemical, (b) electrospinning, (c) electrospray, and (d) sputtering and vapor.

cost [45]. This process is carried out by various methods like potentiostatic, chronoamperometric, cyclic voltammetric, and pulse-potentiostatic deposition [46]. Figure 1.3.6(a) illustrates the electrochemical deposition method schematically.

The electrospinning method is a facile, multipurpose, and direct procedure method to fabricate the biosensor with different nanomaterials such as carbon nanotubes, graphene, TiO_2, Au, Fe_3O_4, and Ag [47–49]. Drug delivery, purification of water, environmental safety, textile industries, and tissue engineering applications are appropriate for electrospinning deposition [47]. The schematic diagram of the electrospinning method is displayed in Figure 1.3.6(b). This method inserts the prepared nanoparticle solution into a stainless-steel syringe with a metallic needle. The needle is connected to the positive electrode, and the collector is connected to the grounded electrode. While the evaporation of the solvent was quick, the polymer solution was evicted to the bench electrode by the voltage of 15 kV. The nanofibers were stabilized in an air atmosphere at a temperature of 280 °C for a duration of 1 h with a heating rate of 5 °C/min. Using the same heating rate it was carbonized at a temperature of 1,100 °C in a nitrogen environment [50].

The electrospray method approaches a technology similar to electrospinning to fabricate the nanostructure, which can be deposited on metallic and nonmetallic surfaces. The schematic diagram of the electrospray method is displayed in Figure 1.3.6(c). The industrial and scientific instruments such as micromachining, microelectronics, nanotechnology, and material technologies are fabricated by this method [51]. This method runs on two power sources: the metallic nozzle is connected to the positive power source, and the electrode is connected to the negative power source. Heat is supplied to the substrate and the electrode, and then heat should be kept at the required temperature. Ensure the temperature to full evaporation of the solvent and keep the working distance between the substrate and nozzle. Before materials deposition, the electrodes should be printed on the substrate. The electric resistance is evaluated by the printed substrates [30].

The thin film of nanomaterials on the surface of the electrode has been fabricated by the vapor deposition method. The vapor deposition or sputtering method is schematically shown in Figure 1.3.6(d). Atomic layer deposition (ALD) and sputter are commonly used techniques for depositing various films by altering high-purity metal foil, evaporation slugs, sputtering targets, and pellets. The use of elevated temperatures in ALD restricts the available options for electrode materials used in film formation. Nevertheless, ALD is used to deposit films of superior quality. Zinc oxide (ZnO) is a highly used nanomaterial for producing thin films with precise thicknesses on various substrates or electrodes. This may be achieved using either ALD or sputtering techniques. In addition, the other variables of the movie may be manipulated. ALD-produced ZnO films have shown to be an effective substrate for the development of biosensors capable of detecting uric acid, grapevine virus A-type proteins, urea, glucose, DNA, and cholesterol [21].

1.3.4.4 Printing techniques

Among the fabrication of biosensor techniques, printing methods are very fascinating deposition methods to deposit a nanomaterial in flexible electrodes. The large-scale and low-cost production industries fabricate the biosensors using this printing technique. In this method, inks are printed on thin films at a low temperature and quick fabrication of soft electronics on different substrates or electrodes. The printing methods are developed by various techniques which are given below, and schematic diagrams of various printing techniques are shown in Figure 1.3.7:
1. Screen printing
2. Inkjet
3. Nozzle jet
4. Laser scribing

The screen-printing method is famous for its great processing efficiency, adaptability to constant roll-to-roll manufacturing, and industrialization, screen printing makes it easy to make inexpensive, portable devices with enormous printing areas [52, 53]. This method uses the patterned mask to spread the ink more easily and quickly than other techniques. The schematic diagram of screen printing is shown in Figure 1.3.7(a).

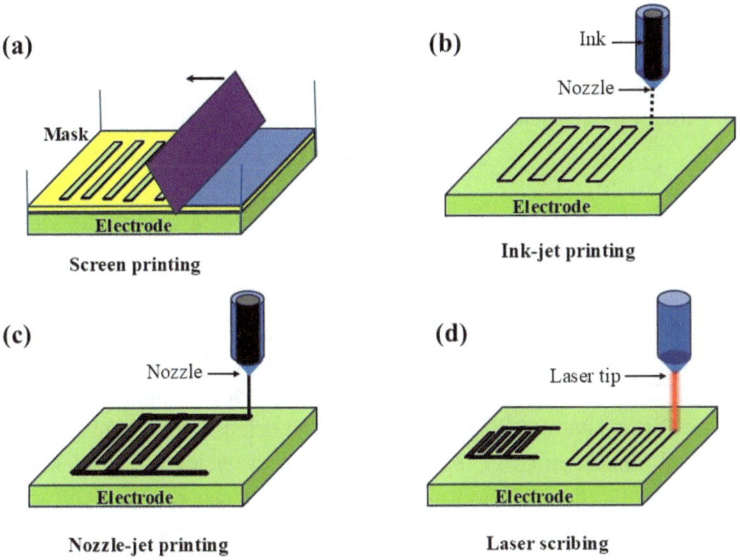

Figure 1.3.7: Systematic diagrams of printing methods on the electrode: (a) screen printing, (b) ink-jet printing, (c) nozzle-jet printing, and (d) laser scribing.

This method utilizes the mesh to transmit the ink on the surface of the electrode. The stencil is used to transfer the designed pattern onto the electrode. The mesh is fabricated from various materials like fabric, metal threads, silk, and synthetic fibers. The mesh does not permit the ink to the nonprinting areas, and it permits the ink to the printable areas only. This printing technique improves the sensor's efficiency due to the nanomaterial inks or paste used for the printing [31, 54–56].

The inkjet is a drop-based noncontact technique compared to the screen-printing technique. High conductive patterns are printed on the electrode or substrate surface by this technique due to its great resolution [57, 58]. In this process, the dropping size flow speed can be controlled, and this method provides a smaller nozzle tip. The inkjet-printing method is shown in Figure 1.3.7(b). The major advantages of this method are eliminating contamination and materials wastage. Mainly, inks are made of nanomaterials such as carbon nanomaterials, metal oxides, and conductive polymers. The sensor, energy storage devices, and transistors are mainly fabricated by this technique [21]. Initially, the silver paste was printed on the reference electrode and then ink was printed to make a counter electrode. Lastly, the insulator was printed as a thin layer [59].

The nozzle-jet printing method is a low-cost and easier method than inkjet printing. It prints a designed pattern on the substrate or electrode surface with a low resolution. This technique uses a big nozzle and high pressure to print the patterns by using high-viscosity inks. The parameters of nozzle tip dimensions, nozzle moving speed, ink pressure, stage temperature, and height between the electrode and nozzle tip are optimized by this printing technique [60]. The schematic diagram of the nozzle-jet printing technique is displayed in Figure 1.3.7(c). Recently, the enzymatic and nonenzymatic biosensors have been fabricated using nanoparticle-based inks, and it is famous for constructing hydrogel-based scaffolds [21]. Initially, silver nanoparticles were printed onto the substrate/electrode surface by nozzle-jet method. Similarly, nanoparticle ink was printed on silver printing. Finally, the printed sensor was dried in an oven at 80 °C for 3 h [61].

The laser-scribing technique fabricates the sensors without a mask. The laser-scribing technique makes it difficult for the growth of biosensors. The additional vital feature of laser scribing is the development of various designed tiny devices [62]. The schematic diagram of the laser-scribing technique is displayed in Figure 1.3.7(d). The separate laser-scribed graphene (LSG) electrodes are prepared by graphene oxide film, and the light scribe DVD label writing method was used to prepare separate LSG electrodes [63]. Initially, GO was layered on the PI substrate and dried at room temperature and pasted on DVD. The LSG-loaded PI foil is printed with a variety of designs five times using a regular LightScribe DVD optical drive. The Nero cover designer software was used to design the pattern. Finally, the PI substrate was taken off the DVD [64].

1.3.5 Conclusions and future perspectives

Advanced nanocomposites and fabrication methods for biosensor applications were discussed. The fabrication techniques utilized to attach nanoparticles to electrode surfaces in the course of biosensor production were covered in this chapter. An overview of all the biosensor manufacturing techniques was provided, with an emphasis on how each process was suitable for achieving efficient sensing performance. The nanomaterial's major role is to enhance the surface area for immobilization of biomolecules. Depending upon the working principle, the fabrication methods of biosensors are classified as follows: (a) direct growth method, (b) coating-based method, (c) direct deposition method, and (d) printing method. Many kinds of literature synthesized the nanoparticles separately to fabricate the biosensors. Usual methods are used to coat the surface of the electrodes like drop casting, screen printing, spin coating, and dip coating. However, some challenges of fabricated electrodes by usual techniques are steadiness but reproducibility is limited. Commonly, usual fabrication methods are easy to use; however, many difficulties occurred during the production of biosensors. However, few of the difficulties are controlled by the deposition of nanomaterials thin films. Currently, the surface of the electrode is coated or grown by many techniques to produce various devices like biosensors, energy storage devices, and photonics. Several issues in usual methods are overcome by deposition methods. This chapter will be useful for researchers who are interested in fabricating biosensors with cost-effectiveness.

References

[1] Madej-Kiełbik L, et al. Biopolymer composites with sensors for environmental and medical applications. Materials. 2022; 15(21): 7493.

[2] Yoon Y, et al. Recent progress in rapid biosensor fabrication methods: Focus on electrical potential application. BioChip Journal. 2024; 18(1): 1–21.

[3] Banica F-G. Chemical Sensors and Biosensors: Fundamentals and Applications, West Sussex, UK: John Wiley & Sons; 2012.

[4] Saha S, Tomar M, Gupta V. Fe doped ZnO thin film for mediator–less biosensing application. Journal of Applied Physics. 2012; 111(10); 102804–102804.

[5] Abhilash M, Thomas D. Biopolymers for Biocomposites and Chemical Sensor Applications. In: Biopolymer Composites in Electronics, Elsevier; Eds: Sadasivuni KK, Ponnamma D, Kim J, Cabibihan J-J, and AlMaadeed MA, editors. Elsevier, Amsterdam, The Netherlands. 2017; pp. 405–435.

[6] Kulkarni MB, Ayachit NH, Aminabhavi TM. Biosensors and microfluidic biosensors: From fabrication to application. Biosensors. 2022; 12(7): 543.

[7] Turner RFB, Sherwood CS. Biocompatibility of perfluorosulfonic acid polymer membranes for biosensor applications, In: Diagnostic Biosensor Polymers. ACS Symposium Series. 1994; Vol. 556, pp. 211–221, Washington, DC, USA.

[8] Lee KM, et al. Chemical design of functional polymer structures for biosensors: From nanoscale to macroscale. Polymers. 2018; 10(5): 551.

[9] Alhadeff E, Bojorge N. Graphite-composites Alternatives for Electrochemical Biosensor. In: John Cuppoletti, editors, Metal, Ceramic and Polymeric Composites for Various Uses, Rijeka, Croatia: IntechOpen; 2011, pp. 597–620.

[10] Gomes NO, et al. Bacterial cellulose-based electrochemical sensing platform: A smart material for miniaturized biosensors. Electrochimica Acta. 2020; 349: 136341.

[11] Ning F, et al. Additive manufacturing of carbon fiber–reinforced plastic composites using fused deposition modeling: Effects of process parameters on tensile properties. Journal of Composite Materials. 2017; 51(4): 451–462.

[12] Singh R, et al. On compressive and morphological features of 3D printed almond skin powder reinforced PLA matrix. Materials Research Express. 2020; 7(2): 025311.

[13] Gürler N, Torğut G. Graphene reinforced potato starch composite films: Improvement of mechanical, barrier and electrical properties. Polymer Composites. 2021; 42(1): 173–180.

[14] Ccorahua R, et al. Starch-cellulose-based triboelectric nanogenerator obtained by a low-cost cleanroom–free processing method. MRS Advances. 2019; 4: 1315–1320.

[15] Bacalzo NP Jr, et al. Controlled microwave–hydrolyzed starch as a stabilizer for green formulation of aqueous gold nanoparticle ink for flexible printed electronics. ACS Applied Nano Materials. 2018; 1(3): 1247–1256.

[16] Prusty G, Das R, Swain SK. Influence of functionalized single-walled carbon nanotubes on morphology, conducting and oxygen barrier properties of poly (acrylonitrile-co–starch). Composites Part B: Engineering. 2014; 62: 236–241.

[17] Sayyar S, et al. 3D printable conducting hydrogels containing chemically converted graphene. Nanoscale. 2017; 9(5): 2038–2050.

[18] Fukushima K, et al. Biodegradation of poly (lactic acid) and its nanocomposites. Polymer Degradation and Stability. 2009; 94(10): 1646–1655.

[19] Kamel S, Khattab TA. Recent advances in cellulose-based biosensors for medical diagnosis. Biosensors. 2020; 10(6): 67.

[20] Thévenot DR, et al. Electrochemical biosensors: Recommended definitions and classification. Biosensors Bioelectron. 2001; 16(1–2): 121–131.

[21] Ahmad R, et al. Deposition of nanomaterials: A crucial step in biosensor fabrication. Materials Today Communications. 2018; 17: 289–321.
[22] Clark LC Jr, Lyons C. Electrode systems for continuous monitoring in cardiovascular surgery. Annals of the New York Academy of Sciences. 1962; 102(1): 29–45.
[23] Kim JY, et al. Tailoring the surface area of ZnO nanorods for improved performance in glucose sensors. Sensors and Actuators, B: Chemical. 2014; 192: 216–220.
[24] Mikani M, et al. Novel I–V disposable urea biosensor based on a dip coated hierarchical magnetic nanocomposite (Fe3O4@ SiO2@ NH2) on SnO2: F layer. Journal of the Chinese Chemical Society. 2017; 64(12): 1446–1459.
[25] Rahmanian R, et al. Development of sensitive impedimetric urea biosensor using DC sputtered Nano-ZnO on TiO_2 thin film as a novel hierarchical nanostructure transducer. Sensors and Actuators B: Chemical. 2018; 256: 760–774.
[26] Pal RK, et al. Micropatterned conductive polymer biosensors on flexible PDMS films. Sensors and Actuators B: Chemical. 2018; 259: 498–504.
[27] Ibrahim AA, et al. Two-dimensional ytterbium oxide nanodisks based biosensor for selective detection of urea. Biosensors Bioelectron. 2017; 98: 254–260.
[28] Cheng X, et al. High performance Cu/Cu_2O nanohybrid electrocatalyst for nonenzymatic glucose detection. Journal of Materials Chemical B. 2016; 4(27): 4652–4656.
[29] Sapountzi E, et al. Gold nanoparticles assembly on electrospun poly (vinyl alcohol)/poly (ethyleneimine)/glucose oxidase nanofibers for ultrasensitive electrochemical glucose biosensing. Sensors and Actuators B: Chemical. 2017; 238: 392–401.
[30] Varea A, et al. Electrospray as a suitable technique for manufacturing carbon-based devices. Journal of Physics D: Applied Physics. 2017; 50(31): 315301.
[31] Suikkola J, et al. Screen-printing fabrication and characterization of stretchable electronics. Scientific Reports. 2016; 6(1): 25784.
[32] Teengam P, et al. Electrochemical paper–based peptide nucleic acid biosensor for detecting human papillomavirus. Analytica Chimica Acta. 2017; 952: 32–40.
[33] Bhat KS, et al. Nozzle-jet printed flexible field–effect transistor biosensor for high performance glucose detection. Journal of Colloid and Interface Science. 2017; 506: 188–196.
[34] Nayak P, et al. Highly efficient laser scribed graphene electrodes for on chip electrochemical sensing applications. Advanced Electronic Materials. 2016; 2(10): 1600185.
[35] Ahmad R, et al. Highly efficient non-enzymatic glucose sensor based on CuO modified vertically-grown ZnO nanorods on electrode. Scientific Reports. 2017; 7(1): 5715.
[36] Ahmad R, Tripathy N, Hahn Y-B. Highly stable urea sensor based on ZnO nanorods directly grown on Ag/glass electrodes. Sensors and Actuators B: Chemical. 2014; 194: 290–295.
[37] Bitsch B, et al. A novel slurry concept for the fabrication of lithium–ion battery electrodes with beneficial properties. Journal of Power Sources. 2014; 265: 81–90.
[38] Kang Z, et al. Direct electrochemistry and bioelectrocatalysis of glucose oxidase in CS/CNC film and its application in glucose biosensing and biofuel cells. RSC Advances. 2017; 7(8): 4572–4579.
[39] Yazdanparast S, et al. Dual-aptamer based electrochemical sandwich biosensor for MCF–7 human breast cancer cells using silver nanoparticle labels and a poly (glutamic acid)/MWNT nanocomposite. Microchimica Acta. 2018; 185: 1–10.
[40] Hussain MM, Rahman MM, Asiri AM. Sensitive L-leucine sensor based on a glassy carbon electrode modified with SrO nanorods. Microchimica Acta. 2016; 183: 3265–3273.
[41] Jang EY, et al. Single walled carbon-nanotube networks on large-area glass substrate by the dip coating method. Small. 2008; 4(12): 2255–2261.
[42] Yang H, Jiang P. Large-scale colloidal self–assembly by doctor blade coating. Langmuir. 2010; 26(16): 13173–13182.

[43] Luo C, et al. Roll-to-roll fabrication of organic nanorod electrodes for sodium ion batteries. Nano Energy. 2015; 13: 537–545.
[44] Wei C, et al. Electrochemical deposition of layered copper thin films based on the diffusion limited aggregation. Scientific Reports. 2016; 6(1): 34779.
[45] Justice Babu K, et al. Three-dimensional dendrite Cu–Co/reduced graphene oxide architectures on a disposable pencil graphite electrode as an electrochemical sensor for nonenzymatic glucose detection. ACS Sustainable Chemistry & Engineering. 2018; 6(2): 1909–1918.
[46] Cheng X, et al. High performance Cu/Cu_2O nanohybrid electrocatalyst for nonenzymatic glucose detection. Journal of Materials Chemistry B. 2016; 4(27): 4652–4656.
[47] Xue J, et al. Electrospun nanofibers: New concepts, materials, and applications. Accounts of Chemical Research. 2017; 50(8): 1976–1987.
[48] Zhang C-L, Yu S-H. Nanoparticles meet electrospinning: Recent advances and future prospects. Chemical Society Reviews. 2014; 43(13): 4423–4448.
[49] Kim JI, et al. A controlled design of aligned and random nanofibers for 3D bi-functionalized nerve conduits fabricated via a novel electrospinning set-up. Scientific Reports. 2016; 6(1): 23761.
[50] Bajaj B, et al. Controllable one step copper coating on carbon nanofibers for flexible cholesterol biosensor substrates. Journal of Materials Chemistry B. 2016; 4(2): 229–236.
[51] Avilez HR, et al. Production of Chitosan Coatings on Metal and Ceramic Biomaterials. In: Amber Jennings J, Bumgardner JD, editors. Chitosan Based Biomaterials, vol. 1, Amsterdam, The Netherlands: Elsevier; 2017, pp. 255–293.
[52] Secor EB, Hersam MC. Emerging carbon and post-carbon nanomaterial inks for printed electronics. The Journal of Physical Chemistry Letters. 2015; 6(4): 620–626.
[53] Kang B, Lee WH, Cho K. Recent advances in organic transistor printing processes. ACS Applied Materials and Interfaces. 2013; 5(7): 2302–2315.
[54] Tong S, Sun J, Yang J. Printed thin–film transistors: Research from China. ACS Applied Materials & Interfaces. 2018; 10(31): 25902–25924.
[55] Xu W, et al. Recent advances of solution-processed metal oxide thin–film transistors. ACS Applied Materials & Interfaces. 2018; 10(31): 25878–25901.
[56] Guzsvány V, et al. Screen–printed enzymatic glucose biosensor based on a composite made from multiwalled carbon nanotubes and palladium containing particles. Microchimica Acta. 2017; 184: 1987–1996.
[57] Gabardo C, Soleymani L. Deposition, patterning, and utility of conductive materials for the rapid prototyping of chemical and bioanalytical devices. Analyst. 2016; 141(12): 3511–3525.
[58] Eshkalak SK, et al. A review on inkjet printing of CNT composites for smart applications. Applied Materials Today. 2017; 9: 372–386.
[59] Carvajal S, et al. Disposable inkjet-printed electrochemical platform for detection of clinically relevant HER–2 breast cancer biomarker. Biosensors and Bioelectronics. 2018; 104: 158–162.
[60] Bhat KS, et al. Low–temperature sintering of highly conductive silver ink for flexible electronics. Journal of Materials Chemistry C. 2016; 4(36): 8522–8527.
[61] Bhat KS, et al. Fully nozzle-jet printed non–enzymatic electrode for biosensing application. Journal of Colloid and Interface Science. 2018; 512: 480–488.
[62] Yi F, et al. Wearable energy sources based on 2D materials. Chemical Society Reviews. 2018; 47(9): 3152–3188.
[63] Griffiths K, et al. Laser–scribed graphene presents an opportunity to print a new generation of disposable electrochemical sensors. Nanoscale. 2014; 6(22): 13613–13622.
[64] Lin S, et al. A flexible and highly sensitive nonenzymatic glucose sensor based on DVD–laser scribed graphene substrate. Biosensors and Bioelectronics. 2018; 110: 89–96.

Ranjan Kumar, Kamlendra Vikram, Sudeepan J., and Sujeet Kumar Mishra*

Chapter 1.4 Evolution of metallic dental implants: historical perspective, needs, and application

Abstract: The aim of contemporary dental practice is to return the patient to standard functionality, communication, wellness, and appearance, irrespective of the deterioration, illness, or trauma affecting the oral and facial structures. Addressing the absence of individual teeth, particularly in the front area, has consistently posed difficulties for dental professionals. As patient expectations rise, the acceptance of removable partial dentures has declined, and numerous patients are now resistant to the idea of preparing healthy teeth for the creation of a fixed partial denture. Dental implant materials include a range of substances that exhibit varying levels of compatibility with the human body. This chapter offers a concise overview and insights into the development of metallic materials used in dental implants, highlighting their historical context and practical applications. The fascination with this field is driven by the diverse range of potential biomedical uses for nanoparticles and nanomaterials, along with the significance of comprehending the materials used in dental implants.

Keywords: Dental implant, metal matrix material, coating, nanomaterial, nanotechnology

1.4.1 Introduction

In recent years, the utilization of dental implants has significantly increased, driven by advancements in medical technology and the growing need for surgical procedures such as dental implants and bone grafting [1]. As the global population ages and the

***Corresponding author: Sujeet Kumar Mishra**, Department of Mechanical Engineering, Birla Institute of Technology, Mesra, Ranchi, Jharkhand, India, e-mail: sujeetmishra@bitmesra.ac.in
Ranjan Kumar, Department of Mechanical Engineering, Birla Institute of Technology, Mesra, Ranchi, India, e-mail: ranjan.mech1990@gmail.com
Kamlendra Vikram, Functional and Biomaterials Engineering Lab, Department of Mechanical Engineering, College of Engineering and Technology, SRM Institute of Science and Technology, Kattankulathur, Kancheepuram, Chennai 603203, Tamil Nadu, India, e-mail: kamlendravikram1@gmail.com
Sudeepan J., Department of Mechanical Engineering, Birla Institute of Technology, Mesra, Ranchi, India, e-mail: sudeepan@bitmesra.ac.in

https://doi.org/10.1515/9783111386027-004

incidence of bone-related issues rises, dental implants have become a critical solution for restoring function and certainly improving the quality of life for many individuals [2, 3]. The potential solution for biomedical implants has been identified in the utilization of biomaterials to serve as replacements for lost or damaged teeth by integrating with the bone. Biomaterials used in dental implants are selected based on key factors such as biocompatibility, mechanical strength, biodegradability, and their physicochemical properties [4, 5]. These materials interact with biological systems in various ways and are classified as inert, active, receptive, or autonomous, depending on their interaction with cellular processes [6, 7]. The careful selection of biomaterials ensures the long-term success of dental implants, which continue to evolve as a vital aspect of modern dental care.

Previously reported data indicate that only in the year 2006, approximately 5.5 million implants were conducted, and soon, by the year 2018, the implant market reached an estimated value of 5 trillion dollars [8]. For decades, the technology has been growing continuously, which further requires the broadening of biocompatible materials and technical frameworks for their viable utilization in dental implants [9]. However, technology always comes with several challenges and requires primary concerns to tackle them. Implantation meets various challenges, such as infections, poor osteointegration, and other adverse effects that create major complications, including root canal complications, gum infections, and biofilm formation [10], and other major issues faced by dentists. A step-by-step procedure of dental surgery and implant is depicted in Figure 1.4.1 as sourced from [11]. Figure 1.4.1 demonstrates the complete procedures carried out during the dental implant, which provides an insight that after drilling, the implants are inserted into the bone using a specialized torquing instrument, followed by the placement of the dental prosthesis [11]. Continuous developing technologies with new engineering advancements are significantly adding value toward the new improved bioimplants with enhanced mechanical properties, biocompatibility, and biomedical efficacy [12–14]. In this regard, nowadays, various biomaterials are being utilized for the restoration and healing of damaged or stressed organs [15]. In today's technological arena, nanotechnology has rapidly evolved into one of the most viable technologies in the field of biomaterial tissue science [16]. On the other hand, 3D technologies based on CAD/CAM are slowly spreading with their improved and huge integration in 3D printing and additive manufacturing technologies [17–19].

Dental implants are nowadays most widely used as replacements for missing teeth, with common causes of tooth loss including gum infection and poor root canal health. Replacing missing teeth with long-lasting dental implants has emerged as a highly effective and advanced solution. Dental implants can be made from a variety of materials, such as ceramics, cobalt-chromium, gold, copper, titanium, and iridio-platinum, offering versatility in their application [12, 20]. Historically, primitive forms of dental implants were used; for instance, ancient Chinese civilizations employed bamboo pins around 4,000 years ago for tooth replacement [21]. In Ancient Egypt, slaves even offered their teeth to pharaohs as substitutes [22]. The development of

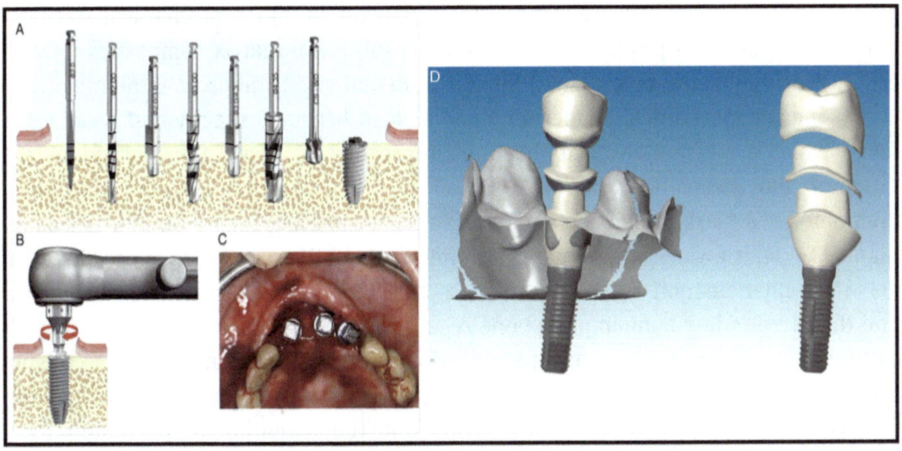

Figure 1.4.1: Steps involved in dental surgery: (A) drilling of appropriate implant diameter, (B) adjusting the implant for insertion, (C) fixing the implant into the hole with a specialized instrument, and (D) insertion of dental prosthesis [11].

modern dental implants gained momentum in the mid-twentieth century, particularly with the introduction of titanium (Ti) as a biocompatible material. In 1952, titanium implants were first tested in rabbit bones, revealing the potential for an osseointegration process where bone adheres to the implant material [20, 23, 24]. This discovery led to the widespread use of titanium in dental implants, as its biocompatibility supports bone integration. However, there are still challenges related to osseointegration, as the bone may not always grow uniformly on the titanium surface, potentially hindering implant stability during healing. To address these issues, surface treatments using ceramics or ceramic-glass materials have been developed to improve osseointegration. Ceramics are also valued for their optical compatibility, which enhances the aesthetics of dental restorations by reducing light scattering [25–27]. The surface structure of implants plays a critical role in their performance, as it can influence the degree of osseointegration. Recent advancements include the use of nanoparticles (NPs) in dental applications, which offer the potential for improving implant surfaces and reducing infection risks. Although metallic dental implants have been used for many years [28], their drawbacks, including limited osseointegration, susceptibility to infection, and mismatched mechanical properties, remain ongoing challenges in the field [29–31].

The increasing demand for medical implants is driven by the aging global population and the rising incidence of bone fractures [2, 3], which necessitate the development of advanced biomaterials for various orthopedic and dental applications [1]. Biomaterials, defined as synthetic or natural substances used to replace or augment parts of living tissues, are critical in medical implants, where they must replicate the function of natural tissue [5]. In particular, the selection of materials for bone implants is governed by key properties such as biocompatibility, mechanical strength, biodegradability, and physicochemical stability [6]. Metal matrix composites (MMCs) are increasingly being investigated for use in dental and orthopedic implants due to their superior mechanical properties, such as load-bearing capacity and wear resistance, which are essential for applications in prostheses, screws, pins, and plates [7]. Titanium alloys, cobalt-chromium (Co-Cr) alloys, and stainless steel have traditionally dominated the market for permanent implants, accounting for 70–80% of all bioimplants [32–34]. However, these materials often face challenges such as stress shielding [35–37], a phenomenon where the stiffness mismatch between the implant material and the surrounding bone impairs bone regeneration and stability, leading to complications like bone resorption and patient discomfort [38, 39]. Furthermore, conventional metallic implants, despite their biocompatibility and corrosion resistance, frequently suffer from poor osseointegration and the potential for toxic metal ion leaching over time, which compromises long-term clinical outcomes [34, 40]. Titanium implants, for instance, are known for their excellent initial biocompatibility, but issues related to weak long-term osseointegration and ion release remain a concern [41]. Additionally, these permanent metal implants often necessitate secondary removal surgeries after healing, which not only increases patient morbidity but also adds to the financial burden of healthcare [24]. To overcome these limitations, there has been a shift toward the development of biodegradable implant materials, which are designed to degrade naturally in the body after fulfilling their function, thus eliminating the need for additional surgery [39]. Magnesium (Mg)-based composites, in particular, have shown great promise as biodegradable implant materials due to their biocompatibility, bioresorbability, and mechanical properties that more closely match those of human bone [42]. Magnesium-based implants are especially attractive for dental applications, as they degrade in the physiological environment and do not require removal, reducing patient discomfort and healthcare costs. These next-generation biomaterials represent a significant advancement in implant technology, combining multifunctional aspects such as biodegradability, enhanced osseointegration, and mechanical stability [40].

In the context of dental implants, MMCs, particularly those incorporating NPs and biodegradable materials like magnesium, hold the potential to address many of the long-standing issues associated with traditional implants. Future research should focus on optimizing the material properties of MMCs to improve their clinical performance and long-term success in dental and orthopedic applications [43, 44]. Magnesium (Mg) is a nontoxic, biocompatible material that is naturally bioresorbable in the human body, making it an attractive option for biodegradable implants. Its ability to degrade in physiological conditions eliminates the need for secondary surgeries, a significant advantage over traditional metallic implants. Magnesium plays a crucial role in the human body, particularly in bone tissue, where it is essential for over 300 enzymatic processes and roughly 50–60% of the body's magnesium is stored in bones. Due to its mechanical properties, such as a density and elastic modulus similar to that of human bone, magnesium has emerged as a promising material for biodegradable medical implants, particularly in orthopedic and dental applications [45].

Despite its advantages, magnesium's rapid corrosion in physiological environments remains a significant challenge, as it can weaken the mechanical integrity of the implant before tissue healing is complete. Proper selection of matrix materials and reinforcements, as well as optimized synthesis processes, is crucial for enhancing the performance of Mg-MMCs in medical applications. Biocompatible reinforcements not only enhance mechanical properties but also ensure that the composite materials are safe for clinical use. The breakdown products of these composites must be safely absorbed or excreted by the body without causing toxicity. Magnesium's high strength-to-weight ratio, ease of casting, and recyclability make it useful in a variety of industries, including aerospace and biotechnology, but improving its corrosion resistance for medical use remains a critical area of research.

The objective of this work is to provide a comprehensive review including the fabrication technique of the biodegradable magnesium matrix composites (Mg-MMCs). Various MMC manufacturing techniques have been incorporated into this study, which further relates the influence of these manufacturing methods with their mechanical behavior. Additionally, the biocompatibility of these composites will be evaluated through degradation studies conducted during corrosion testing. The review aims to offer a clear understanding of how synthesis processes impact the mechanical performance and biocompatibility of biodegradable Mg-MMCs, providing insights into their potential for medical applications.

1.4.2 Materials and methods

Various factors are to be considered for implant materials that contain essential properties useful in clinical applications. The implant materials used from the clinical perspective must have the efficacy and ease of manufacturability, mechanical properties, and compatibility requirements. Figure 1.4.2 indicates the requirement for essential properties that are useful in clinical implants.

1.4.2.1 Metal or metal matrix composites used in bioimplants

Pure magnesium (Mg) and Mg-alloys, such as ZK60A, are commonly used as matrix materials for producing biodegradable Mg-MMCs. Magnesium's biocompatibility, biodegradability, and role in promoting bone formation make it an ideal matrix material. Various reinforcements, including hydroxyapatite (HAP), calcium polyphosphate, β-tricalcium phosphate (β-TCP), zinc (Zn), calcium (Ca), zirconia (Zr), strontium (Sr), graphene platelets, niobium (Nb), and manganese (Mn), are incorporated into the matrix

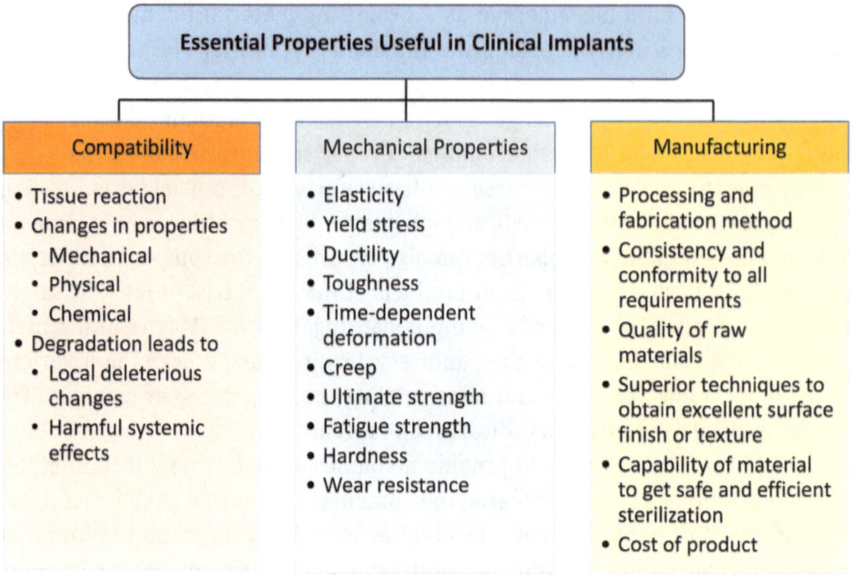

Figure 1.4.2: Essential requirements for implant materials [46, 47].

in varying weight or volume percentages. These composites are typically produced using liquid-state processing or solid-state processing methods, which influence the final properties of the Mg-MMCs. Similarly, titanium (Ti) and its alloys, such as Ti-6Al-4V, are widely used in dental implants and bio-implants due to their excellent biocompatibility, high strength-to-weight ratio, and corrosion resistance. Titanium's ability to osseointegrate, forming a stable bond with bone, makes it a preferred material for long-term implants. Reinforcements, including HAP, calcium phosphate (CaP), bio-glass, zirconia (ZrO_2), and carbon nanotubes (CNTs), are commonly incorporated into titanium alloys to enhance their mechanical properties and bioactivity. These reinforcements improve osseointegration, wear resistance, and the implant's overall durability. The composites are typically fabricated using methods like powder metallurgy, additive manufacturing, and surface coating techniques. These processes help tailor the mechanical properties and biocompatibility of titanium-based implants, making them more suitable for various clinical applications, particularly in dental and orthopedic fields.

1.4.2.2 Synthesis and development of biodegradable MMCs

A composite refers to materials that are engineered by combining two or more distinct with different with chemical or physical elements to achieve enhanced properties [48]. Various methods have been developed to create uniform structures and refine grain sizes, improving the performance of MMCs [49]. Both liquid and solid-state techniques are employed in the production of MMCs, where the matrix is typically a metal or alloy reinforced with fibers, whiskers, or particles [50]. High-performance composites are often developed by hybridizing multiple reinforcement materials. Based on the form of reinforcement, MMCs are categorized into three types: particulate, whisker, and continuous. In biodegradable and biocompatible matrix composites, particulate reinforcements are commonly used, and these composites can be fabricated using various methods such as stir-casting [50], vacuum stir casting [51], stir centrifugal casting [52], high shear solidification, suction casting, disintegrated melt deposition, and powder metallurgy. Each method offers different advantages in achieving the desired mechanical properties and performance. On the other hand, from the perspective of dental implants, various other methodologies for surface treatments are necessary. These surface treatments include anodization, acid etching, alkali surface treatments, acid sol-gel technique, and chemical vapor deposition (CVP).

1.4.2.3 Requirements of nanotechnology and nanomaterials

Nanotechnology and nanomaterials are playing an increasingly vital role in enhancing the functionality of dental implants. The key requirement for these nanomaterials in dental applications is their ability to improve osseointegration, reduce bacterial colonization, and promote rapid healing. Nanotechnology allows the fabrication of nanoscale surface modifications, which mimic the natural extracellular matrix, promoting better cell adhesion and faster tissue regeneration. For instance, nano-coatings like HAP and titanium dioxide enhance bioactivity, creating a favorable environment for bone bonding and integration [53]. Moreover, nanomaterials release functional ions more efficiently than larger particles, ensuring quicker responses and improved bio absorption by surrounding tissues. They can be tailored to release therapeutic agents like antibiotics or growth factors, providing multifunctionality to the implants. NPs like silver, zinc oxide, and magnesium oxide have also shown promise in preventing bacterial infections, thus reducing the risks of peri-implantitis. A challenge in using nanomaterials lies in controlling their degradation rate and ensuring long-term biocompatibility. Bio-inert materials, such as titanium, often require bioactive coatings to establish chemical bonds with bone tissue, which can be achieved through the incorporation of NPs like TCP or polylactic-polyglycolic acid [54]. Through the application of nanotechnology, it is possible to design implants that achieve both rapid tissue integration and reduced risks of postsurgical complications.

1.4.2.4 Advanced nanostructured surface modifications for dental implants

The development of new dental implants aims to reduce failure rates and improve tissue adherence, making surface integration a critical factor. The modification of both surface and bulk material properties influences the interfacial reactions between the implant and surrounding tissue. Nanotechnologies, in particular, have been extensively applied to optimize surface properties. Titanium (Ti) surface nano modifications, for instance, have shown great promise in enhancing bone-implant contact (BIC), osseointegration, and bone development, providing stronger and more stable implants [55]. Three-dimensional nanostructures on implant surfaces have been shown to promote osteogenesis by enhancing cell attachment, growth, and differentiation [56]. Several surface modification techniques have been explored for this purpose. For example, HAP nanocrystalline surfaces produced by plasma spraying improve osteoblast adhesion due to HAP's high protein adsorption capacity among calcium phosphates [57]. However, long-term studies have shown limitations with plasma-sprayed HA coatings due to changes in HA structure at high temperatures, leading to poor adhesion between the coating and the substrate. To address these challenges, incorporating titanium dioxide (TiO_2) into HA coatings has been shown to increase hardness and scratch resistance by up to 36% [58]. Nanocrystalline HAP surfaces fabricated through sputtering deposition methods have demonstrated rapid bone growth and mineral recrystallization, improving biomechanical fixation and BIC in early implantation [59, 60]. Other techniques, such as acid etching, ion-beam-assisted deposition, and grit blasting, have also been shown to enhance bone-to-implant adhesion. Anodization techniques, which create nanoporous surfaces, improve osteoblast-material interactions by increasing surface roughness, lowering the contact angle, and boosting surface energy [61, 62]. Moreover, micro-arc oxidation techniques have been found to enhance bioactivity on Ti surfaces [63]. Additionally, sol-gel-derived HA nanofibers combined with electrospinning facilitate human osteoblast development, with roughness levels increasing with higher calcination temperatures [64]. Nanostructured coatings, such as CVD of nanocrystalline diamond, have also shown bioactive behavior by promoting osteoblastic cell proliferation and mineralized matrix production [65]. Finally, α,ω-diphosphonic acids and alkali hydrothermal treatments have shown promise in improving osteoblast interaction and proliferation, with nanofiber structures demonstrating superior apatite-inductive properties [66]. These advanced nanotechnology-driven surface modifications significantly enhance the performance of dental implants by improving osseointegration, bioactivity, and long-term stability.

1.4.2.5 Nanoscale surface modification

Nanoscale surface modification techniques, such as anodization, acid treatment, alkali treatment, hydrogen peroxide chemical etching, sol-gel processing, and CVD, physical vapor deposition (PVD) are crucial for enhancing the performance of dental implants. These treatments are designed to create nanostructures that improve the surface topography of implants, making them easier to clean and more resistant to bacterial colonization. For instance, they help remove biochemically defective bacteria and oxidize salivary molecules, contributing to better oral hygiene [67, 68]. Surface modifications, such as plasma spraying, acid-etching, anodization, and calcium phosphate coatings, have demonstrated excellent clinical efficacy. Studies show that dental implants with these modifications have a success rate exceeding 95% over 5 years [69]. These nanoengineered surfaces not only enhance the antimicrobial properties of the implant but also promote osseointegration by creating a more favorable interface between the implant and surrounding bone tissue. The use of these advanced techniques has revolutionized implant design, leading to improved long-term outcomes for patients.

1.4.3 Various surface treatment techniques

1.4.3.1 Anodization technique

Anodization is a surface treatment technique widely used to enhance the bioactivity, corrosion resistance, and osseointegration of titanium and titanium alloy dental implants [70]. In this process, an electrochemical reaction occurs, forming a controlled oxide layer on the implant surface, which improves its biocompatibility and mechanical properties [71]. The thickness and porosity of the oxide layer can be adjusted by varying the electrolyte composition, voltage, and duration of anodization. One of the key benefits of anodization is improved osseointegration due to the increased surface roughness and enhanced wettability, which promotes better bone-implant interaction [72]. The results of surface modification on dogs and rabbit models show that bone contact is much higher, with better biomechanical torque removal values for the anodized surfaces than the original surfaces [73]. Furthermore, research has shown that anodized implants with nano-engineered surfaces enhance gingival fibroblast adhesion and alignment, promoting soft-tissue healing. However, challenges with anodization include the risk of poor adhesion between the oxide layer and the titanium substrate, which can lead to surface delamination under mechanical stress. Additionally, achieving uniform oxide coatings on complex geometries and controlling the thickness of the oxide layer pose significant difficulties [74]. To address these issues, researchers have proposed optimizing anodization parameters and incorporating bioactive molecules or NPs into the oxide layer to further enhance osseointegration and reduce bacterial adhesion [75].

1.4.3.2 Surface etching technique

Acid etching is a crucial surface treatment process for enhancing the performance of dental implants by creating specific surface roughness and microstructures that promote osseointegration. This technique exposes the grain boundaries on the implant surface, improving cell attachment and bone integration. The effectiveness of acid etching depends on various factors such as the bulk composition of the material, surface microstructure, type of acid used, etching duration, and surface contamination. Typically, surfaces with roughness (R_a) values between 300 and 1,000 nm are considered minimally rough, which is optimal for dental implants. Studies suggest that acid etching may lead to the formation of a titanium hydride layer, potentially improving implant performance by enhancing surface characteristics (90). Fluorescent histological images have shown increased bone formation around acid-etched titanium implants, especially with calcium-coated surfaces, which outperform other implant types in promoting bone regeneration [76]. The oxide layer formed during acid etching can range from 20 to 100 nm in diameter with a thickness of approximately 10 nm. This layer is typically produced on materials like Ti_6Al_4V and Cr-Co-Mo alloys using high-concentration acids. Implants treated with a combination of sandblasting and acid etching show superior bone anchorage compared to machined or only acid-etched implants due to enhanced torque removal values (91). Additionally, H_2O_2 has been used in surface-etching processes to produce TiO_2, improving the adsorption of RGD(arginyl-glycyl-aspartic acid) cell-adhesive peptides and enhancing surface bioactivity [77, 78]. The use of micro/nanostructured Ti implants, treated with sandblasting and H_2O_2, has been shown to increase surface wettability and promote cellular segregation and gene expression (92). However, while such treatments enhance implant performance, the induced chemical changes on the implant surface due to acid treatments require careful assessment to avoid adverse effects [79].

1.4.3.3 Sol-gel surface modification technique

One of the key benefits of using wet chemical deposition techniques in dental implants is their simplicity, ease of installation, and flexibility in coating implants with complex 3D structures [80]. A common method for enhancing osseointegration is biomimetic modification, where coatings such as calcium phosphate (CaP) are applied. This process typically involves a 14–28-day immersion period in simulated body fluid to develop the desired apatite surface. These biomimetic coatings, characterized by rough and porous calcium-deficient apatite layers, significantly improve cell adhesion and early bone regrowth, making them particularly suitable for dental implants. In vitro studies have demonstrated that sol-gel-sprayed HAP coatings enhance osteoblast proliferation more effectively than plasma-sprayed coatings [81]. During the sintering process, sol-gel techniques allow for the formation of cellular-stage coatings through methods such as dip-coating, spin-coating, or spraying. These techniques enable the deposition of a variety of metal oxides onto metallic and nonmetallic substrates including nanoscale calcium phosphate coatings on dental implants. After the drying process, only the precursor materials remain attached to the implant surface, forming a thin gel-like layer that promotes better bone integration and tissue healing.

This approach is particularly advantageous for dental implants with complex geometries, where the ability to coat intricate surfaces ensures uniform application and enhanced performance [81]. Additionally, biomimetic coatings developed using this method provide better planning for bone formation and improved cellular responses, further enhancing the osseointegration process essential for successful dental implants.

1.4.3.4 Chemical vapor deposition (CVD)

CVD techniques offer a highly efficient method for applying thin film coating on dental implants, particularly when compared to PVD, which requires mechanical forces. The CVD and PVD techniques exist for the synthesis of diamond-like carbon (DLC) coatings. Among all deposition methods, the combination of magnetron sputtering and radiofrequency (RF) plasma processes is the most appropriate for depositing DLC films, owing to the lower substrate temperatures required [82, 83]. The CVD method relies purely on chemical reactions to deposit thin films, making it especially useful for incorporating advanced materials such as graphene in dental implant coatings [84]. In one in vitro study, graphene-coated copper foils produced via CVD were able to induce dental pulp stem cells (DPSC) to deposit mineralized matrices without the need for osteogenic mediums or chemical inducers. This spontaneous secretion of the mineralized matrix by DPSCs highlights graphene's potential to enhance the biological performance of dental implants. CVD techniques can also be applied to titanium (Ti) surfaces in dental implants, particularly through the use of diamond NPs [85]. This method results in ultrahigh hardness, improved strength, and superior adhesion properties. The nanoscale-modified biometal surfaces fabricated using CVD can lead to enhanced osseointegration and improved implant performance [12]. Moreover, CVD enables the deposition of calcium phosphate (CaPO) bio-ceramic nano-coatings on Ti-based implants, which significantly enhance bone integration while improving properties such as abrasion resistance, bond strength, and dissolution rates. Metal-ceramic coatings produced via CVD offer a unique combination of metallic nanocrystalline structures and ceramic toughness, enhancing both the functional and mechanical properties of dental implants. These coatings provide superior durability and improve the overall lifespan of implants, making them highly effective in clinical applications [80].

1.4.3.5 Infections and treatments

Infections remain a significant cause of dental implant failures, and while antibiotics have traditionally been the first line of defense, the growing issue of antibiotic resistance poses a critical challenge to effective treatment. As a promising alternative, mechano-antibacterial effects offer a solution by leveraging physical interactions between the bacterial cell wall and nanoscale structures to combat infections. These nanoscale structures, such as nanopillars and nanowires, stretch and rupture bacterial membranes upon contact, effectively killing bacteria without the need for chemical agents [86]. Sharp-edged nanomaterials, like graphene nanosheets, further enhance this bactericidal action by physically cutting through bacterial membranes.

Research has shown that nanostructures fabricated on titanium supports via plasma etching or hydrothermal treatments exhibit strong antibacterial properties. The sharp nanosheet protrusions formed by these processes create a two-tier hierarchical topography that reduces bacterial attachment and promotes membrane rupture, preventing biofilm formation and minimizing infection risks [80]. Additionally, the release of metal ions and reactive oxygen species from metallic surfaces contributes to bacterial damage by disrupting ATP production, DNA replication, and the integrity of bacterial membranes [87]. Incorporating antibacterial coatings and antimicrobial molecules onto implant surfaces has also shown promise. These coatings can prevent bacterial adhesion and biofilm formation, further reducing the risk of infection and antibiotic resistance [88, 89]. This combination of mechanical and chemical defense mechanisms offers a multifaceted approach to improving the success rate of dental implants.

1.4.4 Conclusion remark

This chapter delves into the compressive characteristics, focusing on the materials and coatings used for dental implants as well as the application aspects of those crafted from MMCs and alloys. We also address potential issues that may occur following the placement of the implant in the oral cavity. By applying the appropriate coating to the surface of the material, the dental implant's qualities may be enhanced, allowing for a more advantageous outcome. The qualities, reflexes, and chemical reactions of the material must be investigated before it can be used in dentistry or on the joints of the human body. In order to get information about the bio-tribological material that was used in the system, this research utilized a variety of different sorts of analytical approaches.

Informed consent statement: Not applicable.

Data availability statement: Not applicable.

Conflicts of interest: The authors declare that there is no conflict of interest.

References

[1] Sumayli A. Recent trends on bioimplant materials: A review. Materials Today: Proceedings. 2021; 46: 2726–2731.
[2] Tong X, Cai W, Lin J, Wang K, Jin L, Shi Z, et al. Biodegradable Zn–3Mg–0.7 Mg2Si composite fabricated by high-pressure solidification for bone implant applications. Acta Biomaterialia. 2021; 123: 407–417.
[3] Kowalski K, Nowak M, Jurczyk M. Mechanical and corrosion properties of magnesium–bioceramic nanocomposites. Archives of Metallurgy Materials. 2016; 61: 1437–1440.
[4] Park JB. Biomaterials Science and Engineering, NY, US: Springer Science & Business Media; 2012. NY, US
[5] Ratner BD. The biocompatibility manifesto: Biocompatibility for the twenty-first century. Journal of Cardiovascular Translational Research. 2011; 4: 523–527.
[6] Gao CC, Kapoor E, Lipford MC, Miller VM, Schroeder DR, Mara KC, et al. Association of vasomotor symptoms and sleep apnea risk in midlife women. Menopause. 2018; 25: 391–398.
[7] Montoya C, Du Y, Gianforcaro AL, Orrego S, Yang M, Lelkes PI. On the road to smart biomaterials for bone research: Definitions, concepts, advances, and outlook. Bone Research. 2021; 9: 12.
[8] Alani A, Kelleher M, Bishop K. Peri-implantitis. Part 1: Scope of the problem. British Dental Journal. 2014; 217: 281–287.
[9] Bhat S, Kumar A. Biomaterials and bioengineering tomorrow's healthcare. Biomatter. 2013; 3: e24717.
[10] Ehrenfest DMD, Coelho PG, Kang B-S, Sul Y-T, Albrektsson T. Classification of osseointegrated implant surfaces: Materials, chemistry and topography. Trends in Biotechnology. 2010; 28: 198–206.
[11] Elias CN, Meyers MA, Valiev RZ, Monteiro SN. Ultrafine grained titanium for biomedical applications: An overview of performance. Journal of Materials Research Technology. 2013; 2: 340–350.
[12] Kandavalli SR, Wang Q, Ebrahimi M, Gode C, Djavanroodi F, Attarilar S, et al. A brief review on the evolution of metallic dental implants: History, design, and application. Frontiers Mater. 2021; 8: 646383.
[13] Andrade AO, Luna AVL, Farias IA, Vasconcelos MG, Vasconcelos RG. Passo a passo clínico dos laminados estéticos: Uma alternativa restauradora em dentes anteriores. Archives of Health Investigation. 2019; 8(9):549-555
[14] Ansarian I, Shaeri M, Ebrahimi M, Minárik P. Tribological characterization of commercial pure titanium processed by multi-directional forging. Acta Metallurgica Sinica. 2019; 32: 857–868.
[15] Bojic S, Volarevic V, Ljujic B, Stojkovic M. Dental stem cells-characteristics and potential. 2014; 29(6): 699–706.
[16] Jin QM, Zhao M, Webb SA, Berry J, Somerman MJ, Giannobile W. Cementum engineering with three-dimensional polymer scaffolds. Journal of Biomedical Materials Research Part A: An Official Journal of the Society for Biomaterials, The Japanese Society for Biomaterials, The Australian Society for Biomaterials, The Korean Society for Biomaterials. 2003; 67: 54–60.
[17] Van Noort Rjdm. The future of dental devices is digital. Dental Materials. 2012; 28: 3–12.
[18] Zandparsa R. Digital imaging and fabrication. Dental Clinics. 2014; 58: 135–158.
[19] Zhao L, Patel PK, Cohen M. Application of virtual surgical planning with computer assisted design and manufacturing technology to cranio-maxillofacial surgery. Archives of Plastic Surgery. 2012; 39: 309–316.
[20] Crabb C. History of dental implants. Dental Nursing. 2006; 2: 398–399.
[21] Misch CE. Contemporary implant dentistry. Dental Nursing. 1999; 8: 90.
[22] Cohen AS, Shen T, Pogrel MA. Transplanting teeth successfully: Autografts and allografts that work. The Journal of the American Dental Association. 1995; 126: 481–485.

[23] Hanawa T. Titanium–tissue interface reaction and its control with surface treatment. Frontiers in Bioengineering Biotechnology. 2019; 7: 170.
[24] Lupi SM, Torchia M, Rizzo S. Biochemical modification of titanium oral implants: Evidence from in vivo studies. Materials. 2021; 14: 2798.
[25] Wang G, Wang S, Dong X, Zhang Y, Shen W. Recent progress in additive manufacturing of ceramic dental restorations. Journal of Materials Research Technology. 2023; 26:1028–1049.
[26] Shenoy A, Shenoy N. Dental ceramics: An update. Journal of Conservative Dentistry Endodontics. 2010; 13: 195–203.
[27] Webster TJ, Siegel RW, Bizios R. Osteoblast adhesion on nanophase ceramics. Biomaterials. 1999; 20: 1221–1227.
[28] Tomsia AP, Launey ME, Lee JS, Mankani MH, Wegst UG, Saiz E. Nanotechnology approaches for better dental implants. The International Journal of Oral Maxillofacial Implants. 2011; 26: 25.
[29] Smeets R, Stadlinger B, Schwarz F, Beck-Broichsitter B, Jung O, Precht C, et al. Impact of dental implant surface modifications on osseointegration. BioMed Research International. 2016; 2016: 6285620.
[30] Přikrylová J, Procházková J, Podzimek Š. Side effects of dental metal implants: Impact on human health (metal as a risk factor of implantologic treatment). BioMed Research International. 2019; 2019: 2519205.
[31] Adhikari J, Saha P, Sinha A. Surface Modification of Metallic Bone Implants – Polymer and Polymer-assisted Coating for Bone In-growth. In: Balakrishnan P, Sreekala MS, Thomas S, editors. Fundamental Biomaterials: Metals, New Delhi, India: Woodhead Publishing, Elsevier. 2018, pp. 299–321.
[32] Khanra AK, Jung HC, Hong KS, Shin KS. Comparative property study on extruded Mg–HAP and ZM61–HAP composites. Materials Science Engineering A. 2010; 527: 6283–6288.
[33] Malhotra R, Dhawan B, Garg B, Shankar V, Nag TC. A comparison of bacterial adhesion and biofilm formation on commonly used orthopaedic metal implant materials: An in vitro study. Indian Journal of Orthopaedics. 2019; 53: 148–153.
[34] Jaiswal S, Kumar RM, Gupta P, Kumaraswamy M, Roy P, Lahiri D. Mechanical, corrosion and biocompatibility behaviour of Mg-3Zn–HA biodegradable composites for orthopaedic fixture accessories. Journal of the Mechanical Behavior of Biomedical Materials. 2018; 78: 442–454.
[35] Parande G, Manakari V, Gupta H, Gupta M. Magnesium-β-tricalcium phosphate composites as a potential orthopedic implant: A mechanical/damping/immersion perspective. Metals. 2018; 8: 343.
[36] Su J, Teng J, Xu Z, Li Y. Corrosion-wear behavior of a biocompatible magnesium matrix composite in simulated body fluid. Friction. 2022; 10: 31–43.
[37] Nakahata I, Tsutsumi Y, Kobayashi E. Mechanical properties and corrosion resistance of magnesium–hydroxyapatite composites fabricated by spark plasma sintering. Metals. 2020; 10: 1314.
[38] Dee KC, Puleo DA, Bizios R. An Introduction to Tissue–Biomaterial Interactions, NJ, US: John Wiley & Sons; 2003.
[39] Shi D. Introduction to Biomaterials, Singapore: World Scientific; 2005.
[40] Del Campo R, Savoini B, Muñoz A, Monge M, Garcés G. Mechanical properties and corrosion behavior of Mg–HAP composites. Journal of the Mechanical Behavior of Biomedical Materials. 2014; 39: 238–246.
[41] Yang J, Guo JL, Mikos AG, He C, Cheng G. Material processing and design of biodegradable metal matrix composites for biomedical applications. Annals of Biomedical Engineering. 2018; 46: 1229–1240.
[42] Feng A, Han Y. The microstructure, mechanical and corrosion properties of calcium polyphosphate reinforced ZK60A magnesium alloy composites. Journal of Alloys Compounds. 2010; 504: 585–593.

[43] Narita K, Tian Q, Johnson I, Zhang C, Kobayashi E, Liu H. Degradation behaviors and cytocompatibility of Mg/β-tricalcium phosphate composites produced by spark plasma sintering. Journal of Biomedical Materials Research, Part B: Applied Biomaterials. 2019; 107: 2238–2253.
[44] Kumar K, Gill R, Batra U. Challenges and opportunities for biodegradable magnesium alloy implants. Materials Technology. 2018; 33: 153–172.
[45] Vikram K, Pramanik S, Paleu V, Bhaumik S. Surface Modification of Magnesium Alloy Employing External Coating for Biomedical Applications. In: Kumar D, Ansari N, editors. Magnesium Alloys for Biomedical Applications, Boca Raton, US: CRC Press. 2025, pp. 109–123.
[46] Long M, Rack H. Titanium alloys in total joint replacement – A materials science perspective. Biomaterials. 1998; 19: 1621–1639.
[47] Hoeppner D, Chandrasekaran V. Fretting in orthopaedic implants: A review. Wear. 1994; 173: 189–197.
[48] Vikram K, Bhaumik S, Pramanik S. Impact of graphite on tribo-mechanical, structural, and thermal behaviors of polyoxymethylene copolymer/glass fiber hybrid composites via Taguchi optimization. Polymer Composites. 2024; 46(2):1768–1787.
[49] Chawla K. Composite Materials: Science and Engineering, NY, US: Springer; 2012.
[50] Saleh B, Jiang J, Fathi R, Xu Q, Wang L, Ma A. Study of the microstructure and mechanical characteristics of AZ91–SiC p composites fabricated by stir casting. Archives of Civil Mechanical Engineering. 2020; 20: 1–14.
[51] Anand N, Ramachandran K, Bijulal D. Microstructural, mechanical and tribological characterization of vacuum stir cast Mg–4Zn/Si3N4 magnesium matrix nanocomposite. Materials Today: Proceedings. 2021; 46: 9387–9391.
[52] Khalili V, Moslemi S, Ruttert B, Frenzel J, Theisen W, Eggeler G. Surface metal matrix nano-composite of magnesium/hydroxyapatite produced by stir-centrifugal casting. Surface Coatings Technology. 2021; 406: 126654.
[53] Zafar MS, Farooq I, Awais M, Najeeb S, Khurshid Z, Zohaib S. Bioactive Surface Coatings for Enhancing Osseointegration of Dental Implants. In: Ed: Kaur G, editors. Biomedical, Therapeutic and Clinical Applications of Bioactive Glasses, New Delhi, India: Woodhead Publishing, Elsevier. 2019, pp. 313–329.
[54] Poon W, Kingston BR, Ouyang B, Ngo W, Chan WC. A framework for designing delivery systems. Nature Nanotechnology. 2020; 15: 819–829.
[55] Salou L, Hoornaert A, Louarn G, Layrolle P. Enhanced osseointegration of titanium implants with nanostructured surfaces: An experimental study in rabbits. Acta Biomaterialia. 2015; 11: 494–502.
[56] Kurella A, Dahotre NB. Laser induced hierarchical calcium phosphate structures. Acta Biomaterialia. 2006; 2: 677–683.
[57] Yang G-L, He F-M, Hu J-A, Wang X-X, Zhao S-f. Effects of biomimetically and electrochemically deposited nano-hydroxyapatite coatings on osseointegration of porous titanium implants. Oral Surgery, Oral Medicine, Oral Pathology, Oral Radiology, Endodontology. 2009; 107: 782–789.
[58] Azari R, Rezaie H, Khavandi A. Investigation of functionally graded HA-TiO2 coating on Ti–6Al–4V substrate fabricated by sol–gel method. Ceramics International. 2019; 45: 17545–17555.
[59] Coelho PG, Granato R, Marin C, Bonfante EA, Janal MN, Suzuki M. Biomechanical and bone histomorphologic evaluation of four surfaces on plateau root form implants: An experimental study in dogs. Oral Surgery, Oral Medicine, Oral Pathology, Oral Radiology, Endodontology. 2010; 109: e39–e45.
[60] Shibli JA, Grassi S, Piattelli A, Pecora GE, Ferrari DS, Onuma T, et al. Histomorphometric evaluation of bioceramic molecular impregnated and dual acid-etched implant surfaces in the human posterior maxilla. Clinical Implant Dentistry Related Research. 2010; 12: 281–288.
[61] Das K, Bose S, Bandyopadhyay A. TiO2 nanotubes on Ti: Influence of nanoscale morphology on bone cell–materials interaction. Journal of Biomedical Materials Research Part A: An Official Journal

of the Society for Biomaterials, the Japanese Society for Biomaterials, the Australian Society for Biomaterials the Korean Society for Biomaterials. 2009; 90: 225–237.
[62] Von Wilmowsky C, Bauer S, Lutz R, Meisel M, Neukam FW, Toyoshima T, et al. In vivo evaluation of anodic TiO2 nanotubes: An experimental study in the pig. Journal of Biomedical Materials Research Part B: Applied Biomaterials: An Official Journal of the Society for Biomaterials, The Japanese Society for Biomaterials,The Australian Society for Biomaterials, The Korean Society for Biomaterials. 2009; 89: 165–171.
[63] Yao Z, Ivanisenko Y, Diemant T, Caron A, Chuvilin A, Jiang J, et al. Synthesis and properties of hydroxyapatite-containing porous titania coating on ultrafine-grained titanium by micro–arc oxidation. Acta Biomaterialia. 2010; 6: 2816–2825.
[64] Bajgai M, Parajuli D, Park S, Chu K, Kang H, Kim H. Hydroxyapatite particulate nanofiber modified titanium: In-vitro bioactivity. Bioceramics Development and Applications. 2010; 1 :1–4.
[65] Amaral M, Dias A, Gomes P, Lopes M, Silva R, Santos J, et al. Nanocrystalline diamond: In vitro biocompatibility assessment by MG63 and human bone marrow cells cultures. Journal of Biomedical Materials Research Part A: An Official Journal of the Society for Biomaterials, The Japanese Society for Biomaterials, The Australian Society for Biomaterials, The Korean Society for Biomaterials. 2008; 87: 91–99.
[66] Wang X, Li Y, Lin J, Yamada Y, Hodgson P, Wen CE. In vitro bioactivity evaluation of titanium and niobium metals with different surface morphologies. Acta Biomaterialia. 2008; 4: 1530–1535.
[67] Hannig M, Kriener L, Hoth-Hannig W, Becker-Willinger C, Schmidt H. Influence of nanocomposite surface coating on biofilm formation in situ. Journal of Nanoscience Nanotechnology. 2007; 7: 4642–4648.
[68] Hannig M, Hannig C. Nanomaterials in preventive dentistry. Nature Nanotechnology. 2010; 5: 565–569.
[69] Le Guéhennec L, Soueidan A, Layrolle P, Amouriq Y. Surface treatments of titanium dental implants for rapid osseointegration. Dent Mater: Official Publication of the Academy of Dental Materials. 2007; 23: 844–854.
[70] Li Y, Li B, Fu X, Li J, Li C, Li H, et al. Anodic oxidation modification improve bioactivity and biocompatibility of titanium implant surface. Journal of Hard Tissue Biology. 2013; 22: 351–358.
[71] Wang C, Bai Y, Bai Y, Gao J, Ma W. Enhancement of corrosion resistance and bioactivity of titanium by Au nanoparticle-loaded TiO2 nanotube layer. Surface Coatings Technology. 2016; 286: 327–334.
[72] Li Y, You Y, Li B, Song Y, Ma A, Chen B, et al. Improved cell adhesion and osseointegration on anodic oxidation modified titanium implant surface. Journal of Hard Tissue Biology. 2019; 28: 13–20.
[73] Sul Y-T, Johansson CB, Röser K, Albrektsson T. Qualitative and quantitative observations of bone tissue reactions to anodised implants. Biomaterials. 2002; 23: 1809–1817.
[74] Yao C, Webster TJ. Anodization: A promising nano-modification technique of titanium implants for orthopedic applications. Journal of Nanoscience Nanotechnology. 2006; 6: 2682–2692.
[75] Vermesse E, Mabru C, Arurault L. Surface integrity after pickling and anodization of Ti–6Al–4V titanium alloy. Applied Surface Science. 2013; 285: 629–637.
[76] Variola F, Yi J-H, Richert L, Wuest JD, Rosei F, Nanci A. Tailoring the surface properties of Ti6Al4V by controlled chemical oxidation. Biomaterials. 2008; 29: 1285–1298.
[77] Mante FK, Little K, Mante MO, Rawle C, Baran GR. Oxidation of titanium, RGD peptide attachment, and matrix mineralization of rat bone marrow stromal cells. Journal of Oral Implantology. 2004; 30: 343–349.
[78] Wang X-X, Hayakawa S, Tsuru K, Osaka A. Bioactive titania gel layers formed by chemical treatment of Ti substrate with a H2O2/HCl solution. Biomaterials. 2002; 23: 1353–1357.
[79] Nazarov DV, Zemtsova EG, Solokhin AY, Valiev RZ, Smirnov VM. Modification of the surface topography and composition of ultrafine and coarse grained titanium by chemical etching. Nanomaterials. 2017; 7: 15.

[80] Bosco R, Van Den Beucken J, Leeuwenburgh S, Jansen J. Surface engineering for bone implants: A trend from passive to active surfaces. Coatings. 2012; 2: 95–119.
[81] Paital SR, Dahotre NB. Calcium phosphate coatings for bio–implant applications: Materials, performance factors, and methodologies. Materials Science Engineering: R: Reports. 2009; 66: 1–70.
[82] Vikram K, Kalita K, Ghadai RK Multi-attribute optimization of diamond-like carbon thin films. AIP Conference Proceedings: AIP Publishing; 2020.
[83] Vikram K, Ragavendran U, Kalita K, Ghadai RK, Gao X-Z. Hybrid metamodel-NSGA-III-EDAS based optimal design of thin film coatings. Computers, Materials Continua. 2021; 66: 1771–1784.
[84] Ahmed W, Sein H, Rajab H, Jackson M. Chemical vapor deposition of diamond coatings onto dental burrs. Journal of Chemical Education. 2003; 80: 636.
[85] Saba T, Saad KSK, Rashid AB. Precise surface engineering: Leveraging chemical vapor deposition for enhanced biocompatibility and durability in biomedical implants. Heliyon. 2024; 10: e37976 (p. 1–25).
[86] Clainche TL, Linklater D, Wong S, Le P, Juodkazis S, Guével XL, et al. Mechano-bactericidal titanium surfaces for bone tissue engineering. ACS Applied Materials Interfaces. 2020; 12: 48272–48283.
[87] Marambio-Jones C, Hoek EM. A review of the antibacterial effects of silver nanomaterials and potential implications for human health and the environment. Journal of Nanoparticle Research. 2010; 12: 1531–1551.
[88] Costerton JW, Stewart PS, Greenberg EP. Bacterial biofilms: A common cause of persistent infections. Science. 1999; 284: 1318–1322.
[89] Heuer W, Elter C, Demling A, Neumann A, Suerbaum S, Hannig M, et al. Analysis of early biofilm formation on oral implants in man. Journal of Oral Rehabilitation. 2007; 34: 377–382.
[90] Palmquist A, Omar OM, Esposito M, Lausmaa J, Thomsen P. Titanium oral implants: surface characteristics, interface biology and clinical outcome. Journal of the Royal Society Interface. 2010; 7 (suppl_5):S515–S527.
[91] Li D, Ferguson SJ, Beutler T, Cochran DL, Sittig C, Hirt HP, Buser D. Biomechanical comparison of the sandblasted and acid-etched and the machined and acid-etched titanium surface for dental implants. Journal of Biomedical Materials Research: An Official Journal of The Society for Biomaterials, The Japanese Society for Biomaterials, and The Australian Society for Biomaterials and the Korean Society for Biomaterials. 2002; 60(2):325–332.
[92] Xie Y, Li J, Yu ZM, Wei Q. Nano modified SLA process for titanium implants. Materials Letters. 2017; 186:38–41.

Kamlendra Vikram, Ranjan Kumar, Avishek Mishra,
and Sumit Pramanik*

Chapter 1.5
Tribological behavior of specific implant materials for dental applications

Abstract: Tribology covers three fundamental topics: friction, wear, and lubrication. As the world advances, tribology plays a crucial role in meeting the requirements for advanced technical innovations, enhancing machine lifespan and elevating the quality of life. The biological application, notably human joints such as knees and hips, continuously move in relation to one another; tribology evolves into bio-tribology. Numerous critical bio-tribological testing events might greatly influence the outcomes of wear concerning implant design and material selection. Tribology is crucial in improving the design and efficacy of biomaterials for medical applications. The joints of the human body, including the hip, knee, jaw, and oral components, must all address issues of wear and lubrication. Dental implant prosthetics represents a rapidly evolving area within contemporary medicine. The rise in social welfare, the aging population, and a significant increase in dental caries contribute to a growing demand for dental implants. In this chapter, we will thoroughly discuss about the wear behavior and impact of biomaterial on the dental application and tribological behaviors of implants and their development in the field of biomedical sciences. In addition, advancements in surface modifications of biomaterials like coatings, functionalization, and the insertion of bioactive molecules are also discussed. The background information and importance of implants materials are followed by a discussion about the role of lubricants in obtaining the ideal features for improved performance.

Keywords: Bio-tribology, biomaterial, dental implant, polymer in orthodontics implant

***Corresponding author: Sumit Pramanik**, Functional and Biomaterials Engineering Lab, Department of Mechanical Engineering, College of Engineering and Technology, SRM Institute of Science and Technology, Kattankulathur, Kancheepuram, Chennai 603203, Tamil Nadu, India, e-mail: sumitprs@srmist.edu.in, ORCID: 0000-0003-0455-1582
Kamlendra Vikram, Functional and Biomaterials Engineering Lab, Department of Mechanical Engineering, College of Engineering and Technology, SRM Institute of Science and Technology, Kattankulathur, Kancheepuram, Chennai 603203, Tamil Nadu, India, e-mail: kamlendravikram1@gmail.com
Ranjan Kumar, Department of Mechanical Engineering, Birla Institute of Technology, Mesra, Ranchi, India, e-mail: ranjan.mech1990@gmail.com
Avishek Mishra, Department of Mechanical and Mechatronics Engineering, Centre for Advanced Materials, Manufacturing and Design, The University of Auckland, Auckland 1023, New Zealand, e-mail: Avishek.mishra@auckland.ac.nz

https://doi.org/10.1515/9783111386027-005

1.5.1 Introduction

Tribology is the study of the relationship between friction, lubrication, and wear that occurs on surfaces that are in relative motion and interact with one another [1, 2]. The term "friction" refers to the physical phenomenon that occurs when one thing rubs or slides against another one [3]. Through the use of compounds that are capable of making the contact surfaces smooth, it is possible to reduce its severity [4]. The progressive erosion of material from surfaces as a consequence of mechanical or chemical impacts is represented by the phenomenon known as wear [4]. Over the last several years, the idea of tribology has gained momentum, mostly due to its connection to biology, which is referred to as bio-tribology. One of the most promising applications of the bio-tribology idea is the use of medicinal biomaterials across the field of dentistry. As a result of their masticatory and phonetic roles, human teeth are a tribological system that is prone to the phenomena of friction, lubrication, and wear. In addition, dental surfaces are subject to deterioration processes such as accidents, cavities, and the loss of tooth tissue, which are often exacerbated by the process of aging [5, 6]. In the field of medicine, the development of materials for the fabrication of biomedical implants, also known as biomaterials (metals, polymers, and ceramics), is directly affected by the qualities and nature of the tissue, organs, and systems that are being replaced or supplemented. The preservation of a complete database of material features is vital for the creation and improvement of existing biomaterials. This information is often more qualitative than quantitative; therefore, it is essential that this database be maintained. Due to the inability to properly define the behavior of biomaterials over a longer length of time and in a variety of human-related environments, it is necessary to analyze implants material while taking into consideration the effect of a large number of system characteristics. It is necessary to have knowledge of a variety of factors in order to understand the complex nature and applications of biomaterials. These variables include biocompatibility, corrosion behavior, mechanical features, stability and durability in adverse environments, wear processes and wear rates throughout exploitation, and many more. In many biological networks or systems, the tribological framework is essential to their functioning. The application of mechanical components to the human body should also get a greater amount of attention in the context of the bio-tribology dealing system. Within the realm of medical research, the application of bio-tribology is intricately linked to the human body. Of particular significance is the apparatus that is used in many kinds of joints, including but not limited to the hip joint, knee joint, synovial joints, artificial joint, and others. It is necessary for the functioning of human joints to improve when injury occurs to these joints as well as in situations that are associated with aging. Trobiology was first used to both live and artificial human joints in the beginning of the late 1960s [7]. The change has been seen in a variety of different places. For instance, the tribological behavior of synovial joints and their replacement, the wear of human dental tissues, the lubrication of red blood cells in narrow capillaries by plasma, the wear of replacement heart valves, the wear of screws and plates in and on bone during fracture repair, the tribology of skin and the friction of hair, biosensors, artificial fingers, and other topics are all examples of things that are being studied. A temporomandibular joint (TMJ) replacement that used polytetrafluoroethylene (PTFE) as the bearing counter face was developed in the latter half of the 1970s. The interpositional implant (IPI), which was made of PTFE, was licensed for use in the market in the year 1983. Due to the fact that PTFE has a low coefficient of friction (COF) and has been used widely as a bearing surface in a variety of technical applications, from a theoretical standpoint, it would seem to be an ideal choice for an implant material. On the other hand, the majority of the more than 25,000 PTFE TMJ implants that patients received were unsuccessful. There was evidence that PTFE was not an acceptable implant material, and this data was gathered as a result of the advancement of research on IPI [8]. According to research that was published in orthopedic medical publications, it is predicted that between 85% and 95% of the nearly 600,000 total joints that are conducted each year in the United States (more than 120,000 artificial hip joints are implanted yearly) would still be operating after 10 years. When designing biomaterials, engineers need to consider the physiologic stresses that will be applied on the implant and ensure that it has appropriate structural integrity. Material selection must also consider biocompatibility with the tissues that are surrounding the implant, the environment, and any corrosion issues that may arise, friction and wear of the articulating surfaces, and implant fixation through either osseointegration (the degree to which bone will grow next to or integrate into the implant) or bone cement. These devices are plagued by a number of significant issues, one of which is entirely attributable to the materials: With complete joint replacements, the polymer cup is subject to wear [7].

Moreover, the tribological behavior of dental implant material and its behavior on the application are also topics that are covered in this chapter. In addition to protective coatings, the influence of various materials, such as polymer-polymer, polymer-metal, and metal-metal interaction, on the surface of dental implant materials is also taken into consideration. The purpose of this chapter is to provide a comprehensive and in-depth review of dental tribology. The chapter will also cover the mechanisms that are responsible for the wear of dental tissues, the experimental methods that are used to quantify tooth wear, the factors that influence oral lubrication and erosion/corrosion, the effects that oral hygiene practice has on dental tissues, and the mechanical properties of dental tissues. Figure 1.5.1 illustrates the specifics of the tribology component of dental implants and the procedure as it relates to application [9].

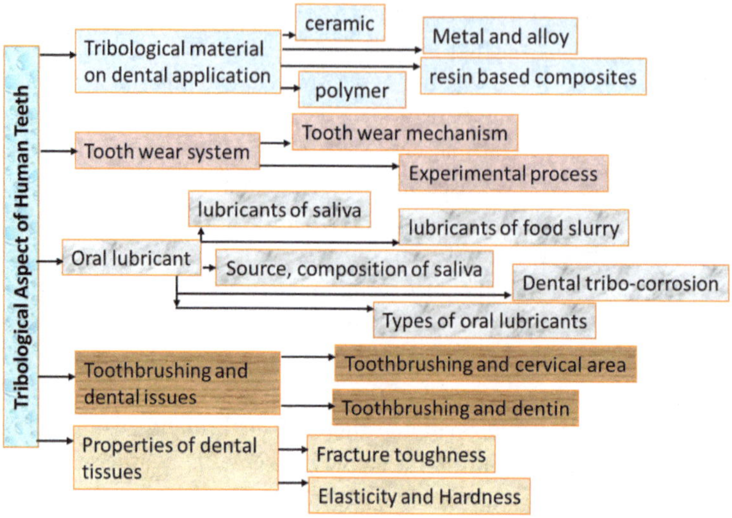

Figure 1.5.1: Details of the tribological aspect of human teeth and properties of the dental.

1.5.2 Recent work of tribology in biomedical

1.5.2.1 Tribology on the dental implant

In the context of the contact between two objects, the term "wear" refers to the process by which the surface of the material is removed. The normal load (N), sliding speed (rpm), and duration time period (min) are the primary factors for effecting the wear, wear loss, and COF [10, 11]. In the case of dental implants, masticatory force is considered to be representative of normal load. The scope of masticatory force is challenging because of the fact that it varies from person to person, depending on factors such as age, eating habits, and the composition of the jaw. When conducting their investigation into the most suitable materials for dental prosthesis, many researchers have used a variety of forces. There are a variety of mastication forces that have been investigated by researchers throughout the course of time and depicted in Table 1.5.1 [12].

Table 1.5.1: In earlier studies, the ranges of masticatory force were evaluated.

S. no.	Envaulted materials for dental implant	Masticatory load	Reference
1	Flat enamel (human tooth)	20 N	[13]
2	Zirconia ceramic, gold-Pt alloy, human tooth	4 N	[14]
3	CoCr alloy	22.24 N	[15]
4	Ti-6Al-4V	3 N	[16]

Table 1.5.1 (continued)

S. no.	Envaulted materials for dental implant	Masticatory load	Reference
5	Commercially pure titanium	20 N	[17]
6	A human tooth with a cusp and a polished flat surface	1.9 N, 3.9 N, 5.88 N	[18]
7	The human teeth	20 N	[19]
8	Composite material based on resin	0.98 N	[20]
9	Cusp and flat (human tooth)	2, 4, 8 N	[21]
10	Flat enamel (human tooth)	30 N	[22]
11	Cast Ti specimen	5 N	[23]

1.5.2.2 Types of wear in dental implants

Four main types of wear mechanisms contribute to dental wear:
(I) abrasive wear (two- and three-body abrasion),
(II) adhesive wear,
(III) fatigue wear, and
(IV) corrosion wear.

The types of wear on dental implants application are depicted in Figure 1.5.2. There are a number of processes that take place simultaneously and sequentially, and dental wear is a complicated phenomenon that is the result of this amalgamation.

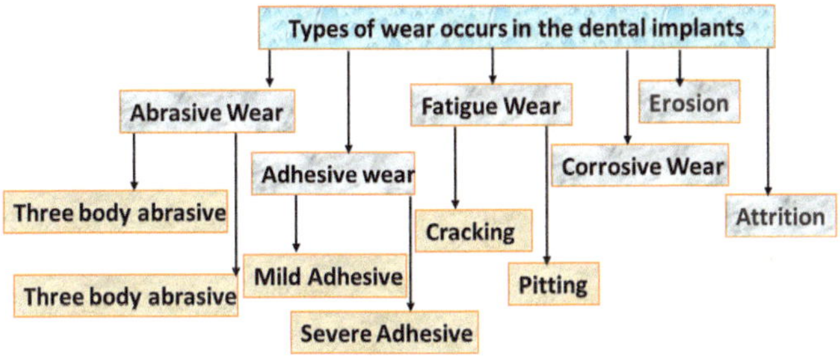

Figure 1.5.2: Types of wear on dental implant application.

1.5.2.2.1 Abrasive wear

Abrasive wear is the kind of wear mechanism that may occur due to the material loss with hard material and softer material. It may be also observed in human teeth, and it is the most common and ubiquitous type of wear mechanism possible. Chewing processes are the sole activities that result in surface-to-surface contact, although this is a common misconception that often arises. Additionally, tooth tightness or involuntary grinding is a significant originator of the condition. This kind of wear is caused by a variety of factors including the force of mastication, the quality of the teeth, the support provided by the bone, and the total number of teeth in the oral cavity. The abrasive wear may be broken down into two distinct groups, namely two-body abrasive wear and three-body abrasive wear.

1.5.2.2.2 Adhesive wear

Adhesive wear starts when two bodies, which are often associated with metal and composite restorations, slide against each other in the oral cavity when there is a large amount of pressure present. This is the beginning of the process of adhesive wear. It is possible for material to be transported from one surface to another depending on the amount of pressure and the sliding motion.

1.5.2.2.3 Fatigue wear

There is a possibility of fatigue loading occurring when a surface is exposed to movement while being subjected to high pressure and cyclic loading. It is possible for the microcracks that are formed to spread to subsurface areas, and this wear starts at the surface. Additionally, enamel fragmentation in the form of debris is the consequence of these microcracks propagating along subsurfaces throughout repeated cycles. Despite the fact that the enamel's prismatic composition prevents the spread of microcracks, the enamel's poor tensile strength speeds up the fragmentation process. On the other hand, this wear may be significantly aggravated in an acidic environment and when subjected to circular loads. The enamel-dentine junction is responsible for preventing the propagation of these microcracks between the enamel and the dentine on the teeth [24].

1.5.2.2.4 Corrosive wear

The term "corrosive wear" refers to the process by which dental molecules are removed or lost when a chemical reaction is present causing the situation. Because of the changes that have occurred in the eating habits of humans, corrosive wear has become one of the most prevalent types of wear. Corrosive wear may be caused by acid interactions with teeth, which can lead to the breakdown of the intermolecular connection between dental tissue molecules, which in turn promotes an increase in the number of other wear mechanisms. No microorganisms are involved in the production of corrosive substances, which may be either extrinsic or intrinsic in nature.

Food items and environmental chemical elements, such as fruits, alcoholic beverages, sodas, and energy drinks, are examples of extrinsic sources [25, 26]. Figure 1.5.3 illustrates the potential locations of a dental implant that are susceptible to tribo-corrosion due to the oral environment impact of the implant when it is subjected to load-bearing activities in the presence of food particles and beverages [27, 28].

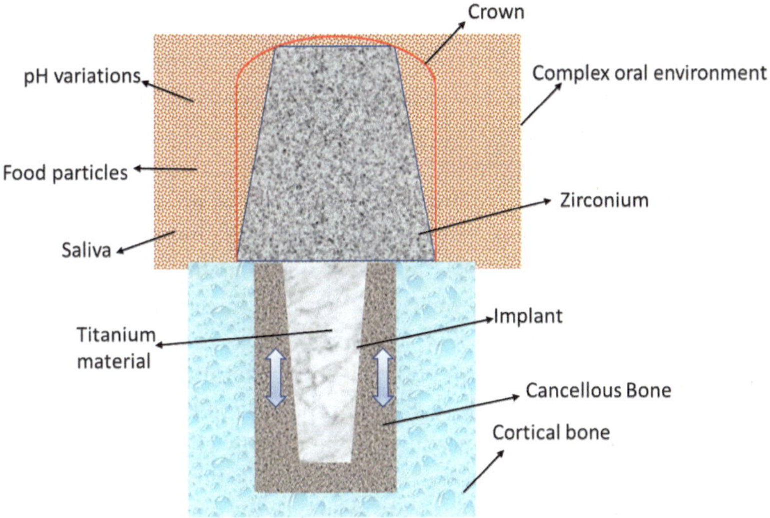

Figure 1.5.3: Potential tribo-corrosion zones in a dental implant influenced by the oral environment during load-bearing activities, particularly in the presence of food particles and beverages.

Due to the fact that implant-supported dental restorations have a high success rate, between 5% and 11% of them fail after 10–15 years and need to be removed [29]. The most common cause of dental implant failure is peri-implantitis, which is an infection of the tissues surrounding the implant, and the loss of supporting bone that is linked with it. One of the most significant factors that contributes to peri-implantitis is the presence of tribo-corrosion at the interface between the implant and the bone as well as at the internal connections between the prosthetic pieces (implant, abutment, and crown). Both the implant and the abutment are often composed of porcelain or zirconia ceramic, while the crown is typically made of titanium or titanium alloy. In spite of the fact that the machining will be quite exact, and there will always be a very little gap between the prosthetic connection and the implant. As a result of micromotion that takes place at the interface between the implant and the abutment and the crown, fretting and crevice corrosion may develop during the process of mastication. Clearly, the condition is comparable to the phenomenon of crevice corrosion caused by mechanical assistance at the taper connections of complete hip replacement systems [30–33]. To enhance the qualities of materials used in dental implant applications, surface coatings are necessary. The coating serves to mitigate material loss, hence providing protection against corrosion and erosion. The surface coating is primarily categorized into two groups: chemical conversion coatings and deposition coatings [34].

1.5.3 Medical application material

The replacement of lost teeth with dental implants has shown to be the most successful treatment option. The correct selection of materials is essential to the success of the implantation process. Surface properties, microstructure, constituent components, and manufacturing techniques of a material may all have an impact on the implant's functionality and dependability. The implant's toughness must be as close as feasible to that of the native bodily component, and the material must be biocompatible [35]. The purpose of the treatments that are used in dentistry is to maintain functionality after the development of chronic caries, wear, and bacterial infections. When teeth are severely damaged as a consequence of wear or traumatic events, treatments are carried out either by preparing the tooth and filling it with restorative materials or by inserting foreign materials into the tooth sockets in order to replace the teeth. It is possible to consider any of these two procedures to be therapy choices. As replacement teeth, dental implants are used in order to restore the functionality, mechanical integrity, and dependability of the teeth that were originally present. During the process of selecting dental materials, several factors are taken into consideration. These include the location of the tooth, its size, biocompatibility, as well as the physical, mechanical, and tribological qualities of the dental material. There are many important considerations to consider when choosing dental materials including aesthetics and cost-effectiveness. Material used in dentistry is often classified into the following categories: metals (or their alloys), polymers, ceramics, and composites. Table 1.5.2 presents a variety of dental materials that have been used, along with the pros and drawbacks that are known about each substance [9, 36]. Figure 1.5.4 depicts the biomaterials for the manufacturing of core and surface coatings of dental implant [37].

Titanium (Ti), which was formerly regarded as a rare metal, is now considered to be one of the most significant metals in the industry and it has now become the implant material that is most often used in dentistry for implants [38]. In the crust of the earth, titanium was the ninth most prevalent element and the fourth most abundant metallic element, after aluminum (Al), iron (Fe), and magnesium (Mg) in terms of abundance. Ti material is capable of being alloyed with a broad range of elements in order to modify its characteristics. This is mostly done for the aim of enhancing its strength, high-temperature performance, creep resistance, reaction to aging heat treatments, and formability [39].

Poly-ether-ether-ketone, often known as PEEK, is an extremely high-performance semicrystalline thermoplastic polymer. It has exceptional thermal and chemical resistance, including resistance to oils and acids, in addition to its very excellent strength and stiffness. PEEK is a material that may be used for the production of dental implants since it is colorless and has an elastic modulus that is comparable to that of bone. On the other hand, PEEK by itself is not biocompatible and does not promote cell attachment. The biological and mechanical characteristics of pure PEEK have been altered as a result of various processing and surface changes that have been suggested in recent research [38, 40]. There is a substantial correlation between the choice of implant material and the choice of implant need and the selection of suitable implant material by dentists. When it comes to endodontic posts, crowns, and bridges, metals are the more popular material option. Combining noble and non-noble (base) metals results in the formation of multifaceted forms. Metals are the preferred option for these dental restorations. Both removable partial dentures and substructures for ceramic fused implant frameworks have been built using cobalt-chromium alloy, which has seen significant usage in both of these applications [41, 42].

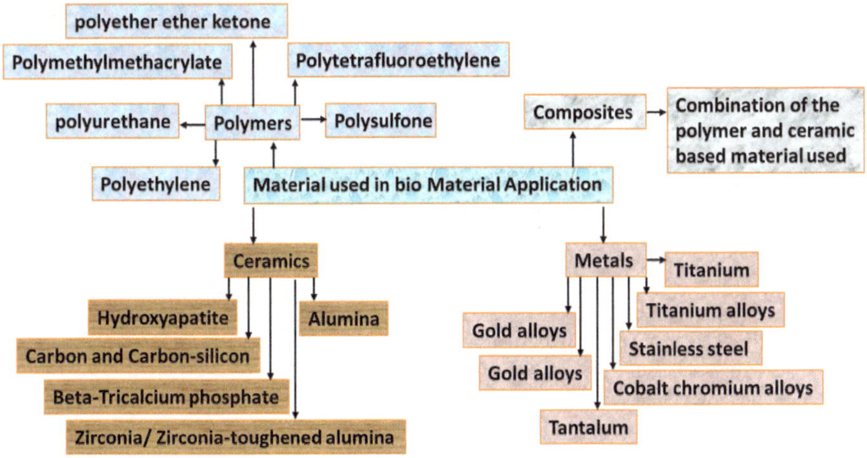

Figure 1.5.4: Biomaterials for the manufacturing of core and surface coatings of dental implant.

Considering its longevity, biocompatibility, minimal abrasiveness against teeth, and aesthetic appeal, gold has been one of the most recommended dental materials for a very long time by dental professionals. In addition to pure gold, gold may also be found in the form of gold alloys. Due to the fact that pure gold is very malleable, it is mostly used in the dentistry field for the purpose of filling minor cavities. On the other hand, gold alloys are utilized for applications that need a high level of mechanical strength such as dental implants. It is possible for the element gold to move from tooth gold to saliva, which may result in contact allergies, particularly for those who suffer from eczema condition [43, 44].

Another dental restorative material that has been used for more than 150 years is dental amalgam, which is also known as silver filling within the dental profession. Dental amalgam has been used for a long time. Typically, it is made up of around 50% elemental mercury, 35% silver, 13% tin, 2% copper, and a trace of zinc. When compared to other dental materials, dental amalgam is the alternative that charges the least amount of money, and it also offers a fairly good degree of reliability [45]. In spite of its many benefits, the use of dental amalgam is contentious owing to the significant amount of mercury it contains. This is due to the fact that the mercury vapor that is released from dental amalgam fillings has the potential to create adverse effects on the organs of humans. Consequently, this has resulted in an ongoing dispute about the safety of utilizing dental amalgam as a material for the purposes of restorative dentistry techniques.

Table 1.5.2: Overview of materials utilized for dental applications.

Material used in dental implant application		**Advantages**	**Disadvantages**
The best materials for dental implant application	Pure gold (Au)	Aesthetics, soft enough as filling materials	Costly and the discharge of elements might cause allergic reactions, a lack of aesthetic appeal
	Gold alloys mixed with Au and Ag	Greater mechanical qualities in comparison to pure Au	The high emission of mercury vapor has the potential to cause a variety of serious illnesses and it is not anesthetized
	Silver amalgam used in dental filling material is also known as silver filling	Inexpensive, good mechanical characteristic	
Metal alloys	Vanadium steel, stainless steel, Ni/Cr-based alloy	Improvements in the mechanical qualities	Excessive release of elements that might cause allergic reactions Less biocompatible and unattractive in appearance
	Titanium alloys for implant	– A high level of resistance to corrosion – Biocompatibility – A higher mechanical quality – Reduced element release rate – Enhanced wear properties after surface treatment	High fretting wear

Table 1.5.2 (continued)

Material used in dental implant application		**Advantages**	**Disadvantages**
Ceramics	Zirconia, Y-TZP, etc.	– It has remarkable resistance to tribo-corrosion – In comparison to metals, biocompatibility is somewhat higher – Enhanced wear properties after surface treatment	High brittleness, and abrasive potential to opposing teeth
	Alumina	– Good compatibility, great wear resistance, and high strength are all characteristics – This material is characterized by its high density, high purity, and resistance to corrosion	Survival rate of alumina implants was lower than that of titanium implants
Resin-based composites	AP-X, Filtek P60, Z350, VITA LC, VITA ZETA composites, etc.	Aesthetics, biocompatibility, chemical inertness, and convenience	Poorer wear, color retention, and poor mechanical properties

1.5.4 Biomaterial characteristics for dental implants

In order for a biomaterial to be considered biocompatible, it must possess its main characteristic of not causing an unfavorable response when it is put into use. This is the most important attribute that a biomaterial must possess. In addition, it is necessary to have strong wear resistance, the ability to osseointegrate, high corrosion resistance, and superior mechanical qualities:

1. Biocompatibility
2. Mechanical properties
3. Osseointegration
4. High corrosion resistance
5. Wear resistance

1.5.4.1 Biocompatibility

The term “biocompatibility” refers to the ability of a biomaterial to perform its intended function in relation to a medical therapy without causing any unfavorable local or systemic effects in the recipient or beneficiary of that therapy, while simultaneously generating the most appropriate and beneficial cellular or tissue response in that particular circumstance and optimizing the clinically representative performance. Both of these characteristics are essential for the biomaterial to be considered biocompatible [46, 47]. To get started, it is essential to consider the many factors that have the ability to influence biocompatibility to some degree. For example, the chemical, physical, and biological properties of the bio-elastomer that comprise biodegradable components are included in these variables. To be more explicit, these concerns often pertain to the following features [46]:

- The material's composition, microstructure, nanoscale structure, and shape
- The structure of the material including its crystallinity and crystallography
- The elastic modulus of the material
- The percentage of water present (water contact angle in degree) in the substance as well as the overall hydrophobic-hydrophilic balance
- The degradation direction, the shape of the degradation product, and the toxicity
- The molecular mobility, chemical gradients, and surface chemical composition, respectively
- The topography of the surface and the energy of the surface
- Surface electrical and electronic characteristics
- Surface toxicity, surface topography, and surface energy
- Leachable, additives, catalysts, pollutants, and the toxicity of these substances
- Use a profile that releases debris

1.5.4.2 Mechanical properties

At the time when the process of mastication is taking place, the restorations or natural teeth are subjected to cyclic stress. The pressures that are imposed on the opposing teeth or restorations as a result of these cyclic stresses create deformation and degrade the durability of these structures over the course of time during the duration of the patient's lifetime. A load or force that is applied to a body results in the body producing a reaction that is equal to and opposite to the load or force that was applied to the body. This is the result of internal tension. Due to the fact that these triggered reactions cause stress and normal load factors, it is of the utmost importance to take them into account. The deformation of materials is caused by a number of different combinations of complicated forces such as compressive, tensile, shear, and bending forces [48].

1.5.4.3 Osseointegration

It is essential for implant stability to have osseointegration, which is described as a direct structural and functional link between organized, living bone, and the surface of a load-carrying implant. Osseointegration is considered a precondition for implant loading and the long-term clinical success of endosseous dental implants. In most cases, all that is required is a person who is both physically and intellectually well, unafflicted by diabetes, or an illness that affects the circulatory system. Patients who are going to have this operation are required to abstain from smoking for a period of 3 months before the procedure, and they will also be prohibited from smoking immediately after the treatment.

1.5.4.4 High corrosion resistance

Due to the fact that mouth fluids and other food cantonment are known to contribute to the occurrence of corrosion on the material that is used in dental implants, it is essential to take notice of the fact that high corrosion resistance materials are required to be used in dental applications. Based on its superior mechanical qualities and resistance to corrosion, titanium and its alloys are used extensively in the manufacturing of dental implants. While dental implants made of titanium have been shown to significantly enhance oral rehabilitation for patients who are either partly or fully edentulous, they also have a pretty good survival rate.

1.5.4.5 Wear resistance

Dental implants need an additional feature, which is the material that is resistant to wear. The term "wear" refers to the capacity of a material to withstand the loss of material from its surface as a result of that substance being subjected to mechanical forces. In addition to the kinds of materials, temperature, humidity, counter body qualities, applied normal load, testing speed, heat treatment, friction products, and other variables, the wear resistance is affected by a wide range of other elements as well.

1.5.5 Wear processes on dental materials

It is possible for restorative surfaces, such as titanium, Zirconia, Noble material, and other material, to become abrasion-prone during the chewing process due to frictional surface contacts with opposing surfaces, such as toothbrush and paste, food bolus, and hard particles originating from dietary sources [49]. It has been suggested that a two-body abrasion occurred when two surfaces were rubbed away from each other by coming into direct touch with their respective asperities [50–52]. Two-body abrasion is a phenomenon that occurs in the oral cavity during a movement of the teeth that is not masticatory. However, it is also possible for this phenomenon to take place on the surface of the prosthetic joint during masticatory tooth movement. It is also important to note that the three-body abrasion is the source of the existence of the "intervening slurry of abrasive particles" in the tribological contact. The wear process that is referred to as fatigue is characterized by the breaking of intermolecular bonds and the formation of a zone of subsurface damage. This disruption is brought about by the movement of surface molecules in response to cyclic stresses. Since this is the case, there is a microcrack forming inside the subsurface that is oblique to the surface. This microcrack has the potential to coalesce to the surface, which may result in the loss of material. It has been often observed that fatigue is linked to the wear of occlusal surfaces. The phenomenon known as adhesive wear is a different kind of wear mechanism that takes place when, after the breakdown of the oxide coating, an attraction is promoted between two surfaces that are moving in relative contact with one another. When surfaces are subjected to friction, wear particles may also attach themselves to them in the form of platelets. It is possible for the microwelds to fracture as a consequence of adhesive wear, which may lead to an increase in the rate of wear. Under conditions of small-amplitude oscillatory movement, another significant wear process that may take place between contacting surfaces is known as fretting [52–54]. The movement may be a result of one of the contacting parts being subjected to cyclic stress, which has the potential to diminish the fatigue strength by 70–80%. The wear of cortical bone against titanium implant surfaces has been linked to the phenomenon known as fretting wear [52].

1.5.6 Concluding remark

In this chapter, we discuss the compressive specifics including the material that is utilized in the dental implant and the problems that may arise after the implant has been placed in the mouth. In order to achieve a higher level of quality in dental implants, it is essential to utilize superior materials, as shown by the findings of a study conducted by a number of researchers in the field of dentistry. The examination reveals that the material's tribological qualities, such as wear resistance and friction coefficient, as well as its mechanical properties, such as fatigue resistance, fracture toughness, and tensile strength, have been improved on many occasions. Regarding the materials that are used for dental restorations, it is also a very significant issue to be considered. These are required to possess a number of fundamental qualities including biocompatibility, noncytotoxicity, resistance to corrosion, and resistance to wear. Dental restorations have been made from a variety of materials including metals and alloys, composites, and ceramics. However, in contrast to human tooth enamel, these materials face a unique set of challenges. In certain instances, it was shown that these materials are cytotoxic, while in other instances, they are characterized by a poor biocompatibility. The wear resistance of dental composites is often less than satisfactory; ceramic, on the other hand, has a more favorable reaction in the oral environment. The substantial amount of research was carried in order to acquire a better understanding of dental tribology and to improve the selection of materials that are used for dental restorations. In the realm of medicine, the term “tribology” or “bio-tribology” is not used very often, which is something that we believe is an intriguing factor to consider.

Informed consent statement: Not applicable.

Data availability statement: Not applicable.

Conflicts of interest: The authors declare that there is no conflict of interest.

References

[1] Spencer ND. Tribology. Faraday Discussions. 2012; 156(1): 435–438.
[2] Vikram K, Pramanik S, Bhaumik S. Thermal kinetic, mechanical and wear properties of glass fiber reinforced polyamide 6 thermoplastic composites. Journal of Applied Polymer Science. 2024; 141(42): e56119.
[3] Heintze S, Zappini G, Rousson V. Wear of ten dental restorative materials in five wear simulators-results of a round robin test. Dent Mater: Official Publication of the Academy of Dental Materials. 2005; 21(4): 304–317.
[4] Kato K. Classification of wear mechanisms/models. Proceedings of the Institution of Mechanical Engineers, Part J: Journal of Engineering Tribology. 2002; 216(6): 349–355.
[5] La Rosa GRM, et al. Application of tribology concept in dental composites field: A scoping review. Journal of Functional Biomaterials. 2022; 13(4): 287.
[6] Zhou Z, Jin Z. Biotribology: Recent progresses and future perspectives. Biosurface Biotribology. 2015; 1(1): 3–24.
[7] Shi B, Liang H. Tribological applications of biomaterials: An overview. Science in China Series A Mathematics Physics Astronomy. 2001; 44(SUPP): 297–306.
[8] Cheryl, R., B., Biomaterials: Body parts of the future, Technology Today, **Fall 1995** issue, Southwest Research Institute.
[9] Zheng Y, et al. Review of dental tribology: Current status and challenges. Tribology International. 2022; 166: 107354.
[10] Vikram K, Pramanik S, Bhaumik S. Effect of hexagonal boron nitride on structural, mechanical, and tribological behavior of polyamide 6/glass fibers (5 wt%) hybrid nanocomposites. Iranian Polymer Journal. 2024; 33(4): 511–530.
[11] Vikram K, Bhaumik S, Pramanik S. Effect of graphite on tribological and mechanical properties of PA6/5GF composites. Journal of Thermal Analysis Calorimetry. 2023; 148(9): 3341–3355.
[12] Saha S, Roy S. Metallic dental implants wear mechanisms, materials, and manufacturing processes: A literature review. Materials. 2022; 16(1): 161.
[13] Zheng J, Zhou Z. Study of in vitro wear of human tooth enamel. Tribology Letters. 2007; 26: 181–189.
[14] Wang L, et al. Friction and wear behaviors of dental ceramics against natural tooth enamel. Journal of the European Ceramic Society. 2012; 32(11): 2599–2606.

[15] Augustyn-Pieniążek J, et al. Selected properties and tribological wear alloys Co-Cr-Mo and Co-Cr-Mo-W used in dental prosthetics. Archives of Metallurgy Materials. 2015; 60(3): 1569–1574.

[16] Xiang D, et al. Comparison of wear properties of Ti_6Al_4V fabricated by wrought and electron beam melting processes in simulated body fluids. Rapid Prototyping Journal. 2020; 26(5): 959–969.

[17] Mathew MT, et al. Influence of pH on the tribocorrosion behavior of CpTi in the oral environment: Synergistic interactions of wear and corrosion. Journal of Biomedical Materials Research, Part B: Applied Biomaterials. 2012; 100(6): 1662–1671.

[18] Eisenburger M, Addy M. Erosion and attrition of human enamel in vitro, part I: Interaction effects. Journal of Dentistry. 2002; 30(7–8): 341–347.

[19] Li H, Zhou Z. Wear behaviour of human teeth in dry and artificial saliva conditions. Wear. 2001; 249 (10–11): 980–984.

[20] Mayworm CD, Camargo SS Jr, Bastian FL. Influence of artificial saliva on abrasive wear and microhardness of dental composites filled with nanoparticles. Journal of Dentistry. 2008; 36(9): 703–710.

[21] Arsecularatne J, Hoffman M. On the wear mechanism of human dental enamel. Journal of the Mechanical Behavior of Biomedical Materials. 2010; 3(4): 347–356.

[22] Borrero-Lopez O, Constantino PJ, Lawn BR. Role of particulate concentration in tooth wear. Journal of the Mechanical Behavior of Biomedical Materials. 2018; 80: 77–80.

[23] Licausi M-P, Muñoz AI, Borrás VA. Influence of the fabrication process and fluoride content on the tribocorrosion behaviour of Ti_6Al_4V biomedical alloy in artificial saliva. Journal of the Mechanical Behavior of Biomedical Materials. 2013; 20: 137–148.

[24] Xu H, et al. Indentation damage and mechanical properties of human enamel and dentin. Journal of Dental Research. 1998; 77(3): 472–480.

[25] Ganss C. Definition of Erosion and Links to Tooth Wear. Monographs in Oral Science, 2006; 20:9–16.

[26] Wang X, Lussi A. Assessment and management of dental erosion. Dental Clinics. 2010; 54(3): 565–578.

[27] Mathew M, et al. Significance of tribocorrosion in biomedical applications: Overview and current status. Advances in Tribology. 2009; 2009(1): 250986.

[28] Ampadi Ramachandran R, et al. Suitability of Ti–Zr alloy for dental implants: Tribocorrosion investigation. Journal of Bio- and Tribo-Corrosion. 2021; 7: 1–15.

[29] Noronha Oliveira M, et al. Can degradation products released from dental implants affect peri-implant tissues? Journal of Periodontal Research. 2018; 53(1): 1–11.

[30] Cruz H, et al. Tribocorrosion and bio-tribocorrosion in the oral environment: The case of dental implants. In: Paulo Davim J, editors. Biomedical Tribology. London, UK: Nova Science Publishers, Inc; 2011, pp. 1–33.

[31] Corne P, et al. Fretting-corrosion behavior on dental implant connection in human saliva. Journal of the Mechanical Behavior of Biomedical Materials. 2019; 94: 86–92.

[32] Alrabeah GO, Knowles JC, Petridis H. Reduction of tribocorrosion products when using the platform-switching concept. Journal of Dental Research. 2018; 97(9): 995–1002.

[33] Fransson C, et al. Extent of peri-implantitis-associated bone loss. Journal of Clinical Periodontology. 2009; 36(4): 357–363.

[34] Vikram K, et al. Surface Modification of Magnesium Alloy Employing External Coating for Biomedical Applications. In: Magnesium Alloys for Biomedical Applications, Boca Raton, USA: CRC Press; 2024, pp. 109–123.

[35] Smith DC. Dental implants: Materials and design considerations. International Journal of Prosthodontics. 1993; 6(2): 106–117.

[36] Zhou Z, Zheng J. Tribology of dental materials: A review. Journal of Physics D: Applied Physics. 2008; 41(11): 113001 (pp.21–22).

[37] Raza SM, et al. Silicon Nitride (Sin): An Emerging Material for Dental Implant Applications. In: Zafar MS, Khurshid Z, García-Gareta E, editors. Dental Implants, Woodhead Publishing; Northwood, UK: Elsevier; 2020, pp. 287–299.
[38] Duraccio D, Mussano F, Faga MG. Biomaterials for dental implants: Current and future trends. Journal of Materials Science. 2015; 50: 4779–4812.
[39] Lautenschlager EP, Monaghan P. Titanium and titanium alloys as dental materials. International Dental Journal. 1993; 43(3): 245–253.
[40] Schwitalla A, Müller W-D. PEEK dental implants: A review of the literature. Journal of Oral Implantology. 2013; 39(6): 743–749.
[41] Makhija SK, et al. Dentist material selection for single-unit crowns: Findings from the national dental practice-based research network. Journal of Dentistry. 2016; 55: 40–47.
[42] Poojar B, et al. Methodology used in the study. Asian Journal of Pharmaceutical and Clinical Research. 2017; 7(10): 1–5.
[43] Knosp H, Holliday RJ, Corti CW. Gold in dentistry: Alloys, uses and performance. Gold Bulletin. 2003; 36(3): 93–102.
[44] Möller H. Dental gold alloys and contact allergy. Contact Dermatitis. 2002; 47(2): 63–66.
[45] Roulet J-F. Benefits and disadvantages of tooth-coloured alternatives to amalgam. Journal of Dentistry. 1997; 25(6): 459–473.
[46] Liu Q, et al. Synthesis, preparation, in vitro degradation, and application of novel degradable bioelastomers – A review. Progress in Polymer Science. 2012; 37(5): 715–765.
[47] Shi R, et al. Recent advances in synthetic bioelastomers. International Journal of Molecular Sciences. 2009; 10(10): 4223–4256.
[48] Suryawanshi A, Behera N. Dental composite resin: A review of major mechanical properties, measurements and its influencing factors. Materialwissenschaft und Werkstofftechnik: Entwicklung, Fertigung, Prüfung, Eigenschaften Und Anwendungen Technischer Werkstoffe. 2022; 53(5): 617–635.
[49] Lambrechts P, et al. How to simulate wear? Overview of existing methods. Dent Mater: Official Publication of the Academy of Dental Materials. 2006; 22(8): 693–701.
[50] Van Houte J, et al. In vitro acidogenic potential and mutans streptococci of human smooth-surface plaque associated with initial caries lesions and sound enamel. Journal of Dental Research. 1991; 70(12): 1497–1502.
[51] Mair LH, Wear in the Mouth: The Tribological Dimension, In: Tooth Wear and Sensitivity, Clinical Advances in Restorative Dentistry, M. Addy, G. Embery, Eds: Edgar WM, Orchardson R; Martin Dunitz Ltd., London, UK; 2000, pp. 181-188.
[52] Souza JC, et al. Wear and corrosion interactions on titanium in oral environment: Literature review. Journal of Bio-and Tribo-Corrosion. 2015; 1: 1–13.
[53] Hiromoto S, Mischler S. The influence of proteins on the fretting–corrosion behaviour of a Ti_6Al_4V alloy. Wear. 2006; 261(9): 1002–1011.
[54] Barril S, Mischler S, Landolt D. Electrochemical effects on the fretting corrosion behaviour of Ti_6Al_4V in 0.9% sodium chloride solution. Wear. 2005; 259(1–6): 282–291.

2 Design

Mohadese Rajaeirad*, Mohammadjavad (Matin) Einafshar, Morad Karimpour, and Nima Jamshidi

Chapter 2.1 Patient-specific implant (PSI) design

Abstract: Patient-specific implants (PSIs) are among the most significant advancements in modern medicine, offering customized solutions tailored to individual anatomical structures. After defining PSIs, this chapter explores their rationale, design process, and clinical applications. It details the integration of advanced medical imaging, computer-aided design (CAD), and 3D printing technologies to enhance the precision and effectiveness of these implants. Furthermore, the chapter examines the role of surgical guides in PSI placement, evaluates design optimization techniques, and discusses challenges such as material selection, regulatory hurdles, and cost considerations. The future of PSIs is also explored, highlighting innovations in artificial intelligence (AI), bioactive materials, and scalable manufacturing techniques. By addressing both current limitations and future directions, this chapter provides a comprehensive understanding of PSIs and their role in personalized medicine.

Keywords: Patient-specific implants, personalized medicine, surgical guides, medical imaging, 3D printing, computer-aided design (CAD), finite element analysis (FEA), biomaterials, regulatory challenges, cost-effectiveness, AI in healthcare, orthopedic implants, cranial implants, maxillofacial reconstruction

2.1.1 Introduction

In the rapidly evolving field of modern medicine, the notion of a "one-size-fits-all" approach is increasingly becoming obsolete. Instead, the focus has shifted toward patient-specific treatments – highly individualized care that takes into account the unique physiological and lifestyle factors of each patient. From personalized medications designed around genetic markers to tailored therapeutic regimens, this trend re-

***Corresponding author: Mohadese Rajaeirad**, Department of Biomedical Engineering, Faculty of Engineering, University of Isfahan, Isfahan, Iran, e-mail: m.rajaeirad@eng.ui.ac.ir
Mohammadjavad (Matin) Einafshar, Department of Material and Production, Aalborg University, Denmark, e-mail: moei@mp.aau.dk
Morad Karimpour, School of Mechanical Engineering, University of Tehran, Tehran, Iran, e-mail: m.karimpour@ut.ac.ir
Nima Jamshidi, Department of Biomedical Engineering, Faculty of Engineering, University of Isfahan, Iran, e-mail: n.jamshidi.academics@gmail.com

https://doi.org/10.1515/9783111386027-006

flects a deeper understanding that every patient is not merely a diagnosis, but an individual with specific needs and characteristics.

A key extension of this personalized approach is found in the development of patient-specific implants (PSIs), which are custom-made medical devices meticulously engineered to meet the unique anatomical needs of individual patients. PSIs are more than just medical tools – they are emblematic of the broader transformation occurring in healthcare, where technology enables clinicians to adapt treatments with unprecedented precision. Just as personalized treatments consider a patient's genetic profile, sensitivities, and lifestyle, PSIs are designed using detailed imaging data to perfectly match the contours and structural requirements of the patient's anatomy.

The emergence of PSIs marks a pivotal moment in the ongoing evolution of personalized medicine, offering not only the promise of enhanced surgical outcomes but also the potential for more efficient, patient-centered care. As we delve deeper into the intricacies of PSIs, their designs, and their clinical applications, it becomes clear that these custom devices are helping to redefine what it means to deliver truly personalized medical care.

Definition of PSIs: PSIs are custom-made medical devices designed to fit the unique anatomical structure of an individual patient. Unlike standard implants, which are manufactured in predetermined sizes and shapes, PSIs are created based on detailed imaging data from the patient, allowing for a precise match to their anatomical features [1, 2].

We define the PSI as follows:

> PSIs are bespoke medical devices precisely engineered to align with an individual's unique anatomical and biomechanical features, providing tailored solutions that enhance integration, functionality, and clinical outcomes.

Implants can be categorized based on their primary function, which reflects the specific needs they address within the body. This classification helps in understanding the diverse roles implants play in medical treatments, ranging from stabilizing structures to enhancing aesthetics. Below is a categorization of implants based on their core functions:

1. **Fixation implants**: These implants are designed to stabilize and support bones or joints, particularly after fractures, surgeries, or reconstructions. They are essential in maintaining the structural integrity of the skeletal system during the healing process.
   - **Examples:**
     - Plates and screws used in orthopedic surgery to fix broken bones.
     - Spinal rods and pedicle screws utilized in spinal fusion surgeries.
     - Cranial fixation devices that secure bone flaps after craniotomy.
2. **Protective implants**: Protective implants serve to shield vulnerable anatomical structures from further injury or damage, or to restore the structural integrity of damaged areas. They play a critical role in protecting vital organs and tissues.

   - **Examples:**
     - Cranial plates that protect the brain after decompressive craniectomy.
     - Titanium meshes used in orbital floor reconstruction to protect the eye and restore the orbital structure.
     - Rib plating systems used to stabilize and protect the thoracic cavity after trauma.
3. **Functional implants**: These implants are designed to restore or enhance the function of a body part, often by replicating the biomechanical properties of the original tissue. They are crucial in restoring mobility, organ function, or other physiological processes.
   - **Examples:**
     - Hip and knee replacements in joint arthroplasty.
     - Cardiac implants like pacemakers or heart valves.
     - Dental implants that replace missing teeth and restore chewing function.
4. **Cosmetic implants**: Cosmetic implants are used primarily to improve the aesthetic appearance of a body part, often in reconstructive or elective surgeries. They are particularly important in restoring or enhancing the visual symmetry and appearance of the body.
   - **Examples:**
     - Breast implants used in augmentation or reconstruction postmastectomy.
     - Facial implants (e.g., chin, cheek, and jaw implants) used in reconstructive surgery to restore facial symmetry and contour.
     - Custom cranial implants used to restore the normal contour of the skull after trauma or surgery.
5. **Protective-functional implants**: These implants combine both protective and functional aspects, ensuring that the integrity and functionality of the anatomical area are maintained or restored. They are particularly useful in cases where both protection and restoration of function are critical.
   - **Examples:**
     - Cranioplasty implants, which not only protect the brain but also restore the cranial contour and support normal neurological function.
     - Orbital implants that protect the eye while restoring normal eye movement and appearance.

Recent advancements in PSIs have been driven by groundbreaking technologies such as 3D printing and additive manufacturing (AM). These innovations allow for the creation of implants that are uniquely designed to conform to a patient's anatomical features, improving both precision and clinical outcomes. Materials such as polyetheretherketone (PEEK) and titanium mesh are now being utilized and customized using computer-aided design (CAD) and 3D printing techniques [3, 4]. These innovations have particularly enhanced the design and application of implants in complex areas like the craniomaxillofacial region, where the demand for accuracy is especially critical [3].

Compared to standard implants, PSIs offer numerous advantages, particularly regarding fit and customization. Research has shown that PSIs achieve high dimensional accuracy and morphological similarity, which are essential for reconstructive procedures in areas with complex anatomy [4]. Clinical outcomes with PSIs have demonstrated reduced operative times, fewer complications, and improved postoperative aesthetics. The use of PSIs has been associated with a lower risk of complications such as infections and implant displacement, in contrast to standard implants [5]. However, PSIs come with higher costs and longer manufacturing times, which present challenges to their widespread adoption.

Personalized medical treatments, including the integration of PSIs, have been shown to improve recovery rates and enhance overall clinical outcomes by tailoring interventions to each patient's specific needs. In fields such as oncology, personalized approaches have resulted in more effective and less toxic treatments, contributing to better outcomes for patients with complex diseases such as multiple myeloma [6]. Similarly, personalized strategies in stroke rehabilitation have shown improved functional recovery by addressing the unique characteristics of each patient's condition [7].

PSIs are particularly vital in cases where standard implants do not provide an adequate fit or function due to the patient's unique anatomical structure. Conditions such as maxillofacial deformities, cranial defects, and various orthopedic issues benefit from the precision and customization of PSIs, ensuring better anatomical conformity and functional outcomes [1, 3]. Furthermore, challenges such as implant resorption, infection, and displacement, which are common with prefabricated implants, can be mitigated with PSIs that are specifically designed to match the patient's anatomy [3].

Importance and necessity of PSIs: The importance and necessity of personalized surgical implants are underscored by their role in enhancing medical treatment and patient outcomes. Specifically, these advances can be categorized into three key areas:

1. **Personalized medical treatment:** PSIs represent a significant advancement in personalized medicine, providing tailored solutions that align with the specific anatomical needs of each patient. This high level of personalization contributes to better surgical outcomes by minimizing complications and enhancing the precision of the intervention.
2. **Improved patient outcomes:** PSIs have been associated with improved patient outcomes, particularly in terms of both functionality and aesthetics. The precise fit and anatomical conformity of these implants reduce the likelihood of postoperative complications and lead to faster recovery times, contributing to enhanced patient satisfaction [8].

3. **Adaptation to unique anatomical structures:** PSIs are particularly advantageous in cases involving complex or altered anatomy, whether due to trauma, disease, or congenital conditions. By accommodating irregularities and asymmetries, PSIs offer a superior match to the patient's specific anatomical needs, which off-the-shelf solutions cannot provide.

2.1.2 Why we need PSIs

PSIs have become increasingly vital in modern medicine due to the vast anatomical variations among individuals. These implants are custom-designed to fit the unique characteristics of each patient, ensuring a more precise and effective treatment outcome compared to standard implants. PSIs are particularly valuable in addressing anatomical variability, reducing surgical complications, and enhancing both functional and aesthetic outcomes.

2.1.2.1 Addressing anatomical variability

The human body displays significant anatomical diversity, which can vary greatly due to factors such as genetics, nationality, race, and individual development. Differences in bone structure, tissue density, and organ size are observed not only between individuals but also between populations of different ethnic backgrounds. For example, craniofacial anatomy often differs across races, influencing the shape and contours of the skull, jaw, and face [1]. These variations make it challenging to use standard implants universally, as a "one-size-fits-all" approach can lead to suboptimal outcomes.

Anatomical variability is particularly evident in patients with congenital deformities, traumatic injuries, or previous surgeries. For instance, in maxillofacial defects, the bone structure can vary significantly from patient to patient due to trauma, congenital anomalies, or prior surgical interventions. In these cases, PSIs are essential for creating a customized solution that matches the patient's specific bone contours and unique anatomical features [1]. This tailored approach allows surgeons to restore function and aesthetics more effectively.

2.1.2.1.1 Examples of common anatomical variations necessitating PSIs

PSIs are often essential due to specific anatomical variations that render standard implants unsuitable across several medical fields (Figure 2.1.1).

In cranial surgery, complex defects – such as those caused by craniosynostosis or traumatic injuries – frequently require PSIs for precise contouring and functional restoration. Standard implants may fail to provide the exact fit needed to correct cranial asymmetries, making PSIs crucial for optimal outcomes [9]. For instance, in cases like craniosynostosis, PSIs can correct contour deformities, such as supraorbital retrusion and bitemporal hollowing, leading to both aesthetic and functional improvements.

In orthopedic surgery, anatomical variations such as hip dysplasia or complex spinal deformities like hemivertebrae are prevalent. Standard implants often do not accommodate these unique bone structures, complicating proper alignment and stability. PSIs are tailored to conform precisely to the patient's skeletal structure, enhancing stability and ensuring long-term success. For example, in congenital scoliosis with hemivertebrae, PSIs can be customized to fit the spine's unique curvature, providing stability and improving overall quality of life [10, 11].

Similarly, in maxillofacial and dental surgery, significant variations in jaw structure necessitate custom implants, especially for patients undergoing procedures like hemimandibulectomy or maxillectomy. These custom solutions restore both functional movement and aesthetics, crucial for proper chewing, speech, and facial symmetry. In cases involving extensive jaw tumors requiring mandibular resections, PSIs help restore jaw continuity and support dental rehabilitation, enabling patients to regain function and aesthetic balance [12, 13].

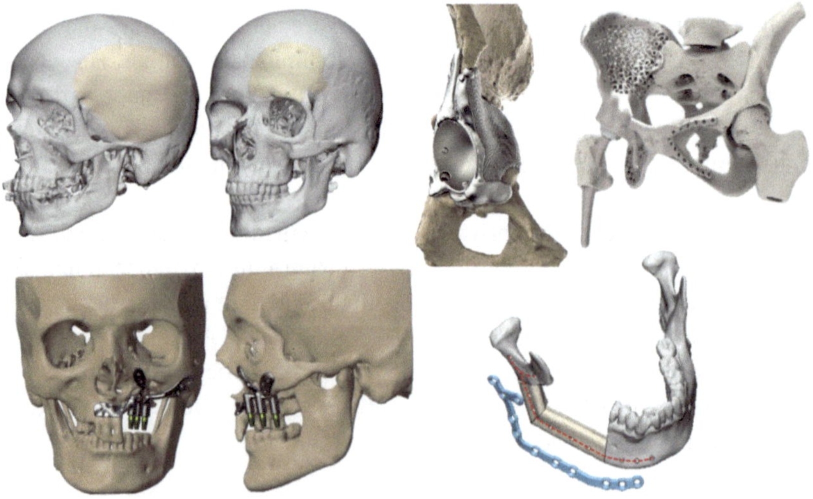

Figure 2.1.1: Common patient-specific implants.

2.1.2.2 Customization and precision in surgery

One of the most significant benefits of PSIs is the customization they offer, providing a tailored fit to the patient's unique anatomy. This precision leads to a more secure and stable implant, minimizing the need for intraoperative adjustments and reducing the risk of complications.

- **Reduction in surgery time**: PSIs are predesigned to fit the patient's anatomy precisely, which significantly reduces the time required for adjustments during surgery. For example, in complex hip surgeries, PSIs can streamline procedures, leading to shorter operating times and lower risks of intraoperative complications [11]. The precise fit also minimizes the need for real-time surgical improvisation, contributing to smoother, faster procedures.
- **Reduction in postoperative complications**: By fitting perfectly to the patient's anatomy, PSIs reduce the likelihood of complications such as implant displacement, infection, or poor integration. For instance, in orbital reconstructions, customized implants have been shown to lower the risk of postoperative issues like diplopia (double vision) and hardware extrusion, ensuring better functional outcomes and patient satisfaction [12].

2.1.2.2.1 Enhanced functionality and aesthetics

PSIs not only provide functional benefits but also significantly improve aesthetic outcomes, which is particularly crucial in surgeries involving the face, skull, and visible parts of the body.

- **Aesthetic improvements**: In cranial and maxillofacial surgeries, PSIs allow for precise contouring that matches the patient's natural anatomy. This level of precision leads to superior aesthetic results compared to standard implants, particularly in regions where appearance plays a critical role in patient satisfaction [9].
- **Functional improvements**: PSIs enhance functional outcomes by ensuring proper alignment and stability. For instance, in surgeries for hip dysplasia, custom implants have resulted in improved joint function, pain reduction, and higher patient satisfaction, as evidenced by postoperative assessment scores [11].

2.1.2.2.2 Customization benefits: case studies and examples

Numerous case studies have demonstrated the effectiveness of PSIs in various surgical applications. In cranial reconstructions, for example, PSIs have been used to restore both the structure and appearance of the skull following trauma or congenital deformities [14, 15]. Similarly, in complex facial repairs, PSIs have been successful in restoring both function and aesthetic appeal, leading to improved patient outcomes

[1, 2]. These cases underscore the significant advantages of PSIs in modern surgical practice. By addressing anatomical variability, reducing surgical complications, and enhancing both functional and aesthetic outcomes, PSIs have become an indispensable tool in personalized medicine.

Clinical cases involving PSIs highlight their advantages across various fields of medicine. In orthopedics, 3D printing technology has been employed to create highly accurate PSIs. While specific cases are not detailed, the use of 3D-printed implants likely results in a more precise fit to patient anatomy, which can enhance surgical outcomes and overall patient recovery [16]. In maxillofacial surgery, a notable case involves the use of presurgical nasoalveolar molding (PNAM) for treating labial-alveolo-palatal clefts. This technique uses custom-made devices to prevent nasal deformities and improve surgical outcomes, demonstrating the value of tailored approaches in addressing complex craniofacial conditions [17]. PNAM also has psychological benefits, aiding in the reintegration of patients post-surgery by enhancing both functional and aesthetic results [17].

The advantages of PSIs are further seen in their ability to address complex medical conditions. While specific cases of failed standard implants transitioning to PSIs are not provided, the use of PNAM for cleft lip and palate defects illustrates how custom solutions can overcome challenges that might be difficult to manage with traditional methods. This approach underscores the potential of PSIs to handle congenital anomalies and other intricate conditions [17].

Some key lessons inferred from these cases include the importance of interdisciplinary approaches, particularly in managing complex cases like cleft lip and palate. Additionally, the development of manufacturing techniques such as 3D printing is crucial for advancing the precision and effectiveness of PSIs in various fields [16]. Presurgical planning, coupled with the use of custom devices, clearly improves surgical outcomes, as shown by the PNAM technique [17].

Long-term success in the use of PSIs often involves ongoing follow-up procedures. In orthopedic cases, this typically includes regular radiographic imaging and physical examinations to assess implant performance over time [18]. While detailed long-term data on PSIs were not provided, the PNAM cases suggest that achieving satisfactory surgical results and psychological reintegration requires sustained follow-up [17]. Although the search results offer limited specific information on PSIs, the examples point to the broader potential of patient-specific approaches in improving both functional and aesthetic outcomes in various medical fields.

2.1.3 How we design PSIs

The design of PSIs is a complex, multistep process that utilizes cutting-edge technology, advanced medical imaging, and precise computational tools to create custom implants tailored to the unique anatomical needs of individual patients. This intricate process begins with the acquisition and processing of detailed patient data, progresses through segmentation, and culminates in the design, material selection, and validation of the implant. The end result is an implant that optimally fits the patient's anatomy and provides enhanced functional and aesthetic outcomes.

2.1.3.1 Predesign considerations

The success of PSI design hinges on understanding patient-specific requirements. This necessitates close collaboration between the surgical team and engineers to ensure all anatomical and functional needs are addressed. Surgeons provide critical insights into clinical requirements, while engineers focus on the technical aspects, ensuring that the implant meets the necessary specifications. Effective collaboration is vital for achieving a design that not only fits the patient's anatomy but also fulfills the functional and aesthetic goals of the procedure.

2.1.3.2 Advanced medical imaging techniques

The foundation of PSI design lies in advanced medical imaging techniques, which are crucial for creating accurate anatomical models and ensuring precise implant fitting. Computed tomography (CT) scans are invaluable in producing detailed cross-sectional images, particularly in orthopedic and cranial surgeries where high spatial resolution is critical. CT scans excel in visualizing hard tissues, such as bone, making them indispensable for applications where structural accuracy is paramount [19, 20].

Magnetic resonance imaging (MRI) complements CT by providing superior imaging of soft tissues. This is particularly beneficial in procedures where the preservation and reconstruction of soft tissues are essential, such as in craniofacial or joint surgeries. In some applications, like total knee arthroplasty, MRI has demonstrated enhanced bone resection accuracy and improved PSI stability compared to CT [21]. The integration of these imaging modalities with 3D modeling techniques allows for the creation of precise anatomical replicas, which are integral to the successful design and implementation of PSIs.

2.1.3.3 Data processing and segmentation

The collection and processing of imaging data are critical steps in PSI design. Segmentation, which involves isolating specific anatomical structures, defines the boundaries of tissues and bones, ensuring that the digital model accurately reflects the patient's anatomy. Advances in machine learning algorithms have significantly enhanced segmentation accuracy and efficiency, directly impacting the precision of the 3D model and, consequently, the fit of the implant. Accurate segmentation is essential for better implant design, improving surgical outcomes, and reducing the likelihood of complications. After the segmentation and design process, software such as MIMICS is often used to check the fit of the implant, ensuring that the bone and implant do not intersect, which could lead to complications during surgery.

2.1.3.4 Design process

The design process for PSIs begins with a detailed evaluation of the patient's anatomy, typically using CT scans for bone assessment. Specialized software, such as MIMICS or 3D Slicer, is employed during this phase to generate a detailed model of the affected area. The involvement of the surgeon is crucial throughout this process, as ongoing discussions ensure that the implant design aligns with the specific treatment requirements Figure 2.1.2 (a).

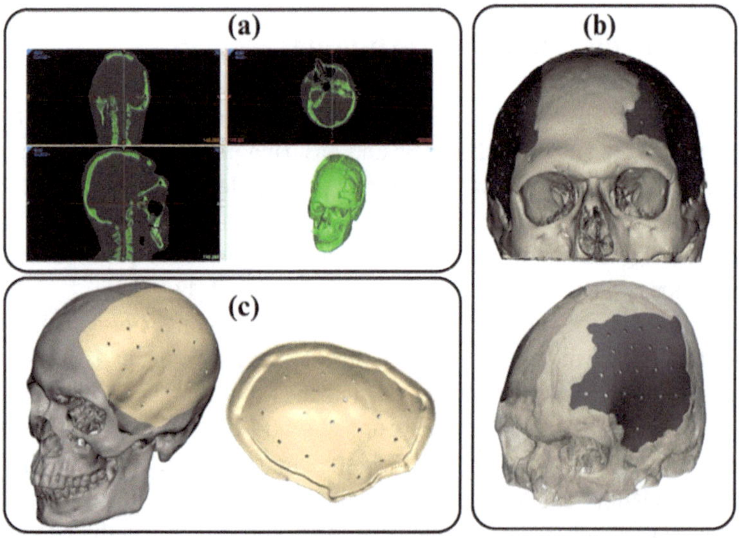

Figure 2.1.2: (a) 3D models generated using CT scan images in Mimics software; (b) customized cranial plate designed using the 3D model from (a) for a specific patient; (c) another example of a customized cranial plate, viewed from the inside and outside. The undercut area, critical for better fit, must be removed based on fracture site and placement orientation to optimize implant fit and function.

In the case of cranial implants, for example, it is essential to avoid relying on weakened bone at the fracture site Figure 2.1.2 (b) and (c) and instead extend the implant to more stable regions for support. For maxillofacial implants, which involve inherently weaker bones, special attention is given to minimizing pressure on vulnerable areas to prevent secondary fractures Figure 2.1.3. In complex cases involving soft tissue, MRI may complement the CT scan to provide a comprehensive understanding of the anatomical structures, particularly when the implant involves arteries or other delicate tissues. The design process carefully balances function and aesthetics, ensuring that the PSI provides adequate strength, does not irritate the skin, and meets both functional and cosmetic goals.

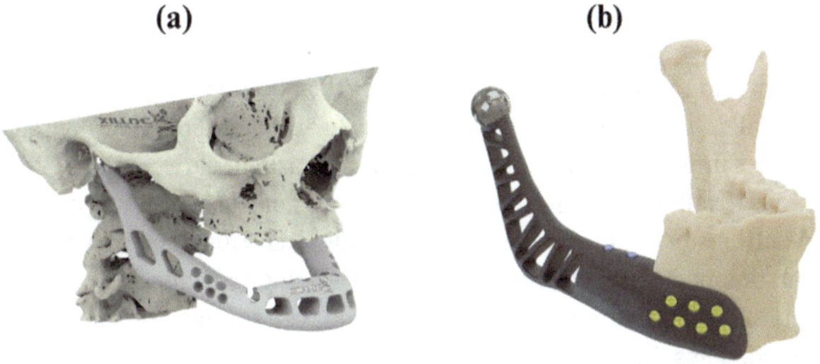

Figure 2.1.3: (a) Full mandibular implant produced by Xilloc Medical, Sittard-Geleen, The Netherlands; and (b) partial mandibular implant produced by Fanavaran Jarahyar Sharif (FJSharif), Tehran, Iran.

In regions where the soft tissue covering the implant is thin, such as the chin, special consideration is given to the tactile interaction between the skin and the implant. The number and placement of screws are determined not only by biomechanical requirements but also by ensuring minimal irritation or discomfort to the patient. Additionally, load-bearing characteristics must be analyzed to guarantee that the implant can withstand physiological forces without compromising structural integrity. To verify these biomechanical properties, finite element analysis (FEA) simulations are often conducted using software such as Abaqus or Ansys, allowing for virtual testing under various load conditions. These simulations can save time and costs compared to physical prototype testing.

2.1.3.5 Fitting, simulation, and finalization

Once the design is finalized, virtual simulations are conducted to evaluate the contact points between the implant and the bone. A prototype may be printed to physically verify the fit. Throughout this process, the incision technique and implant orientation are continuously discussed with the surgeon to ensure optimal surgical outcomes. Upon approval of the final design by the surgical team, the implant is manufactured using suitable materials, such as PEEK, selected for their mechanical properties and biocompatibility.

2.1.3.6 Validation and verification

Design verification and validation are critical steps that ensure the PSI meets the required specifications and functions as intended. Verification confirms that the design and its digital models accurately represent the design input requirements, essentially answering the question, "Are we building the implant right?" This involves simulations and testing procedures that verify the integrity of the design based on predefined standards.

Validation, on the other hand, ensures that the final product meets the needs of the end user – the patient and the medical team. It addresses the question, "Are we building the right implant?" Validation requires testing the device in real-world or clinically relevant environments to confirm its suitability for actual use in patient care.

In PSI design, verification often includes virtual simulations using FEA to predict how the implant will perform under physiological conditions. This approach allows for thorough evaluation of stress distributions and load-bearing capacities before clinical implementation. For validation, clinical trials or cadaver studies are commonly employed to assess the PSI's fit, functionality, and long-term stability, ensuring patient safety and improved surgical outcomes.

2.1.3.7 Mechanical and biological properties of PSI materials

The selection of materials for PSIs is crucial, as the chosen material must exhibit both mechanical and biological properties that are suitable for the specific clinical application. Mechanically, the material must be strong enough to withstand the physical forces exerted on the implant during normal activity. Durability and resistance to wear are essential to ensure the longevity of the implant. Biologically, the materials must be biocompatible to prevent adverse reactions such as inflammation or rejection by the body.

Titanium and its alloys, such as ceramic-titanium (CeTi), are commonly used in PSIs due to their excellent biocompatibility and mechanical strength. Titanium is particularly effective in applications requiring structural support, such as cranial and orthopedic implants. However, the primary drawbacks of titanium include its higher cost and the potential for producing imaging artifacts during MRI scans, which can obscure postoperative evaluations [19].

PEEK is another material used for its flexibility, lightweight nature, and radiolucency, which allows for clear postoperative imaging. PEEK's good biocompatibility makes it suitable for a variety of applications, though it is not as strong as metals such as titanium and may require surface modifications to enhance its osseointegration capabilities [19, 21].

2.1.3.8 Pros and cons of common PSI materials

PSIs are designed to fit the unique anatomical features of each patient, offering tailored solutions for various medical applications. The selection of material is crucial, as it directly impacts the implant's performance, biocompatibility, and long-term success. Table 2.1.1 provides a clear comparison of each material's strengths and limitations, helping guide the decision-making process when selecting the appropriate material for specific applications. This ensures that the chosen material aligns with the mechanical demands and biological environment of the implant site, ultimately contributing to improved patient outcomes.

Table 2.1.1: The most commonly used materials in PSIs, along with their respective advantages and disadvantages.

Material	Pros	Cons
Titanium and titanium alloys	– High biocompatibility, reducing the risk of rejection and inflammation – Excellent mechanical strength and durability, suitable for load-bearing applications – Promotes bone growth around the implant (osseointegration)	– Expensive – Can cause metallic artifacts in CT imaging, complicating postoperative assessments
Polyetheretherketone (PEEK)	– Radiolucency allows for clear postoperative imaging – Well-tolerated by the body, making it versatile in various applications – Can be tailored to mimic bone's mechanical properties	– Lower strength compared to metals, limiting its use in high-load areas – May have higher bacterial adhesion compared to titanium
Hydroxyapatite	– Promotes bone integration due to its similarity to the bone mineral – Enhances biological response to the implant	– Brittle, often used as a coating rather than a bulk material – Complex and costly application process
Composite materials	– Can be engineered for specific mechanical and biological properties – Versatile, used in a range of applications from dental to orthopedic implants	– Designing and manufacturing can be challenging and costly

Table 2.1.1 (continued)

Material	Pros	Cons
Polymers and polymer composites	– Reduces imaging artifacts, improving postoperative care – Easily molded and customized for specific patient needs	– Generally lower mechanical strength compared to metals, limiting structural application
Silicone	– Highly customizable for complex shapes, ideal for cosmetic and reconstructive uses – Generally well-tolerated by the body	– Limited load-bearing capacity due to lower mechanical strength

2.1.3.9 Advancements in CAD and simulation tools

CAD tools have revolutionized PSI design by enabling the creation of highly detailed and customizable 3D models tailored to a patient's specific anatomical features. Recent advancements in CAD technology allow for more intricate and precise designs, incorporating fine details that ensure a better fit and function of the implant. Simulation tools such as FEA are employed to predict the mechanical behavior of the implant under physiological conditions, helping optimize the design before manufacturing.

Some CAD systems are integrated with robotic systems, such as the Sculptor Robotic Guidance Arm, which assist surgeons in the precise placement of implants. This integration enhances accuracy during surgery, leading to improved patient outcomes [19].

2.1.3.10 Case study: cranial implants

Cranial implants made from materials such as PEEK and titanium represent significant advancements in PSI technology, particularly in cranial reconstruction Figure 2.1.4. Titanium cranial meshes are widely used for their exceptional strength, durability, and ability to integrate with the bone tissue, making them suitable for cases requiring substantial structural support, such as large or irregular bone defects. PEEK cranial plates, on the other hand, offer a unique advantage due to their ability to closely mimic the natural thickness and mechanical properties of the bone, making them ideal for calvarial reconstructions following decompressive craniectomies, tumor resections, or trauma.

PEEK, being radiolucent, does not interfere with postoperative imaging techniques such as CT, MRI, or radiation therapy, enhancing its utility for ongoing patient care. PEEK implants also feature innovative technologies such as InterFix and XSu-

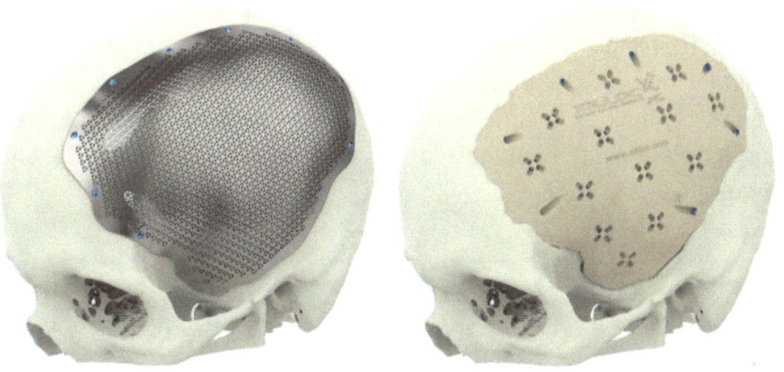

Figure 2.1.4: Cranial defect treated by titanium mesh (left) and PEEK implant (right) (all the images belong to Xilloc Company).

ture, which optimize the fixation process and soft tissue attachment. These technologies contribute to better surgical processes and improved postoperative outcomes.

2.1.4 Surgical guides in PSI design

2.1.4.1 Types of surgical guides

Before delving into the specific evaluation of the design process, it is important to distinguish between two main categories of surgical guides: **dissection guides** and **implantation guides**. These guides serve different purposes in surgical procedures and have distinct design considerations (Figure 2.1.5).

2.1.4.1.1 Design and application

- **Dissection guides**: These guides are designed to provide clear and accurate pathways for dissection, with a focus on achieving optimal accuracy in tissue removal while minimizing damage to the surrounding structures. By navigating and mapping anatomical structures during surgery, dissection guides help surgeons identify critical anatomical landmarks and pathways, such as blood vessels or nerves, to avoid during dissection. For instance, in breast reconstruction surgery, 3D-printed vascular models are employed to map perforator vessels, enhancing the precision of dissection and reducing operative times [22].
- **Implantation guides**: In contrast, implantation guides are designed to ensure the accurate placement of implants within the prepared surgical site, with a primary focus on achieving precise alignment and positioning according to the patient's

unique anatomy. This precision is crucial for optimizing both functional outcomes and aesthetic results. Typically created using CAD/CAM (computer-aided manufacturing) technology, these guides are tailored to fit the patient's specific anatomical features, making them essential in fields like dental implantology, where the correct placement of implants is critical to fulfilling both aesthetic and functional requirements [23, 24].

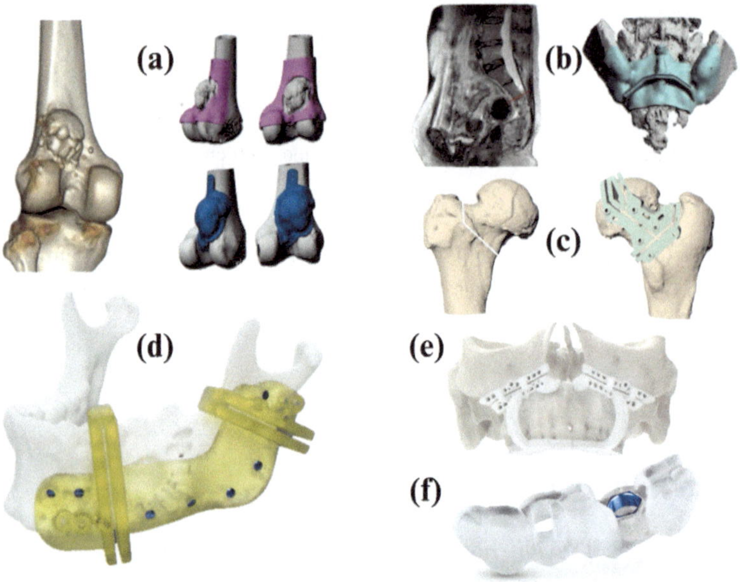

Figure 2.1.5: Surgical guides used for (a) distal femur, (b) pelvis, (c) femoral head, and (d) mandible tumor resection, as well as for (e) orthognathic osteotomy and (f) dental implantation.

2.1.4.1.2 Critical steps in the design and fabrication of surgical guides

1. **Imaging and data acquisition**: High-resolution imaging techniques, such as CT or MRI, are employed to capture detailed anatomical data.
2. **Segmentation and 3D modeling**: The acquired images are processed to create a 3D model of the relevant anatomy. Accurate segmentation is critical for defining the structures of interest.
3. **Design using CAD software**: The surgical guide is designed using CAD software, tailored to the patient's anatomy and the surgical plan.
4. **Simulation and validation**: The design is often tested using simulation software to ensure it meets the surgical objectives.
5. **Fabrication**: The guide is manufactured, typically using 3D printing technology, allowing for precise and rapid production.

2.1.4.1.3 Advantages and challenges of surgical guides

The use of surgical guides in tumor resection and reconstruction offers several distinct advantages:

- **Increased precision**: Surgical guides enhance the accuracy of surgical cuts and implant placement, reducing the risk of damaging the surrounding healthy tissues.
- **Reduced operating time**: By providing a clear surgical roadmap, these guides help streamline procedures, decreasing the time spent in the operating room (OR).
- **Improved safety**: Surgical guides assist surgeons in avoiding critical structures, thereby minimizing the risk of complications and improving overall patient safety.

However, there are challenges associated with their use:

- **Complexity of design**: The development of accurate guides requires detailed imaging and precision manufacturing, which can be both time-consuming and resource-intensive.
- **Surgical integration**: Successful implementation of surgical guides demands proper training and seamless integration into the surgical workflow. Surgeons must be familiar with these tools and skilled in their use for the guides to effectively enhance surgical outcomes.

For example, in orthopedic surgery, 3D-printed guides have been demonstrated to enhance surgical precision, reduce operative time, and decrease blood loss in the treatment of complex fractures, such as displaced middle-third clavicle fractures [25]. Similarly, in maxillofacial surgery, the use of 3D-customized implants and surgical guides during mandibulectomy and reconstructive procedures has significantly improved the accuracy of bone resections and implant placements, resulting in enhanced functional and aesthetic outcomes [26].

The role of surgical guides in complex surgeries: Surgical guides are integral to complex surgical procedures, as they enhance precision, mitigate the risk of errors, and ultimately improve patient outcomes. By providing well-defined pathways for surgical interventions, these guides significantly reduce the likelihood of complications while enhancing both functional and aesthetic results. For instance, in maxillofacial surgeries, surgical guides facilitate precise bone cutting and accurate implant placement in the jaw and facial bones, which is critical for effective functional restoration and optimal aesthetic outcomes. Similarly, in orthopedic surgeries, these guides assist in the accurate alignment and positioning of implants, both of which are essential for joint functionality and longevity. Moreover, they help minimize intraoperative radiation exposure by decreasing the need for repeated imaging assessments.

The role of surgical guides in tumor resection and reconstruction: Within the context of personalized medicine, patient-specific treatments now extend beyond custom implants to encompass surgical guides, which play a significant role in improving precision during tumor resection and subsequent reconstruction. These guides are particularly valuable in complex surgical scenarios where anatomical structures are delicate, and accuracy is of paramount importance. Developed using advanced imaging techniques and 3D printing technologies, surgical guides are specifically tailored to the unique anatomical features of each patient, thereby transforming the approach to surgical planning and execution across various medical disciplines and ultimately enhancing surgical outcomes.

2.1.4.1.4 Applications of surgical guides in various specialties

Neurosurgery: In neurosurgery, achieving a balance between maximizing tumor resection and preserving healthy brain tissue is critical. Surgical guides, developed using multimodal imaging and 3D printing technologies, facilitate this delicate balance by providing precise navigation through complex neuroanatomy. These guides enable the accurate removal of tumors while minimizing the risk of neurological deficits, a technique often referred to as the "onco-functional balance." This approach ensures that critical brain regions responsible for functions such as language and motor skills are preserved, leading to better patient outcomes [27].

Orthopedic oncology: In orthopedic oncology, 3D-printed surgical guides play a pivotal role in the precise removal of bone tumors and subsequent reconstruction of the affected areas. These guides assist surgeons in making accurate cuts and positioning implants with greater precision, which is particularly important when dealing with complex anatomical structures. Their use has been shown to enhance both the planning and execution of procedures, contributing to improved clinical outcomes and reduced complications [28].

Maxillofacial surgery: Maxillofacial surgery, often involving intricate reconstructions of facial structures, greatly benefits from patient-specific surgical guides. These guides, created through 3D imaging and printing, enable highly customized surgical approaches tailored to the patient's unique anatomy. The precision offered by these guides is essential not only for restoring function but also for achieving optimal aesthetic results. By aligning surgical actions with the exact contours of the patient's facial structure, surgical guides help surgeons achieve more accurate and satisfactory outcomes.

Dermatology: In dermatology, patient-specific 3D biomodels have emerged as valuable tools for guiding the surgical treatment of nail tumors. These models allow surgeons to visualize the affected area with greater clarity, facilitating more precise and minimally invasive procedures. For instance, 3D models have been effectively used in surgeries for glomus tumors and subungual keratoacanthomas, reducing operating times and improving patient understanding of the procedure [29].

Spinal tumor surgery: Spinal tumor surgery has also seen significant advancements with the use of patient-matched 3D-printed guides. These guides, created from CT scans, assist in the accurate placement of cervical pedicle screws – a procedure that demands extreme precision to avoid complications such as nerve damage. By ensuring the correct placement of these screws, surgical guides reduce the need for intraoperative fluoroscopy, streamline the procedure, and enhance patient safety, leading to better overall outcomes in spinal surgeries [30].

2.1.5 Evaluation of the design process

- **Key considerations for designing a PSI**
  The design of a PSI necessitates meticulous collaboration between surgeons and engineers, focusing on several critical aspects:
  - **Anatomical fit**: The implant must be precisely tailored to match the patient's unique anatomy. This precision is crucial for both functional outcomes and aesthetic considerations.
  - **Material selection**: It is essential to select materials that are biocompatible, durable, and appropriate for the specific anatomical location and intended function of the implant. The choice of material directly influences the implant's integration with the host tissue and its longevity.
  - **Mechanical properties**: The implant must possess the necessary mechanical stability and load-bearing capacity to withstand the physiological forces it will encounter. This requires a detailed assessment of the implant's structural integrity under various stress conditions.
  - **Biological compatibility**: Evaluating the potential for osseointegration is vital, as it determines how well the implant will bond with the surrounding bone and tissue. Additionally, minimizing the risk of adverse biological reactions is essential to ensure long-term implant success.
  - **Surgical approach**: The implant design must consider the planned surgical procedure, ensuring that the implant can be effectively and easily placed during surgery. This requires careful planning to accommodate the constraints of the surgical environment.

2.1.5.1 Clinical feedback and adaptation

Incorporating feedback from surgeons and patients post-surgery is essential for the continuous improvement of PSI designs:

This comprehensive design process, from predesign considerations to clinical feedback and adaptation, ensures that PSIs are meticulously crafted to meet the highest standards of functionality, safety, and patient satisfaction.

- **Incorporating surgeon and patient feedback**: After the implant has been used in surgery, feedback is collected to assess its performance. This feedback is invaluable for making further refinements, ensuring future implants are even more effective.
- **Adjustments based on clinical experiences**: Clinical experiences may necessitate adjustments to the implant's shape, size, or material to better meet the patient's needs and improve outcomes.
- **FEA and computational modeling:** FEA plays an essential role in both the verification and optimization of PSIs. Software tools such as **ANSYS**, **Abaqus**, and **COMSOL Multiphysics** are extensively used to simulate the mechanical behavior of the implant under various conditions. These simulations help engineers identify potential weaknesses, such as areas prone to stress concentration or deformation, allowing them to optimize the design for improved performance. **Computational modeling**, using platforms like **AnyBody** and **OpenSim**, extends the analysis by predicting how the implant will interact with biological tissues and the patient's unique biomechanics. This combined approach ensures that the implant is not only structurally sound but also functional within the body.

Simulation of mechanical performance and biological interactions: Through advanced simulation techniques, the mechanical performance of a PSI can be evaluated under realistic conditions. Using FEA software like **ANSYS**, **Abaqus**, and **COMSOL**, engineers simulate scenarios such as load-bearing capacities, fatigue under repetitive stress, and impact resistance. These simulations help verify that the implant will withstand the mechanical demands it will face once implanted.

In parallel, **biomechanical simulations** using platforms like **AnyBody** and **OpenSim** allow for the analysis of the implant's integration with the surrounding tissues. These tools help model the dynamic interactions between the implant and the musculoskeletal system, ensuring proper load distribution, joint functionality, and overall biocompatibility. By assessing these interactions, engineers can further validate the implant's design for long-term stability and functionality within the patient's body.

In clinical practice, once the fitting of a PSI is validated and verified by the surgeon—particularly in relatively straightforward cases—computational simulations are often deemed unnecessary. However, in complex scenarios where muscle dynamics and biomechanical integration play a critical role, or when the design presents structural challenges or multiple design alternatives, FEA or specialized simulation software may be employed to optimize implant performance and ensure functional efficacy.

2.1.5.2 Prototyping and iterative design

Prototyping plays a crucial role in the PSI design process. By creating physical prototypes, designers can test and refine the implant before the final version is manufactured. This step is essential for verifying that the implant meets all design specifications and performs as intended.

The design process for PSIs is inherently iterative, involving multiple cycles of feedback and refinement. Initial prototypes undergo rigorous testing and simulation, generating valuable insights that inform subsequent design adjustments. This iterative approach ensures that each version of the implant is progressively improved, ultimately achieving the best possible fit and function for the patient. By incorporating feedback from both prototype testing and simulations, the final design is optimized to meet the highest standards of precision and performance Figure 2.1.6.

2.1.5.3 Shape and topology optimization

Lattice structures represent a significant advancement in the design of implants created through AM. These structures offer several advantages that enhance the performance of the implants.

One of the primary benefits of lattice structures is improved osseointegration. By mimicking the porous nature of the bone, lattice structures promote bone ingrowth, facilitating better integration of the implant with the surrounding tissue. This leads to increased stability and longevity of the implant. Additionally, lattice structures help mitigate the issue of stress shielding, which is common with solid implants. Stress shielding can lead to bone resorption and implant failure by unevenly distributing mechanical loads. Lattice structures address this issue by allowing for a more uniform distribution of stresses, thus reducing the likelihood of such complications.

Another advantage of lattice structures is their contribution to weight reduction. By incorporating lattice designs, the weight of the implant can be significantly reduced without compromising its strength. This not only makes the implant more comfortable for the patient but also eases handling during surgery. Furthermore, lattice structures offer enhanced mechanical properties due to their design flexibility. This flexibility enables the optimization of properties such as stiffness and strength to closely match those of the surrounding bone, thereby improving the overall performance of the implant.

In the design process, shape optimization techniques are employed to ensure that the implant conforms precisely to the patient's anatomy. This process reduces the

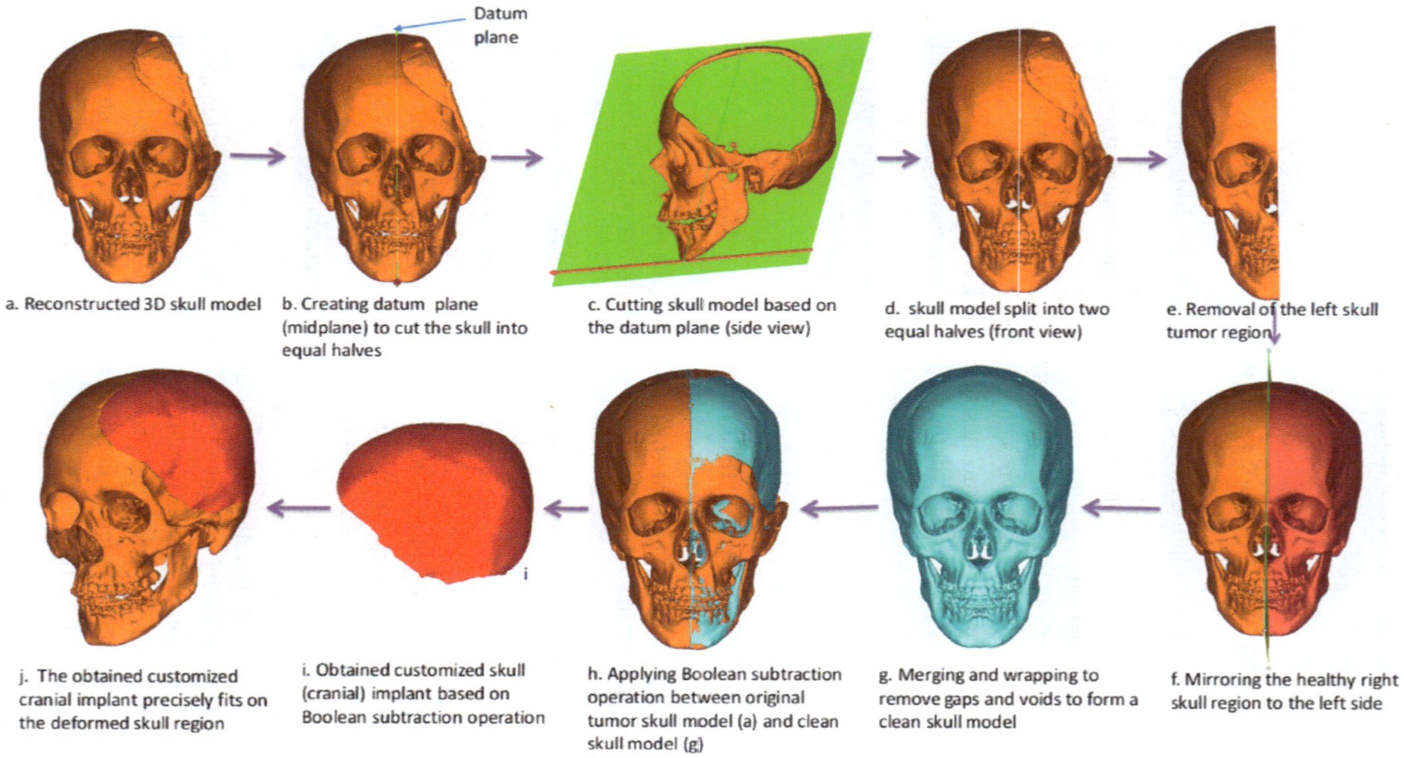

Figure 2.1.6: Steps involved in the design of customized cranial implant in 3-Matic®: (**a**) Reconstruction of the 3D skull model; (**b**) generation of the midplane; (**c**) cutting of the skull model; (**d**) splitted model; (**e**) removal of the defected side; (**f**) mirroring of the right side; (**g**) merging of two sides; (**h**) application of Boolean subtraction; (**i**) customized implant; and (**j**) skull with fitted implants [31] (Open access).

amount of material used, resulting in a lighter and more comfortable implant. Similarly, topology optimization is used to enhance the internal structure of the implant. This technique ensures that the implant is robust enough to withstand physiological forces while being as lightweight as possible.

2.1.5.4 Patient perspectives on PSIs

PSIs are becoming an increasingly important option in personalized medical treatments, and understanding patient perspectives on these implants is crucial for their successful implementation. An analysis of patient preferences and experiences with PSIs compared to standard implants reveals several key insights into what patients value and the factors that influence their decision-making.

1. **Patient values and decision-making**: Patients tend to prioritize improvements in their quality of life when considering PSIs. For instance, studies on mini-implants for mandibular overdentures indicate that patients experience better oral health-related quality of life (OHRQoL) with PSIs compared to standard implants. Specifically, treatments involving four mini-implants were found to provide greater satisfaction than those using two standard implants, highlighting the value patients place on personalized approaches that meet their specific needs [32]. Additionally, having alternative treatment options is important to patients, especially in cases where standard implants may not be feasible due to factors like insufficient ridge width, chronic health conditions, fear of surgery, or financial constraints [32].
2. **Quality of life improvements and concerns:** Patients generally perceive PSIs as offering significant improvements in their quality of life, particularly in dental applications. The study on mini-implants showed that PSIs could lead to comparable or even superior OHRQoL outcomes when compared to standard implants [32]. However, there are concerns associated with the use of PSIs, particularly regarding the longevity and survival rates of the implants. For example, mini-implants have shown lower survival rates (89% and 82%) compared to standard implants (99%), which could raise concerns among patients about the long-term reliability of this newer technology [32].
3. **Common themes in patient feedback**: Trust and expectations play a significant role in patient decision-making when it comes to PSIs. For example, a study on esophageal cancer treatment preferences found that patients who opted for active surveillance tended to have greater trust in their bodies and held positive expectations for their health outcomes. In contrast, patients who preferred standard surgery often sought to minimize uncertainty [33]. In dental applications, patient feedback on mini-implants indicated high satisfaction rates, suggesting that personalized treatments like PSIs can lead to positive patient experiences [32].

4. **Cultural and socioeconomic factors**: Although the search results do not explicitly address cultural and socioeconomic factors, the influence of cost on treatment choices is evident. The fact that cost can limit access to standard implants suggests that socioeconomic status plays a role in the decision-making process. Patients with limited financial resources may be more likely to seek alternative treatments like mini-implants, which could be more affordable or better suited to their specific needs [32].
5. **Role of patient education and communication**: Effective communication and patient education are critical for the successful adoption of PSIs. While not explicitly covered in the search results, the NOSANO study on esophageal cancer treatment preferences underscores the importance of understanding patient mindsets and preferences in making treatment decisions. A conceptual model developed in the study emphasizes the need for healthcare professionals to tailor their communication to align with patient preferences, thereby facilitating informed decision-making and increasing patient satisfaction with their chosen treatment options [33].

2.1.5.5 Additional insights

In specific medical applications, the outcomes of PSIs can vary depending on factors such as the surgeon's experience and the complexity of the procedure. For example, in total ankle arthroplasty, PSIs showed similar outcomes to standard referencing in terms of implant positioning and operative time, suggesting that the benefits of PSI may be more pronounced in some areas than others [34]. Additionally, in orbital reconstruction, PSIs demonstrated high accuracy in achieving planned outcomes, which could be a significant factor in building patient trust and satisfaction with this technology [35].

2.1.6 Challenges and future directions in PSI design

2.1.6.1 Technological limitations in PSI design and production

Despite the significant advancements in PSI design and manufacturing, several technological challenges persist. A primary issue lies in **manufacturing precision**. Although 3D printing has advanced substantially, achieving the necessary precision for complex anatomical structures, particularly at the microscale, remains difficult. The **resolution of 3D printers**, especially when working with biocompatible materials, often falls short of the exacting standards required for certain PSIs.

Moreover, **material limitations** are a significant challenge. Identifying materials that possess the optimal balance of mechanical properties, biocompatibility, and printability is an ongoing concern. The range of materials suitable for medical applications is limited compared to traditional manufacturing methods, complicating the production of high-quality PSIs.

Another critical issue is **production time**. The process of designing, manufacturing, and sterilizing PSIs is time-consuming, which can be problematic in cases requiring urgent intervention. Additionally, **quality control** presents challenges, particularly in ensuring consistent quality across different production batches and maintaining sterility throughout the manufacturing process.

2.1.6.2 Regulatory hurdles in PSI approval

The regulatory landscape for PSIs varies significantly across regions. In the United States, the Food and Drug Administration (FDA) typically classifies PSIs as Class II or III medical devices, necessitating rigorous premarket approval processes. The FDA has introduced a "custom device exemption" for certain PSIs, but this exemption is limited to specific cases and does not address broader regulatory challenges.

In contrast, the European approach, governed by the Medical Device Regulation (MDR), is generally more flexible. PSIs in Europe often fall under the MDR, which, while still stringent, can be less time-consuming than the FDA process. However, both regions require manufacturers to demonstrate safety and efficacy, although the specific pathways and requirements can differ considerably.

2.1.6.3 Ethical concerns

Ethical concerns surrounding personalized medical devices, particularly those involving predictive and advisory functionalities in brain implants, are numerous and complex. One significant issue is patient autonomy. There are apprehensions regarding how these advanced technologies might influence a patient's decision-making capacity and sense of control. As real-time monitoring of brain activity becomes integral to these devices, questions arise about who ultimately exercises control over the patient's cognitive processes, leading to potential feelings of "loss of control" among patients [36, 37].

Additionally, fairness and accessibility present critical ethical dilemmas. The high costs associated with personalized systems may lead to disparities in access to these innovative treatments. This raises broader questions of justice and fairness in the application of artificial intelligence (AI)-driven healthcare technologies, which could encompass all aspects of personalized medical implants [38, 39].

Privacy and data protection also remain paramount concerns. The design of personalized medical devices often requires the use of personal health data, which raises significant privacy issues, especially with the increasing integration of AI and machine learning. This is further complicated by the prospect of long-term monitoring of these implants, which involves continuous data collection and poses challenging questions about data ownership, usage, and patient privacy [37, 39].

Informed consent is another complex issue related to the ethical implications of personalized systems. There is a growing concern about the adequacy of information provided during the consent process for these sophisticated technologies, as patients may not fully understand the implications of what they are agreeing to [40]. Furthermore, ethical dilemmas are anticipated in light of potential long-term failures or complications associated with personalized systems, necessitating careful consideration of how these issues will be managed [41].

Regulatory bodies play a crucial role in addressing these ethical challenges, particularly concerning current regulatory processes. Agencies such as the FDA typically classify personalized medical systems as Class II or III medical devices, imposing rigorous premarket approval processes. However, there is an expressed need for regulatory frameworks to evolve to adequately address the unique characteristics of personalized medical devices.

Safety measures are also an essential component of the regulatory landscape. Emphasizing clinical trials and post-marketing surveillance is vital to ensure long-term safety and efficacy. Moreover, the implementation of CAD/CAM technology can enhance precision and minimize risks associated with these devices [38].

The challenge for regulatory bodies lies in balancing innovation and safety. As technological advancements occur at an unprecedented pace, it is imperative that regulatory processes adapt to maintain patient safety while accommodating the personalized nature of these systems. Flexible approval processes are necessary to ensure that safety standards are rigorously upheld without stifling innovation.

Privacy and data concerns are particularly critical in the context of personalized medical systems. The necessity for robust data protection measures is underscored by the reliance on personal health data in device design. The integration of AI and machine learning exacerbates these privacy concerns, necessitating careful examination of how personal health data is used and protected [39]. Furthermore, the long-term monitoring inherent in some of these devices raises questions about data ownership and patient privacy, particularly in implants with predictive capabilities [36, 37].

Ethical dilemmas also extend to the long-term use of personalized medical systems. A significant question concerns responsibility for potential failures or complications arising from these implants. Identifying whether liability rests with the manufacturer, healthcare provider, or the patient is critical for ethical accountability [41]. Additionally, there is a pressing need for clear protocols regarding long-term monitoring, maintenance, and potential revision or removal of these systems.

The psychological impact of having a personalized implant, particularly one with predictive or advisory capabilities, warrants thorough examination. The long-term effects on patients' mental well-being must be recognized and addressed [36, 37].

While personalized medical systems offer significant potential benefits, they also pose intricate ethical challenges that require ongoing dialogue among stakeholders, including patients, healthcare providers, manufacturers, and regulatory bodies. Key issues such as balancing innovation with safety, ensuring equitable access, protecting patient privacy, and maintaining patient autonomy must be critically addressed as the field of personalized medical devices continues to progress.

2.1.6.4 Cost-benefit analysis of PSIs

The integration of PSIs into clinical practice presents a unique set of cost-related challenges that can limit their widespread adoption despite their potential benefits. Understanding the various cost factors, as well as the long-term economic impact of PSIs, is essential for evaluating their overall value in healthcare.

2.1.6.4.1 Cost factors and accessibility of PSIs

The higher costs associated with PSIs are primarily driven by several key factors. First, the design and engineering of PSIs require significant time and expertise, as each implant must be custom-made to fit the patient's unique anatomy. This level of personalization is labor-intensive, especially in complex cases, which makes PSIs more expensive compared to standard implants [42, 43]. The manufacturing process further adds to the cost, as PSIs are typically produced using advanced techniques such as 3D printing. These methods, while offering high precision and quality, are more costly than traditional mass production, particularly when producing intricate designs [43, 44].

Moreover, the materials used in PSIs must meet stringent medical standards for biocompatibility and mechanical strength. These high-quality materials, such as titanium and PEEK, are expensive, contributing to the overall cost of the implants [45]. Regulatory compliance also plays a significant role in driving up costs. The development and production of PSIs involve navigating complex regulatory requirements, which include rigorous testing and detailed documentation to ensure the safety and efficacy of the implants. Meeting these standards is both time-consuming and costly, further increasing the financial burden of PSIs [46, 47].

2.1.6.4.2 Efforts to address cost challenges

To address these cost challenges, ongoing innovations are focused on making PSIs more cost-effective. One approach involves the use of AI and machine learning to streamline the design process. By partially automating the creation of complex implants, these technologies can reduce labor costs and potentially lower the overall expense of producing PSIs [44, 48]. Additionally, advancements in 3D printing technology and other manufacturing methods aim to improve production efficiency, which could help reduce costs associated with small-batch production. The development of new materials that are both biocompatible and cost-effective is also a key area of focus, with the potential to lower material costs without compromising the quality of the implants [45, 49].

2.1.6.4.3 Long-term healthcare cost impact

While PSIs involve high upfront costs, they offer the potential for long-term savings in several areas. For instance, the precise fit and alignment provided by PSIs can reduce the need for revision surgeries, which are often required with standard implants due to poor fit or alignment. Although not always directly addressed in the literature, this reduction in revision surgeries could lead to significant cost savings over time by decreasing the frequency of additional procedures [42, 43]. Furthermore, the improved fit and alignment offered by PSIs may also lead to shorter hospital stays and reduced rehabilitation times, contributing to overall healthcare cost savings [46].

2.1.6.4.4 Cost savings from improved outcomes

PSIs can also lead to substantial cost savings through enhanced surgical outcomes. For example, the use of PSIs has been shown to significantly reduce operative times, which translates into lower costs and improved efficiency in the OR. In total ankle arthroplasty, for instance, the use of PSIs reduced operative time from 168 min to 137 min, illustrating the potential for cost savings in this context [43, 44]. Additionally, the precise sizing of implants with PSIs minimizes the need for intraoperative adjustments and reduces the likelihood of complications. In one study, PSIs achieved 100% accuracy for tibial components and 83% accuracy for talar components, highlighting their potential to reduce surgical complications and associated costs [43, 48].

2.1.6.4.5 Economic models for cost-effectiveness

The cost-effectiveness of PSIs can vary depending on the economic model used to evaluate them. For example, using traditional OR accounting methods, one study found that PSIs reduced costs by an average of $7,597 per procedure in total ankle arthroplasty, demonstrating their potential for significant cost savings [43]. However, when evaluated using time-driven activity-based costing, PSIs were found to be $836 more expensive on average. This discrepancy underscores the importance of the chosen economic model in determining the perceived cost-effectiveness of PSIs [43]. Additionally, the quality-adjusted life year methodology, while not specifically applied to PSIs in the studies reviewed, is often used to assess the cost-effectiveness of medical interventions. This approach could provide valuable insights into the long-term value of PSIs by considering improvements in patient's quality of life [47].

2.1.6.4.6 Upfront costs versus long-term benefits

While PSIs generally involve higher upfront costs due to the personalized nature of their design and production, they may offer significant long-term benefits. The initial investment required for PSIs is higher compared to standard implants, but this investment may be justified by the potential for improved outcomes [42, 43]. For instance, PSIs could lead to long-term cost savings through reduced complications, fewer revision surgeries, and shorter hospital stays. In total knee arthroplasty, the use of customized components resulted in similar outcomes at a lower cost, suggesting that PSIs could offer similar benefits in other contexts [47].

Although specific research on PSIs for chronic conditions is limited, their potential to improve outcomes and reduce complications could be particularly beneficial in managing long-term conditions, offering additional cost savings over time [46, 48]. Overall, while PSIs generally incur higher upfront costs, they hold the potential for significant long-term cost savings. These savings could arise from reduced operative time, improved surgical outcomes, and fewer complications. However, the cost-effectiveness of PSIs varies depending on the economic model used and the specific medical application. More research is needed to fully quantify these benefits across different medical fields and patient populations.

2.1.6.4.7 Technical challenges in PSI manufacturing

Even with the notable progress in PSI manufacturing, various technical challenges continue to hinder the widespread use and efficiency of these implants. The difficulties stem from the accuracy and clarity of production procedures, compatibility of materials, and postprocessing needs for achieving the desired quality and function.

A major obstacle in PSI production is reaching the extremely precise level needed to duplicate intricate anatomical structures, especially on a small scale. Current 3D printers, especially when used with biocompatible materials, often do not meet the high standards required for medical purposes in terms of resolution capabilities. This restriction may impact the precision of the implants, possibly resulting in less than ideal fits and performance.

Another important hurdle is ensuring compatibility with materials. Choosing materials that are biocompatible and suitable for advanced manufacturing processes is crucial, yet continues to pose a persistent challenge. The variety of materials for 3D printing in medical use is still limited, restricting the customization and functionality of PSIs. This difficulty is increased by the requirement for these substances to satisfy stringent regulatory criteria while also having the mechanical characteristics required to function effectively as implants.

Many 3D-printed PSIs need thorough postprocessing in addition to addressing precision and material obstacles to achieve desired precision and surface quality standards. Postprocessing involves tasks like finishing, sterilizing, and further machining, which can impact the implant's final size and surface properties. This not only raises both the manufacturing time and cost but also brings in extra factors that can affect the consistency and dependability of the end product.

2.1.6.5 Challenges in complex geometries and multi-material constructs

Developing PSIs requires tackling various specific obstacles associated with intricate shapes and combinations of materials. The intricate internal structures or highly detailed external surfaces of certain implants can challenge the capabilities of current 3D printing technologies due to their geometric complexity. Maintaining uniform quality in different shapes of the same implant is difficult, as inaccuracies can cause structural issues or imperfect fits.

Multimaterial printing is also a challenging aspect of PSI manufacturing. Even though multimaterial 3D printing technology is available, developing PSIs that combine multiple materials and adhere to medical regulations remains a complex endeavor. The bonding and integration of various materials in one implant is intricate due to each material's unique properties and behaviors during manufacturing. Maintaining the structural integrity and biocompatibility of the implant while achieving a smooth integration poses a continuous challenge in the industry.

Moreover, the research and development of functional gradients within implants are crucial, as material properties gradually transition to emulate natural tissue structures. This method focuses on developing implants that mimic the behavior of natural tissues more accurately and effectively. Yet, it is still difficult to achieve these gradients using current manufacturing methods because of the precise control needed over the implant's material composition and distribution.

2.1.6.6 Innovations needed for scaling PSI production

Increasing the production of PSIs comes with numerous challenges that require important innovations throughout different phases of the design and manufacturing process. One important area that needs innovation is the automation of design procedures. Creating more sophisticated AI and machine learning algorithms is essential for automating and enhancing the development of PSIs. These technologies can simplify the design of intricate implant shapes, cutting down on the time and skill required for making personalized implants. Furthermore, it is crucial to have enhanced software integration among imaging, design, and manufacturing phases to guarantee a smooth workflow from acquiring initial patient data to producing final implants.

Another vital area for creativity lies in high-speed, high-precision production. Although current 3D printing methods provide great customization options, they often lack in terms of speed and scalability for production. Advancements are critically needed to speed up production while maintaining the necessary precision for medical implants. This might entail creating innovative manufacturing methods that blend the customization features of 3D printing with the speed and effectiveness of traditional manufacturing techniques.

Quality assurance and uniformity are equally important when expanding PSI manufacturing. It is crucial to have advanced in situ monitoring systems in place to ensure that every implant produced meets the necessary specifications through real-time quality control during manufacturing. Furthermore, it will be crucial to establish uniform protocols for PSI production to ensure consistency and reliability at various manufacturing locations. These protocols would make it easier for PSIs to be more widely used by ensuring that implants made in various places meet the same rigorous standards.

2.1.6.7 Impact of material properties on PSI success

The properties of the materials used in constructing PSIs play a significant role in determining their success. Strength and durability are among the most crucial material characteristics. The PSIs require materials that can handle the mechanical strains of the implant location and maintain stability over time. It is crucial to prioritize attributes like flexibility or biodegradability alongside mechanical properties for certain applications that demand a combination of strength and adaptability.

Another vital factor is biocompatibility, which guarantees the implant's ability to integrate well with surrounding tissues and not trigger a negative immune reaction. Ensuring long-term biocompatibility is essential for avoiding rejection and facilitating the successful integration of the implant within the body. Current research is dedicated to creating materials that not only prevent immune responses but also encourage tissue integration or regeneration, improving the overall effectiveness of PSIs.

Biodegradability is considered advantageous for certain PSIs, especially for implants that are meant to be temporary or stimulate tissue regrowth. Nevertheless, managing the rate at which degradation occurs and guaranteeing the safety and non-toxicity of degradation by-products pose major obstacles. Study in this field aims to develop materials that break down at a specific pace and generate substances that can be readily absorbed or expelled by the body.

2.1.6.8 Current research and development efforts

Research and development in PSI design and production are quickly progressing, with various main areas of concentration that offer to greatly advance the field. An example of this is the advancement of superior materials. Scientists are developing biocompatible materials with enhanced mechanical properties and improved printability. Furthermore, there is an increasing interest in smart materials that have the ability to alter their characteristics based on physiological factors, which could result in implants that are more adaptable and effective.

Hybrid manufacturing methods are another area of progress to watch. By merging 3D printing with conventional manufacturing techniques, it could be feasible to utilize the advantages of both methods. This might include employing multiprocess manufacturing systems that smoothly combine various fabrication techniques, enabling the production of more intricate and personalized implants in a more efficient way.

Bioprinting is making great progress as well, as technological advancements allow for the development of implants containing living cells. This method could enhance implant integration and functionality by closely imitating natural tissue structures. In the same way, ongoing research is being conducted on in situ manufacturing technologies that could enable the direct creation or alteration of implants in the OR. This could allow surgeons to customize implants to fit the patient's requirements instantly, potentially enhancing results and minimizing the necessity for further surgeries.

2.1.6.9 Current challenges

The most common complications associated with PSIs, reported by literature include:
1. **Screw loosening**: Implant stability can be a concern, as reported in cases of screw loosening [50].
2. **Postoperative dissatisfaction**: Managing patient expectations is crucial, as postoperative dissatisfaction related to aesthetic outcomes or functional issues has been reported [50].
3. **Postoperative infection**: Maintaining sterility and proper wound care is essential to avoid postoperative infections [50].
4. **Implant breakdown, exposure, or displacement**: Although not observed in some studies [51], these potential complications are known risks associated with PSIs.
5. **Proximal migration and subluxation**: Common complications in proximal humerus reconstruction may also occur with PSIs in orthopedic applications [52].
6. **Nonunion and aseptic loosening**: These complications, noted in humerus reconstruction, could potentially occur with PSIs [52].
7. **Glenoid component failure**: This is a known complication in shoulder arthroplasty and may occur with PSIs [53].

8. **Instability, wear, and loosening**: Inaccurate component positioning can result in these issues, affecting the long-term survival of the implant (2017).
9. **Compromised peri-implant health**: Particularly in dental implants [53], the risk of compromised peri-implant health is noted, especially in medically compromised patients [54].
10. **Fracture complications**: Fracture-related complications, such as infection, can affect patient outcomes, particularly in cases of multiple fractures or elderly patients.

2.1.7 Future trends and innovations in PSIs

The future of PSIs is being shaped by several key trends and innovations, particularly in the areas of AI integration, emerging materials, manufacturing techniques, and advancements in imaging and modeling technologies. These developments hold the potential to significantly enhance the design, production, and clinical application of PSIs, ultimately improving patient outcomes and broadening the scope of personalized medicine.

– **Integration of AI and machine learning in PSI design:** The integration of AI and machine learning into the design of PSIs is revolutionizing the field by introducing several significant benefits. One of the primary advantages is the use of machine learning algorithms for automated segmentation, which enhances both the accuracy and speed of image segmentation – a critical step in the creation of PSIs. By improving this process, AI helps to ensure that the implants are tailored precisely to the patient's unique anatomical features, leading to better-fitting and more effective implants.

AI is also being employed to optimize implant designs based on patient-specific data and biomechanical principles. This allows for the development of more effective and customized solutions that are fine-tuned to the patient's specific needs. Predictive modeling through machine learning further enhances the design process by anticipating patient outcomes and implant performance. This capability enables designers to make more informed decisions, potentially reducing the risk of complications and improving the overall success rate of surgeries involving PSIs.

Moreover, AI-powered systems are being developed to improve quality control processes during the manufacturing of PSIs. These systems ensure higher levels of consistency and precision, which are essential for producing reliable implants that meet strict medical standards. Collectively, these advancements in AI and machine learning are transforming the landscape of PSI design, making it more efficient, accurate, and responsive to patient needs.

– **Emerging materials and manufacturing techniques:** The development of emerging materials and innovative manufacturing techniques is set to significantly enhance the future capabilities of PSIs. Bioactive materials, which actively promote tissue integration and regeneration, are at the forefront of these advancements. These materials have the potential to improve long-term outcomes for patients by facilitating better integration of the implant with surrounding tissues, thus reducing the likelihood of complications such as implant rejection or failure.

Smart materials represent another promising area of development. These materials can change their properties in response to environmental stimuli, such as temperature or pH levels, which could lead to more adaptive and functional implants. For instance, an implant might become more rigid or flexible depending on the mechanical demands placed upon it, providing dynamic support tailored to the patient's needs.

Nanomaterials are also being explored for their ability to enhance both the mechanical and biological properties of PSIs. By incorporating nanomaterials, implants can be made stronger, more durable, and more biocompatible, improving their effectiveness in clinical applications. Furthermore, the advent of 4D printing technology, which enables the creation of implants that can change shape or function over time in response to physiological conditions, offers dynamic solutions that can adapt as the patient's condition evolves.

Hybrid manufacturing techniques, which combine 3D printing with traditional machining methods, present another exciting opportunity. This approach allows for the production of more complex and precise implants, thereby improving their functionality and durability. The integration of these advanced materials and manufacturing techniques represents a pivotal step forward in the development of PSIs, offering new possibilities for personalized medicine.

– **Advances in imaging and modeling technologies:** Advancements in imaging and modeling technologies are crucial for enhancing the design and effectiveness of PSIs. Improved imaging resolution, achieved through high-resolution CT and MRI, enables more precise anatomical mapping. This level of detail allows for better delineation of complex structures, which is essential for fitting PSIs to the unique anatomies of individual patients.

Moreover, improvements in imaging techniques, such as motion artifact reduction, contribute to the accuracy of PSI design. For certain applications, such as assessing bone resection and implant stability, MRI may offer superior results compared to CT, providing more accurate data for designing implants. The integration of real-time data processing into imaging systems further enhances the design process by enabling dynamic imaging that captures anatomical changes during surgery. This capability allows for on-the-fly adjustments to the implant design, improving surgical outcomes.

Advanced 3D modeling techniques also play a vital role in the design and production of PSIs. These techniques allow for the custom 3D printing of models tailored to the patient's specific needs, enhancing the fit and functionality of implants. Additionally, these models enable simulations that improve surgical planning, helping surgeons to anticipate challenges and plan for the most effective outcomes.

The incorporation of AI and machine learning into imaging and modeling processes is another significant advancement. Predictive analytics, driven by AI, can analyze previous surgical data to optimize implant designs, while automated segmentation improves the efficiency and accuracy of creating 3D models. However, these advancements also present challenges, including the management of large datasets produced by imaging technologies, compliance with regulatory standards for safety and efficacy, and the potential increase in costs associated with these advanced technologies. Balancing these costs with the benefits in patient outcomes will be essential for healthcare providers.

2.1.7.1 Future implications of PSIs in personalized medicine

The rise of PSIs marks a significant step forward in personalized medicine. These advancements have the potential to transform how we approach medical care. As PSIs continue to evolve, their impact is expected to extend far beyond current applications, driving innovation across various medical fields and setting new standards for individualized healthcare.

2.1.7.2 Latest innovations in patient-specific implants

Recent breakthroughs in PSI technology are revolutionizing personalized treatments. From new materials to cutting-edge manufacturing methods, these advancements are delivering more accurate and effective medical solutions. Some notable innovations include:

- **Precision medicine:** The same principles used to create PSIs – such as tailored fit and functionality – could be applied to drug delivery systems and treatment plans. This would enable therapies to be fine-tuned to each patient's unique characteristics, improving overall effectiveness.
- **Regenerative medicine:** Custom-designed scaffolds, made possible through PSI technology, are opening new possibilities in tissue engineering. This could be game-changing for organ transplants and wound healing, offering patient-specific solutions that improve integration and tissue function.

2.1.8 Conclusion

In this chapter, we have explored the multifaceted realm of PSI design, elucidating its pivotal role in advancing personalized surgical practices. Our discussion began with an overview of the critical necessity for PSIs, outlining their evolution from conceptualization to their practical applications. Through rigorous examination of the design process – encompassing predesign considerations, advanced medical imaging, data processing, and simulation – we have highlighted the intricate methodologies involved in crafting these bespoke implants.

The evaluation of PSI design emerges as a cornerstone of effective implementation, ensuring that these tailored solutions not only meet but exceed the requirements for precision and functionality. As detailed in the chapter, the design process involves several crucial steps, including the acquisition of patient-specific data, CAD modeling, and the development of surgical guides. The evaluation techniques, such as FEA and prototyping, play an indispensable role in validating the efficacy, safety, and durability of the implants.

Despite the promising advancements in PSI technology, the current landscape reveals both opportunities and challenges. Recent studies indicate that while PSIs have achieved notable improvements in surgical accuracy, the clinical outcomes and patient satisfaction have not consistently surpassed those of conventional methods. Addressing this gap requires ongoing research and innovation, particularly in enhancing manufacturing precision, exploring new materials, and optimizing quality control processes.

Looking ahead, the future of PSIs holds significant potential for transforming surgical practices. Continued advancements in AI, machine learning, and new manufacturing techniques such as 4D printing are poised to refine PSI design and broaden its applications. However, for PSIs to realize their full potential, a concerted effort is needed to address cost factors and ensure that these innovations are accessible and practical for widespread use.

In conclusion, the evaluation of PSI design remains a dynamic field, with its success largely dependent on iterative improvements and technological advancements. By integrating patient feedback, leveraging cutting-edge tools, and addressing both technical and economic challenges, we can pave the way for PSIs to deliver on their promise of enhanced surgical precision and patient outcomes. The future of PSIs will undoubtedly continue to evolve, driving forward the boundaries of personalized medicine and surgical excellence.

References

[1] Alasseri N, Alasraj A. Patient-specific implants for maxillofacial defects: Challenges and solutions. Maxillofacial Plastic and Reconstructive Surgery. 2020; 42(1): 15. doi: 10.1186/s40902-020-00262-7.

[2] Vivigdha V, Krishnan M, Santhosh Kumar MP, Senthil Murugan P, Rajamanickam P. Multidisciplinary approach to patient-specific implants (PSIs): A case report and review of literature. Cureus. 2023; 15 (6): e41238. doi: 10.7759/cureus.41238.

[3] Mowar A, Dubey P, Parveen S. The role of patient-specific implants in craniomaxillofacial reconstructive surgery: A scoping review. International Journal of Creative Research Thoughts (IJCRT). May 2022;10(5):i772–i778.

[4] Neha S, Aghlmandi S, Dalcanale F, Seiler D, Zeilhofer H-F, Honigmann P, Florian MT. Quantitative assessment of point-of-care 3D-printed patient-specific polyetheretherketone (PEEK) cranial implants. International Journal of Molecular Sciences. 2021; 22(16): 8521. doi: 10.3390/ijms22168521.

[5] Sun H-HB, Fallah H. Efficacy and cost of maxillary patient-specific implants in orthognathic surgery: A review of three patient cases. The Permanente Journal. 2022; 26(1): 137–142. doi: 10.7812/TPP/21.054.

[6] Reiman LT, Walker ZJ, Babcock LR, Forsberg PA, Mark TM, Sherbenou DW. A case for improving frail patient outcomes in multiple myeloma with phenotype-driven personalized medicine. Aging and Cancer. 2021; 2(1–2): 6–12. doi: 10.1002/aac2.12022.

[7] Chang WH. Personalized approaches to stroke: One step forward for functional recovery of stroke patients. Journal of Personalized Medicine. 2022; 12(5): 822. doi: 10.3390/jpm12050822.

[8] Benignus C, Buschner P, Kristin Meier M, Wilken F, Rieger J, Beckmann J. Patient specific instruments and patient individual implants – A narrative review. Journal of Personalized Medicine. 2023; 13(3): 426. doi: 10.3390/jpm13030426.

[9] Yehuda C, Alabdulkarim A, Mirko SG. Patient-specific implants and fat grafting for contour deformities post craniosynostosis reconstruction: A therapeutic approach. The Journal of Craniofacial Surgery. 2023; 34(3): 959–963. doi: 10.1097/SCS.0000000000009178.

[10] Tajrian A, William CHP, Natarajan P, Lennox A, Koinis L, Ralph JM. Anterior lumbar interbody fusion (ALIF) for lumbar hemivertebra in an adult using three-dimensional-printed patient-specific implants and virtual surgery planning: A technical report. Surgical Techniques Development. 2023; 12(4): 199–210. doi: 10.3390/std12040019.

[11] Petohazi A, Musbahi O, Akilapa O, Gee M, Harnett P. 392 outcomes of anatomical custom patient-specific implants after 3D CT pre-operative planning for complex hip conditions. British Journal of Surgery. 2022; 109(1): znac039.268. doi: 10.1093/bjs/znac039.268.

[12] Desai DD, Alwani M, Sheen D, Narayanan A, Gordin E. The use of patient-specific orbital reconstruction implants during maxillectomy reconstruction. Facial Plastic Surgery & Aesthetic Medicine. 2023; 25(5): 403–408. doi: 10.1089/fpsam.2022.0201.

[13] Rahimov CR, Ahundov AA, Hajiyeva GI, Ch R, Rahimli DAS, Farzaliyev IM. Treatment of extensive tumors of the jaws by hemimandibuloectomy with simultaniouse reconstruction of the mandible, arthroplasty of temporomandibular joint, orthopedic rehabilitation supported by dental implants. Head and Neck Tumors (HNT). 2020; 10(3): 97–110. doi: 10.17650/2222-1468-2020-10-3-97-110.

[14] Koper D, Ter Laak-poort M, Lethaus B, Yamauchi K, Moroni L, Habibovic P, Kessler P. Cranioplasty with patient-specific implants in repeatedly reconstructed cases. Journal of Cranio-Maxillofacial Surgery. 2019; 47(5): 709–714. doi: 10.1016/j.jcms.2019.01.034.

[15] Mukul SK, Mishra M, Singh G, Singh A, Kumar Das M. Patient-specific implants in maxillofacial reconstruction: A case report. The Traumaxilla. 2019; 1(2–3): 76–80. doi: 10.1177/26323273211069665.

[16] Dong X-P, Zhang Y-W, Pei Y-J, Wang Z, Zhang X-X, Xing-Liang Y, Zi-Zheng A, Mei Y-X, Jing-Na L. Three-dimensional printing for the accurate orthopedics: Clinical cases analysis. Bio-Design and Manufacturing. 2020; 3(2): 122–132. doi: 10.1007/s42242-020-00070-y.

[17] Sabir S, Azhari M, Rokhssi H, Merzouk N, Bentahar O. Presurgical nasoalveolar molding: An advantageous adjunctive neonatal therapy for cleft lip and palate defects in 2 clinical cases. Journal of Pediatric and Neonatal Individualized Medicine. 2021; 10: e100249. doi: 10.7363/100249.

[18] García-González M, María Muñoz Guzón F, González-Cantalapiedra A, Manuel González-Fernández P, Otero Pérez R, Asunción Serra Rodríguez J. Application of shark teeth–derived bioapatites as a bone substitute in veterinary orthopedics. Preliminary clinical trial in dogs and cats. Frontiers in Veterinary Science. 2020; 7: 574017. doi: 10.3389/fvets.2020.574017.

[19] Barrett A, Andrews B, Cobb J. Accuracy of truly patient-specific UKA procedures using sculptor RGA. Journal of Bone and Joint Surgery – British Volume. In Orthopaedic Proceedings (Vol. 95, No. SUPP_15, pp. 11–11). Bone & Joint.

[20] Troitino I, White TE, Lozo J. A comparison of the accuracy of WATCHMAN and WATCHMAN FLX device sizing between CT, TEE, and patient specific 3D models. Proceedings of IMPRS. 2021; 4(1): doi: 10.18060/25779.

[21] Kwon O-R, Kang K-T, Son J, Choi Y-J, Suh D-S, Koh Y-G. The effect of femoral cutting guide design improvements for patient-specific instruments. BioMed Research International. 2015; 2015: 978686. doi: 10.1155/2015/978686.

[22] DeFazio MV, Arribas EM, Ahmad FI, Le-Petross HT, Liu J, Chu CK, Santiago L, Clemens MW. Application of three-dimensional printed vascular modeling as a perioperative guide to perforator mapping and pedicle dissection during abdominal flap harvest for breast reconstruction. Journal of Reconstructive Microsurgery. 2020; 36(5): 325–338. doi: 10.1055/s-0039-1700991.

[23] Sedov YG, Avanesov AM, Saleev RA, Saleeva GT, Yarulina ZI. A classification of surgical guides application for dental implantation. Stomatologiia. 2021; 100(1): 84–88. doi: 10.17116/stomat202110001184.

[24] Unsal G-S, Turkyilmaz I, Lakhia S. Advantages and limitations of implant surgery with CAD/CAM surgical guides: A literature review. Journal of Clinical and Experimental Dentistry. 2020; 12(4): e409–e417. doi: 10.4317/jced.55871.

[25] Gobbato B, Moya D, Menor Fusaro F, Valente S. Practical guide to 3D printing and planning in orthopedics and traumatology. Rev Asoc Argent Ortop Traumatol. 2022; 87(1), 122–130.

[26] Kim MK, Ji Ham M, Rae Kim W, Kim HG, Jun Kwon K, Gon Kim S, Wook Park Y. Investigating the accuracy of mandibulectomy and reconstructive surgery using 3D customized implants and surgical guides in a rabbit model. Maxillofacial Plastic and Reconstructive Surgery. Springer. 2023; 45(1): doi: 10.1186/s40902-023-00375-9.

[27] Dadario NB, Brahimaj B, Yeung J, Sughrue ME. Reducing the cognitive footprint of brain tumor surgery. Frontiers in Neurology. Frontiers Media SA. 2021; 12. doi: 10.3389/fneur.2021.711646.

[28] Terry G, Rivlin M, John AA. Three-dimensional printing technology and role for custom implants in orthopedic oncology. Techniques in Orthopaedics. 2018; 33(3): 166–174. doi: 10.1097/BTO.0000000000000292.

[29] Nakamura RC, Leverone A, Carolina Avila DA, Werner H, Canella C. Advantages of nail surgical procedures using a 3D biomodel. Skin Appendage Disorders. 2024; 10(4): 312–320. doi: 10.1159/000537871.

[30] Heyde C-E, Osterhoff G, Uja S, Völker A, Von der Höh NH, Jarvers JS. Patient-matched 3-D-printed guides for the insertion of cervical pedicle screws in tumor surgery. Case Reports in Surgery. 2021; 2021: 8135996. doi: 10.1155/2021/8135996.

[31] Moiduddin K, Hammad Mian S, Umer U, Alkhalefah H. Fabrication and analysis of a Ti6Al4V implant for cranial restoration. Applied Sciences. Multidisciplinary Digital Publishing Institute. 2019; 9(12): 2513. doi: 10.3390/app9122513.

[32] Souza RFD, Ribeiro AB, Della Vecchia MP, Costa L, Cunha TR, Reis AC, Albuquerque RF. Mini vs. standard implants for mandibular overdentures: A randomized trial. Journal of Dental Research. 2015; 94(10): 1376–1384. doi: 10.1177/0022034515601959.

[33] Hermus M, Berend JVDW, Rebecca THC, Collee G, Noordman BJ, Peter-Paul LOC, Jan Willem TD. et al., Patient preferences for active surveillance vs standard surgery after neoadjuvant chemoradiotherapy in oesophageal cancer treatment: The NOSANO-study. International Journal of Cancer. 2023; 152(6): 1183–1190. doi: 10.1002/ijc.34327.

[34] Heisler L, Vach W, Katz G, Egelhof T, Knupp M. Patient-specific instrumentation vs standard referencing in total ankle arthroplasty: A comparison of the radiologic outcome. Foot and Ankle International. 2022; 43(6): 741–749. doi: 10.1177/10711007221077100.

[35] Rana M, Moellmann HL, Schorn L, Lommen J, Rana M, Wilkat M, Hufendiek K. Primary orbital reconstruction with selective laser melting (SLM) of patient-specific implants (PSIs): An overview of

96 surgically treated patients. Journal of Clinical Medicine. 2022; 11(12): 3361. doi: 10.3390/jcm11123361.
[36] Cook M, Gilbert F. Are predictive brain implants an indispensable feature of autonomy?. Bioethica Forum. 2015; doi: 10.24894/BF.2015.08038.
[37] Gilbert F. A threat to autonomy? the intrusion of predictive brain implants. AJOB Neuroscience. 2015; 6(4): 4–11. doi: 10.1080/21507740.2015.1076087.
[38] Joshi S, Javed Salema H, Pawar S, Sunil Nair V, Koranne V, Dilip Sane V. Patient-specific implants in maxillofacial reconstruction – a case report. Annals of Maxillofacial Surgery. 2023; 13(2): 258–261. doi: 10.4103/ams.ams_126_23.
[39] Fan L, Ruijs N, Yuan L. Ethics & AI: A systematic review on ethical concerns and related strategies for designing with AI in healthcare. AI. 2022; 4(1): 28–53. doi: 10.3390/ai4010003.
[40] Bluhm R, Sipahi ED, Achtyes ED, McCright AM, Cabrera LY. Stakeholders' ethical concerns regarding psychiatric electroceutical interventions: Results from a US nationwide survey. AJOB Empirical Bioethics. 2024; 15(1): 11–21. doi: 10.1080/23294515.2023.2224592.
[41] Manahan MA. what do clinicians and organizations owe patients with recalled implanted devices or materials?. AMA Journal of Ethics. 2021; 23(9): E679–684. doi: 10.1001/amajethics.2021.679.
[42] Sackeyfio A, Samyshkin Y, Riemer J, Blake M, Mhishi S, Prett M, Elliot L, Refoios Camejo R. #4140 Daprodustat versus recombinant human erythropoietin for treating anaemia of chronic kidney disease: Cost effectiveness model methodology and findings. Nephrology Dialysis Transplantation. 2023; 38(1): gfad063a_4140. doi: 10.1093/ndt/gfad063a_4140.
[43] Savage-Elliott I, Wu VJ, Isabella W, Timothy Heffernan J, Rodriguez R. Comparison of time and cost savings using different cost methodologies for patient-specific instrumentation vs standard referencing in total ankle arthroplasty. Foot & Ankle Orthopaedics. 2019; 4(4): 2473011419884278. doi: 10.1177/2473011419884278.
[44] Al-Kibsi T, Fma A-R, Al-Shamahy Ha AL-HK, Oai A-D, Bhh A-S. The effect of dental implants on secretory Iga level in saliva. International Journal of Dentistry and Oral Health. 2024; 10(1): doi: 10.16966/2378-7090.411.
[45] Carmon I, Zobrab A, Alterman M, Tabib R, Cohen A, Kandel L, Greenberg A, Reich E, Casap N, Dvir-Ginzberg M. Chondrocytes supplemented to bone graft-containing scaffolds expedite cranial defect repair. Scientific Reports. 2023; 13: 19192. doi: 10.1038/s41598-023-46604-z.
[46] Gvaramia D, Fisch P, Flégeau K, Huber L, Kern J, Jakob Y, Hirsch D, Rotter N. Evaluation of bioprinted autologous cartilage grafts in an immunocompetent rabbit model. Advanced Therapeutics. 2024; 7(6): 2300441. doi: 10.1002/adtp.202300441.
[47] Monu J, Clement N, Phil JW. All-polyethylene tibia components have the same functional outcomes and survival, and are more cost-effective than metal-backed components in patients 70 years and older undergoing total knee arthroplasty. Bone & Joint Open. 2022; 3(12): 969–976. doi: 10.1302/2633-1462.312.BJO-2022-0063.R1.
[48] Garside R, Pitt M, Anderson R, Rogers G, Dyer M, Mealing S, Somerville M, Price A, Stein K. The effectiveness and cost-effectiveness of carmustine implants and temozolomide for the treatment of newly diagnosed high-grade glioma: A systematic review and economic evaluation. Health Technology Assessment (Winchester, England). 2007; 11(45): iii–iv. doi: 10.3310/hta11450.
[49] Pruett JE, Everman S, David Torres Fernandez E, Davenport K, Romero DG, Yanes Cardozo LL. Androgen receptor blocker improves the cardiometabolic profile in a rat model of polycystic ovary syndrome, but at what cost?. Journal of the Endocrine Society. 2021; 5(1): A803–A804. doi: 10.1210/jendso/bvab048.1634.
[50] Sung-Ho H, Choi J-Y. Restoration of facial bone defect using patient-specific implant. The Journal of Craniofacial Surgery. 2022; 33(7): 2063–2066. doi: 10.1097/SCS.0000000000008551.

[51] Saponaro G, Todaro M, Barbera G, Scivoletto G, Foresta E, Gasparini G, Moro A. Patient-specific facial implants in polyetheretherketone and their stability: A preliminary study. Annals of Plastic Surgery. 2023; 90(6): 564–567. doi: 10.1097/SAP.0000000000003527.
[52] Tilton M, Tucker S, Manogharan G, Edward JF, Lewis G (2018): Patient-Specific Prosthesis for Load-Bearing Reconstruction of Proximal Humerus after Tumor Resection.
[53] Sanchez-Sotelo J, Sperling JW. Preoperative planning and patient-specific instrumentation improve accuracy of shoulder arthroplasty. Orthopedic Surgery Update, 2017;11(1). Mayo Clinic. Corpus ID: 209379757.
[54] Vissink A, Spijkervet F, Raghoebar GM. The medically compromised patient: Are dental implants a feasible option?. Oral Diseases. 2018; 24(1–2): 253–260. doi: 10.1111/odi.12762.

Hossein Sahbafar*

Chapter 2.2
Modeling techniques of bone tissue scaffolds

Abstract: In regenerative medicine, bone tissue engineering is essential for resolving the problems associated with bone injuries and anomalies. In order to facilitate cell attachment, proliferation, and differentiation, scaffolds are crucial to this process because they provide a three-dimensional framework that resembles the extracellular matrix that exists in nature. In bone tissue engineering, computational and numerical studies are essential because they allow the performance and design of scaffolds to be optimized before they are physically implemented. Numerical techniques for assessing bone tissue scaffolds include finite element analysis, computational fluid dynamics, representative volume element analysis, and fluid-structure interaction. With the help of these computational techniques, scaffolds that are specifically designed for a certain purpose may be designed more effectively, increasing biocompatibility, encouraging vascularization, and eventually leading to better healing results in clinical settings. Their importance resides in cutting down on the expenses and duration of experiments while accelerating the development of efficient bone regeneration techniques. This chapter provides an in-depth review of the numerical and computational techniques used to model and simulate bone tissue scaffolds.

Keywords: Bone tissue engineering, scaffold, modeling, simulation, finite element analysis (FEA), computational fluid dynamics (CFD), representative volume element (RVE), fluid-structure interaction (FSI)

2.2.1 Introduction

In regenerative medicine, bone tissue engineering is essential for resolving the problems associated with bone injuries and defects [1–4]. To encourage bone regeneration, it combines cells, biomaterials, and bioactive substances [5, 6]. In order to facilitate cell attachment, proliferation, and differentiation, scaffolds are essential to this process because they provide a three-dimensional (3D) framework that resembles the extracellular matrix found in nature [7–10]. Various types of scaffolds have been developed, including natural polymers (e.g., collagen and chitosan) [11–14], synthetic polymers (e.g., polylactic acid and polycaprolactone (PCL)) [15–18], and composite materials that combine both [19–21]. Enhancing their potential to promote vasculariza-

***Corresponding author: Hossein Sahbafar,** School of Mechanical Engineering, College of Engineering, University of Tehran, Tehran, Iran, e-mail: hosseinsahba@gmail.com, hsahbafar@ut.ac.ir

https://doi.org/10.1515/9783111386027-007

tion and bone integration, these scaffolds may be designed with precise mechanical properties, porosity, and biodegradability. The development of these scaffolds has greatly improved patient outcomes and opened the door for novel orthopedic and dental therapies [22–25].

In bone tissue engineering, computational and numerical studies are essential because they allow scaffold performance and design to be optimized before physical implementation [26, 27]. Methods such as finite element analysis (FEA), computational fluid dynamics (CFD), representative volume element (RVE) analysis, and fluid-structure interaction (FSI) provide insights into mechanical behavior, fluid flow, and biological interactions within scaffolds [28–34]. While CFD mimics the movement of nutrients and waste products, which are necessary for cell survival, FEA aids in the prediction of stress distribution and structural integrity under load. RVE analysis facilitates the microscale knowledge of material characteristics, whereas FSI investigates the relationship between biological fluids and scaffold mechanics [28–34]. With the use of these computational techniques, scaffolds that are specifically designed for a certain purpose may be designed more effectively, increasing biocompatibility, encouraging vascularization, and eventually leading to better healing results in clinical settings. Their importance resides in cutting down on the expenses and duration of experiments while speeding up the development of efficient bone regeneration strategies [35–37].

In this chapter, computational and numerical methods used for modeling and simulating bone tissue scaffolds are discussed in detail.

2.2.2 Finite element analysis and computational fluid dynamics and their combinations with other approaches for modeling bone tissue scaffolds

2.2.2.1 Finite element analysis (FEA)

A computational technique called FEA is used to predict how structures will react to forces. It simulates the mechanical properties of scaffolds in bone tissue engineering, guaranteeing the best possible design for load-bearing and biological integration. Using FEA and mechanical testing, Wang et al. examined the mechanical properties and bone ingrowth of a unique "Hexanoid" scaffold design in comparison to four traditional scaffold designs. In comparison with the other designs, the Hexanoid scaffold performed better in terms of bone volume percentage and demonstrated mechanical strength that was equivalent to that of human compact bone. The study's weaknesses were pointed out by the authors, who recommended combining the findings with clinical rehabilitation procedures to optimize new scaffold designs in further studies [38].

Tabatabaee et al. conducted a study with the objective of simulating a bone scaffold composed of poly(2-hydroxyethyl methacrylate) (PHEMA), gelatin, and graphene oxide (GO) under compression stress using the FEA in SOLIDWORKS and ABAQUS. The results demonstrated that GO might promote osteogenic differentiation by stabilizing the scaffold and lowering the chance of failure. The lack of experimental validation is one of the study's limitations; thus, future work should focus on refining scaffold parameters and confirming the computational model via in vitro and in vivo investigations [39]. Additionally, using FEA (Figure 2.2.1), Scocozza et al. examined the mechanical properties of a 3D-printed hybrid scaffold composed of PCL and an alginate-based hydrogel containing lyosecretome, a freeze-dried formulation of mesenchymal stem cell secretome. The findings demonstrated that although the addition of the hydrogel did not always affect the mechanical performance, the scaffold's mechanical properties might be enhanced by adjusting the infill pattern. It was discovered that the computational framework offered a reliable method for predicting the scaffolds' mechanical performance, which is well reflected in the upcoming clinical uses in the maxillofacial domain. The lack of a long-term in vivo assessment, which ought to be the focus of further research, was the study's limitation [40].

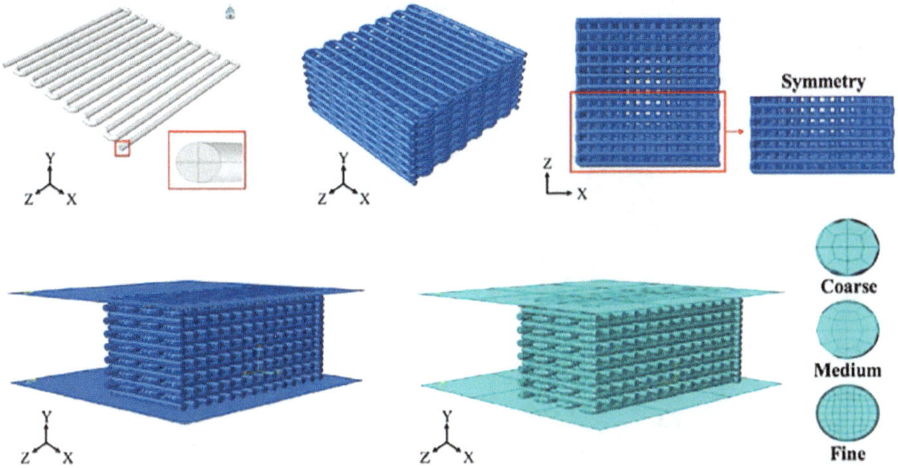

Figure 2.2.1: Bone scaffold model developed in [40] for studying its mechanical properties. Reprinted with permission under CC BY 4.0.

In addition, a numerical model was provided by Batoni et al. to examine the degrading behavior of 3D-functionalized scaffolds made of polyester for bone regeneration. The release of ICOS-Fc, a bioactive protein, was examined in relation to two distinct scaffold designs: (i) a scaffold featuring an exterior surface functionalization without macroporosity and (ii) a scaffold including an interior-functionalized macroporous architecture. The objective of the model was to maximize scaffold design in order to regulate protein release and degradation over time and space. The model did not take into consideration all degradation pathways, but the findings offered guidance for designing scaffolds to enable the regulated distribution of bioactive compounds [41]. Furthermore, Perier-Metz et al. examined the effect of scaffold architecture on major bone defect regeneration using an in silico model. The model indicated that lower scaffold surface-area-to-volume ratios and decreased stiffness encouraged bone formation and that material stimulation of progenitor cells and scaffold surface guiding of cellular processes were important contributors. The scientists demonstrated the potential of in silico techniques to methodically assess scaffold design parameters for bone regeneration by comparing model predictions to experimental data in sheep; however, further validation is required [42]. Using a novel multispiked connecting scaffold (MSC-Scaffold), Uklejewski et al. aimed to develop a validated computational model for biomimetic fixation of resurfacing endoprostheses (Figure 2.2.2). A strong correlation between simulated and experimental findings was achieved by the researchers by performing a micro-CT (computed tomography) study of bone density distribution during the embedding of the MSC-Scaffold in periarticular bone and modifying the numerical model appropriately. With the use of the verified finite element (FE) model, the biomimetism of the MSC-Scaffold prototype was verified. Although further research is required, the study verified the numerical model for creating the material and geometric aspects of a novel form of resurfacing endoprosthesis attachment and offered insights into the mechanical behavior of the implant-bone system [43].

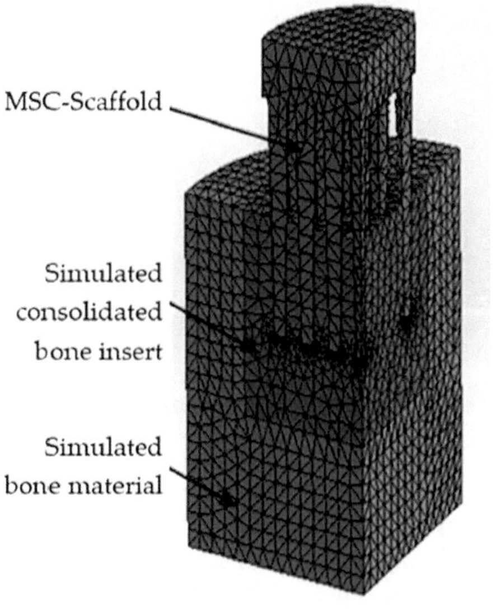

Figure 2.2.2: Modified FE model of the MSC-Scaffold [43]. Reprinted with permission under CC BY 4.0.

Furthermore, Sampson et al. used a combination of thermally induced phase separation (TIPS) and 3D-plotting (3DP) to create hierarchical 3D scaffolds for bone tissue engineering consisting of poly(lactic-co-glycolic acid) (PLGA) and nanohydroxyapatite (nHA). They compared the cell growth in a perfusion bioreactor and a static culture plate to a commercial scaffold, performed a comprehensive factorial design of experiments, investigated the internal structures, and predicted the fluid flow. Though significant diversity was noted, the findings showed the 3DP/TIPS scaffolds' promise. Further research might focus on refining the scaffold architecture and bioreactor parameters to augment cellular growth and tissue regeneration [44]. The objective of Pais et al. was to use a remodeling algorithm inspired by biology to develop an optimal bone scaffold. A gyroid infill with bone tissue was subjected to mechanical properties and density field comparison using an FE approach and radial point interpolation meshless method. The results demonstrated the gyroid scaffold's ability to simulate the characteristics of titanium alloy and bone, indicating its potential use in bone tissue engineering applications. Future research might examine the use of the bioinspired remodeling method to create more complicated scaffold geometries since the current work was limited to a homogenized model of the gyroid infill [45]. Additionally, Lowe et al. utilized FEA to assess the changes in material stiffness and stress distributions in a polymer-assisted 3D PCL/β-tricalcium phosphate (β-TCP) scaffold for the repair of mandibular defects (Figure 2.2.3). The computational simulations demonstrated that the majority of the stress occurred in the mandibular bone, and the scaffold showed a compressive modulus that was similar to cortical bone. At first, the scaffold only supported a small amount of loading, but as new bone infiltration increased, it could support a greater amount. The authors recommended further investigation into the mechanical performance of the scaffold throughout the process of degeneration and under dynamic loading conditions [46].

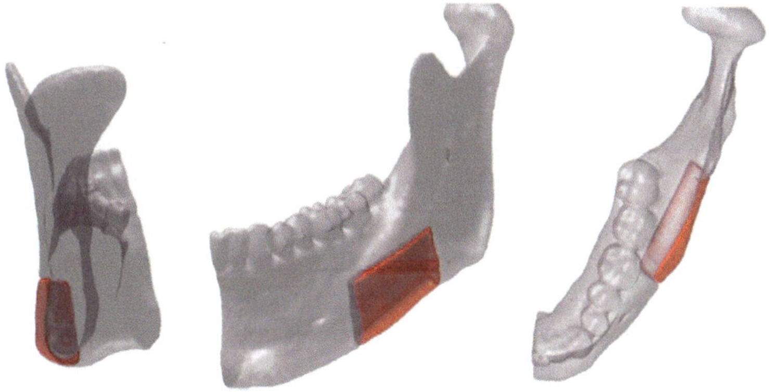

Figure 2.2.3: Schematic of the model developed for evaluating the material stiffness changes and stress distributions in a polymer-assisted 3D PCL/β-TCP scaffold. Reprinted with permission from [46]. Copyright (2021) Elsevier.

Furthermore, the objective of Baneshi et al. was to create the perfect bone scaffold by combining freeze-drying and 3D printing technologies. A multicomponent scaffold covered with a bionanocomposite material was constructed and characterized, and its mechanical, biological, and simulation capabilities were assessed. The produced scaffold exhibited appropriate porosity, wettability, bioactivity, and biocompatibility for bone tissue engineering applications, as shown by the findings. One of the limitations was that the scaffold's performance was not tested in vivo. However, long-term in vivo investigations and scaling up manufacturing for clinical trials should be explored in future studies [46]. In addition, by examining the functions of bone grafting and scaffold construction, Perier-Metz et al. aimed to determine if mechanobiological principles of bone regeneration may explain scaffold-supported bone defect repair. In order to compare predictions with in vivo data, they created an in silico model that included scaffold-guided migration/deposition and bone graft stimulatory effects. The results demonstrated that the key elements influencing bone defect filling in a 3D-printed titanium scaffold were graft osteoconductive qualities and scaffold surface-driven extracellular matrix deposition. With this information, more efficient 3D scaffold constructs might be created, and their preclinical optimization for scaffold-based bone defect repair could be facilitated [47]. Maslov carried out a numerical investigation of tissue regeneration within a porous scaffold and developed a biomechanical model. The stress-strain state of a poroelastic medium was examined, and the effective elastic moduli of the "bone-implant" system were evaluated using FEA. The findings demonstrated correlations between implant porosity, external stimulatory load parameters, and regeneration features. Higher-frequency dynamic impacts, which are known to have a more significant influence on bone callus regeneration, were not taken into account in the research, which was limited to static and low-frequency cyclic loads [48]. In another study, the objective of Esmaeili et al. was to use 3D printing to create and evaluate a porous polymeric-HA scaffold for the treatment of femur fractures. They carried out FEA under various stress scenarios and created a femur model with varied HA compositions. The results demonstrated that replacing a femur with the appropriate strength and stability is possible when using the porous scaffold model. The authors spoke about three scaffold architectures that may be created using current bioprinting methods: zonal, hybrid, and mono. One of the limitations was not assessing the scaffold's biological performance and long-term mechanical properties, which might be investigated in follow-up studies [49].

Table 2.2.1 presents the summary of the studies that used the FEA method to model and analyze bone tissue scaffolds.

Table 2.2.1: Summary of the literature on the use of the FEA approach for numerical analysis of bone tissue scaffolds.

Objectives	Simulation or modeling method	Software used	Scaffold composition	Results	Limitations	Future recommendations	Reference
Simulate bone ingrowth in a novel scaffold design	FEA and bone remodeling theory	ABAQUS	Ti-6Al-4V metal scaffold	Novel Hexanoid scaffold design showed higher bone ingrowth compared to conventional designs	–	Integrate results with clinical rehabilitation to optimize novel scaffold designs	[38]
Simulate PHEMA-gelatin scaffold with GO for bone tissue engineering	FEA	SOLIDWORKS and ABAQUS	PHEMA, gelatin, and GO	GO can support scaffold stability and reduce failure, beneficial for osteogenic differentiation	–	Experimental validation and further optimization of scaffold composition and structure	[39]
Predict mechanical response of 3D-printed hybrid scaffold	Structural FEA	Autodesk Inventor and ABAQUS	PCL and alginate-based hydrogel	Infill pattern affects scaffold stiffness; alginate inclusion does not always impact mechanical performance	–	Further investigate the impact of scaffold design and composition on mechanical properties	[40]
Computational model for polyester-based scaffold degradation	FEA	COMSOL	Functionalized polyester-based 3D scaffold	Optimized scaffold design to control degradation and protein release	Focused on hydrolytic degradation only, no experimental validation	Experimental validation, and consideration of other degradation mechanisms	[41]

(continued)

Table 2.2.1 (continued)

Objectives	Simulation or modeling method	Software used	Scaffold composition	Results	Limitations	Future recommendations	Reference
Predict the impact of scaffold design on large bone defect regeneration	FEA and agent-based modeling techniques	ABAQUS/CAE 6.18	–	Identified key factors: (1) scaffold surface guidance of cellular migration and tissue formation, and (2) scaffold material stimulation of progenitor cell activity. Lower scaffold surface-area-to-volume ratio and reduced scaffold stiffness favored bone formation	–	Experimentally validate model predictions and investigate additional scaffold design parameters	[42]
Develop numerical model for biomimetic fixation of resurfacing endoprostheses	Micro-CT analysis and FEA	–	MSC-Scaffold	Validated numerical model for MSC-Scaffold implant-bone interaction	Focused on one specific scaffold design	Explore other scaffold designs and optimization	[43]
Develop hierarchical 3D scaffolds for bone tissue engineering	FEA	COMSOL	PLGA and nHA	Optimal scaffold showed improved cell proliferation in perfusion bioreactor compared to static culture and a commercial scaffold	Substantial variability in results	Further optimize scaffold design and bioreactor culture conditions	[44]

Develop optimized bone scaffold using bioinspired remodeling	FEA and radial point interpolation meshless method	–	Gyroid infill with PLA, and comparison to bone tissue	Gyroid scaffold exhibits similar density and mechanical properties as bone	Only one scaffold material considered	Explore other scaffold materials, and in vivo testing of optimized scaffolds	[45]
Evaluate material stiffness changes of a PCL/β-TCP scaffold	FEA	SolidWorks, MATLAB R2018b, and NASTRAN	PCL and β-TCP	Scaffold achieved compressive modulus close to cortical bone. FEA showed stress distribution in scaffold and mandible	Limited data on progression of bone regeneration in porous biomaterial systems	Further explore mechanical performance of scaffold during bone regeneration	[46]
Develop a novel multicomponent scaffold for bone tissue engineering	FEA	ABAQUS	Alginate, polyvinyl alcohol, bioceramic-titanium nanoparticles, and electroconductive polylactic acid (PLA)	Scaffold with 78% porosity, 27% apatite formation, good wettability, and 99% biocompatibility	–	Further in vivo and clinical studies to validate the scaffold's performance	[50]
Investigate scaffold effects on bone defect healing	FEA	ABAQUS	3D-printed honeycomb titanium structure	Graft osteoconduction and scaffold-guided matrix deposition essential	Validation only against one in vivo dataset	Optimize 3D scaffold design for bone regeneration	[47]

(continued)

Table 2.2.1 (continued)

Objectives	Simulation or modeling method	Software used	Scaffold composition	Results	Limitations	Future recommendations	Reference
Modeling bone regeneration within a porous scaffold	FEA	–	Regular or chaotic internal structure	Parametric analysis of bone regeneration under external mechanical stimulation	Static and low-frequency loads only, and high computational resources required	Investigate dynamic high-frequency mechanical loads, and extend to heterogeneous scaffold structures	[48]
Fabricate and analyze porous polymeric-HA scaffold for femur fractures	FEA	ABAQUS	PLA polymer and HA nanoparticles	Porous scaffold model showed proper strength and mechanical stability for femur replacement	Only static analysis performed, and no dynamic testing	Investigate dynamic loading, optimize scaffold design, and test in vivo	[49]

2.2.2.2 Computational fluid dynamics (CFD)

Transport and fluid flow phenomena are simulated using CFD. It evaluates waste disposal and nutrition supply in scaffolds, which is important for tissue regeneration and cell viability in bone tissue engineering. In a perfusion bioreactor, Channasanon et al. examined the impact of scaffold geometry on the osteogenic differentiation of MC3T3-E1 pre-osteoblasts. Using stereolithography, three oligolactide-HA scaffold geometries were produced and assessed. The impact of flow behavior on cell response was predicted via the use of CFD simulation. The highest performance was achieved by the LC-1000 scaffold because of its ideal ratio of permeability to flow-induced shear stress. The lack of investigation into other scaffold materials and cell types is one of the study's weaknesses. Future studies should look at more intricate scaffold designs and bioreactor operation settings [51]. In addition, the objective of Karaman et al. was to ascertain how bone regeneration was affected by network solid and sheet solid structures in three distinct triply periodic minimum surface (TPMS) architectures: diamond, gyroid, and I-WP. To assess the scaffolds' permeability, wall shear stress, and velocity distribution, they used CFD simulations (Figure 2.2.4). The findings suggested that sheet solid TPMS structures are more beneficial for biomechanical settings since they had higher surface area, lower wall shear stress, and comparable permeability when compared to network solid TPMS structures at identical porosities. The work was limited to computational studies; experimental validation and examination of other TPMS structures may be part of further research [52]. In another investigation, Liu et al. used a computer model that combined species transport models with volume of fluid (VOF) to predict the distribution of glucose and oxygen in a 3D osteochondral scaffold. The model demonstrated that the scaffold's linked structure enhanced nutrient transfer by comparing well with in vivo tissue regeneration patterns. Although experimental validation is still a barrier, the scientists stated the

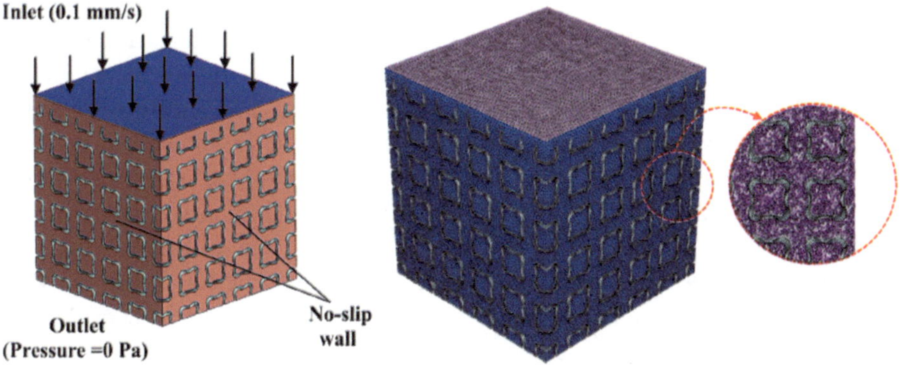

Figure 2.2.4: Schematic of the model and boundary conditions for CFD analysis in [52]. Reprinted with permission from [52]. Copyright (2023) Elsevier.

model might be used to better understand nutrition transport and tissue creation inside the scaffold. Future studies should focus on including cell migration, proliferation, and differentiation in the model [53].

Table 2.2.2 presents the summary of the studies that used the CFD method to model and analyze bone tissue scaffolds.

2.2.2.3 Combination of FEA and CFD

Some investigations, based on their objectives and nature, have employed both FEA and CFD methods. In contrast to traditional diamond, gyroid, and I-WP scaffolds, Li et al. aimed to thoroughly assess the biomechanical and biological properties of biomimetic Voronoi bone scaffolds. To examine the scaffolds' functionality, the researchers conducted in vitro and in vivo tests, mechanical testing, and computational modeling. They discovered that, in contrast to the other scaffold designs, the Voronoi scaffolds had bionic mechanical behavior and an adequate hydrodynamic response, which promoted cell proliferation and osteogenesis. The research did not examine the impacts of altering pore size and shape; however, it was limited to a particular material. In order to better enhance Voronoi scaffolds' characteristics for bone tissue engineering applications, future studies should examine how well they work when designed with various parameters and materials [54].

Table 2.2.3 presents the summary of Li et al., who used the combination of FEA and CFD methods to model and analyze bone tissue scaffolds [54].

2.2.2.4 Representative volume element (RVE) and FEA

A small, statistically representative portion of a substance is called an RVE. It helps FEA in bone tissue engineering by defining scaffold characteristics, improving mechanical simulation accuracy, and optimizing design. Using the space holder approach, Liang et al. aimed to develop and describe a wollastonite-titanium porous scaffold for use in pharmaceutical applications. The results demonstrated that the scaffold's porosity and compressive strength had both increased with the addition of TiO_2 nanoparticles. The mechanical properties were predicted using Abaqus simulations, with an inaccuracy of 18% in comparison to experimental data. According to the study's findings, the sample containing 10% TiO_2 demonstrated promise for use in pharmaceutical and maxillofacial applications; nevertheless, further research is required to fully optimize the scaffold's properties [55]. In addition, Li et al.'s aim for femoral applications was to develop and evaluate porous bone scaffolds made from alginate-HA composites. In addition to performing mechanical, biological, and computational analyses – which included the use of RVE and FEA methods – they also employed 3D printing and freeze-drying procedures to produce the scaffolds

Table 2.2.2: Summary of the literature on the use of CFD approach for numerical analysis of bone tissue scaffolds.

Objectives	Simulation or modeling method	Software used	Scaffold composition	Results	Limitations	Future recommendations	Reference
Investigate the effect of scaffold geometry on osteogenic differentiation	CFD	Ansys CFX	Oligolactide-HA	LC-1400 scaffold showed highest cell proliferation; LC-1000 scaffold had optimal balance of permeability and flow-induced shear stress for osteogenesis	–	Explore other scaffold geometries and compositions for improved osteogenic differentiation	[51]
Determine the effect of TPMS scaffold architecture on permeability, wall shear stress, and velocity	CFD	SolidWorks and ANSYS Fluent	Diamond, gyroid, and I-WP TPMS architectures, network solid and sheet solid structures, 60–80% porosity	Network solid TPMS scaffolds have higher permeability than sheet solid. Wall shear stress decreases with increasing porosity. Sheet solid structures have greater surface area, lower wall shear stress, and similar permeability	Only compares three TPMS architectures, and no experimental validation	Experimentally validate, investigate other TPMS architectures, and optimize scaffold design for bone regeneration	[52]
Predict oxygen and glucose distribution in 3D osteochondral scaffold	CFD, coupled VOF, and species transport model	ANSYS Fluent	3D osteochondral scaffold	Oxygen distribution correlates with tissue regeneration; scaffold has good interconnected structure	No experimental validation	Integrate cell growth and tissue formation in the model	[53]

Table 2.2.3: Summary of Li et al. who employed the combination of FEA and CFD approaches to model and analyze bone tissue scaffolds [54].

Objectives	Simulation or modeling method	Software used	Scaffold composition	Results	Limitations	Future recommendations	Reference
Integrated evaluation of biomimetic bone scaffold properties	FEA and CFD	Python for formulating the algorithm, Grasshopper and Rhinoceros3D 7.0 for designing the scaffold, and Fluent 6.3 for CFD simulation	Titanium alloy and 3D printed	Voronoi scaffold exhibited biomimetic mechanics and osteogenesis	Only one biomaterial (titanium alloy) evaluated	Investigate biomimetic scaffolds with other biomaterials	[54]

(Figure 2.2.5). The PLA-alginate/20 wt% HA scaffold had the most desirable mechanical properties, according to the results; however, the 30 wt% HA scaffold had fewer biological upsides. The study was limited by its emphasis on short-term biological assessment; further studies should look at long-term in vivo performance and further refine the characteristics and composition of the scaffold [56]. Additionally, Jiang et al. used polyvinylpyrrolidone (PVP), alginate, and titanium nanoparticles (Ti-NPs or TiO_2 nanoparticles) using a freeze-drying process to produce a porous polymeric-based scaffold for dental tissue regeneration. Multiscale FEA and artificial neural networks (ANNs) were used to predict the parameters of the scaffold, which demonstrated improved mechanical and biological capabilities. The authors recommended further study to improve the scaffold's composition and investigate its potential in tooth tissue regeneration in light of the findings, which demonstrated the dependability of the established FEM and ANN models [57].

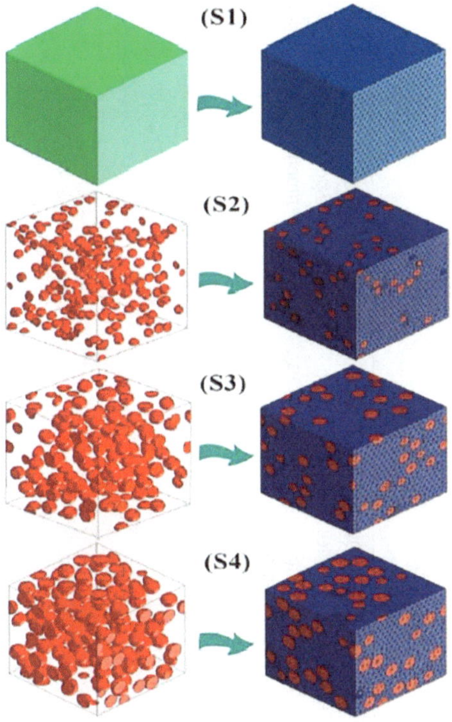

Figure 2.2.5: Schematic of RVE-generated model and their corresponding mesh model of alginate/HA scaffolds in different weight percentages. Reprinted with permission from [56]. Copyright (2021) Elsevier.

Table 2.2.4 presents the summary of the studies that used the RVE/FEA methods to model and analyze bone tissue scaffolds.

Table 2.2.4: Summary of the literature on the use of the RVE/FEA approach for numerical analysis of bone tissue scaffolds.

Objectives	Simulation or modeling method	Software used	Scaffold composition	Results	Limitations	Future recommendations	Reference
Fabrication and characterization of wollastonite-titanium porous scaffold for pharmaceutical application	RVE and FEA	ABAQUS	Wollastonite, titanium nanoparticles and NaCl as space holder	Increased compressive strength and porosity with increasing titanium content. Prediction of mechanical properties using ABAQUS showed 18% error compared to experimental data	Evaluation limited to 28 days of testing	Longer-term in vitro and in vivo studies to assess degradation and biocompatibility	[55]
Fabricate and characterize porous bony scaffold for femoral applications	RVE and FEA	ABAQUS	PLA, alginate, and HA	Improved mechanical strength and biological properties with the addition of HA nanoparticles	–	Investigate long-term in vivo performance, and optimize scaffold design and composition	[56]
Fabrication and evaluation of a porous scaffold for dental tissue repair	FEA, RVE, and artificial neural network (ANN)	ABAQUS/ Explicit and VUSDFLD subroutine	PVP, alginate, and titanium nanoparticles (Ti-NPs or TiO_2)	Improved mechanical and biological properties, and good agreement between experimental and simulation results	Limited to specific scaffold composition and properties	Investigate other scaffold materials and compositions, and in vivo studies	[57]

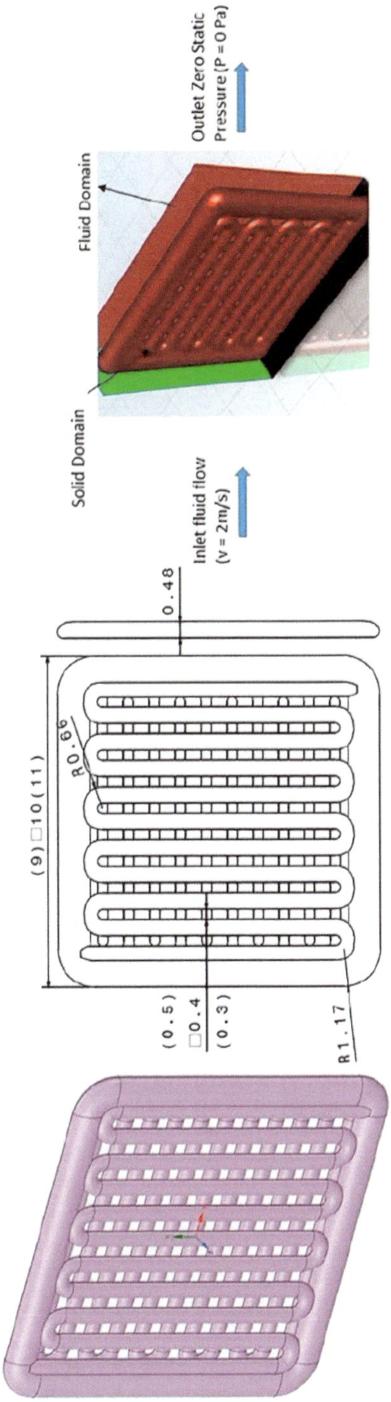

Figure 2.2.6: Schematic illustration of the model created in [58] for FSI analysis of the PEI-PLA bone scaffold. Reprinted with permission under CC BY 4.0.

Table 2.2.5: Summary of the literature on the use of FSI/CFD methods for numerical analysis of bone tissue scaffolds.

Objectives	Simulation or modeling method	Software used	Scaffold composition	Results	Limitations	Future recommendations	Reference
Simulate mechanical stimulation of cells on porous hydrogel scaffold	FSI	SOLIDWORKS and ANSYS 2020R2	3D-structured porous hydrogel scaffold seeded with human mesenchymal stem cells (MSCs)	Compression amplitude affects cell differentiation into bone, cartilage, and fibrous tissue	–	Explore effects of different scaffold designs and stimulation parameters on cell differentiation	[59]
Comparative study of turbulence methods for 3D scaffold design	CFD and FSI	Ansys Fluent and Ansys Mechanical APDL	PEI, ULTEM 1010, and PLA	DES model shows better performance for 3D scaffolds	Difficulty in maintaining constant pore and layer size	Explore other turbulence models and materials for 3D scaffold design	[58]

2.2.2.5 Fluid-structure interaction (FSI) and CFD

The study of FSI looks at how fluid flow and structural deformation interact. It optimizes designs for nutrition delivery and mechanical stability during tissue regeneration in bone tissue engineering by combining CFD with scaffold mechanics. In order to develop a 3D-printed scaffold and select the best numerical technique to simulate the scaffold for tissue engineering, Suffo et al. carried out a comparison research of turbulence approaches. On scaffolds composed of polyetherimide (PEI) and PLA, they evaluated several turbulence models using CFD methods and the FSI approach (Figure 2.2.6). While there were no notable differences across the materials, the detached eddy simulation (DES) model outperformed the others for the typical operating regime of 3D scaffolds. Unpredictable variability in pore and layer size was shown by PLA scaffolds. The research emphasizes how crucial it is to take geometric characteristics and scaffold architecture into account when selecting and developing a scaffold [58]. Furthermore, Azizi et al. predicted cell differentiation on the surface of a 3D-structured hydrogel scaffold seeded with cells as a result of dynamic compressive force stimulation by simulating the FSI. The influence of compression amplitude on cell differentiation was examined in the research, and it was discovered that although increasing compression initially increased cartilage differentiation before lowering it, it decreased the bone tissue development. The authors came to the conclusion that, despite some drawbacks, such as the need for experimental validation, FSI simulation is a flexible method in computational mechanobiology for studying the effects of scaffold design and stimulation on cell differentiation [59].

Table 2.2.5 presents the summary of the studies that used the FSI/CFD methods to model and analyze bone tissue scaffolds.

2.2.3 Other modeling and simulation techniques for modeling bone tissue scaffolds

Some investigations have employed other types of simulation, modeling, and mathematical methods to study bone tissue scaffolds. A computational model was developed by Marquez-Florez et al. to mimic the in vivo regeneration of articular cartilage caused by scaffold implantation. The scaffold, subchondral bone, and articular cartilage were all represented as poroelastic materials in the model. The findings clarified how the scaffold provided a mechanical environment that was favorable to cartilage regeneration and how subchondral bone remodeled. Before performing in vivo research, the scientists proposed that the computational model would be a helpful tool for predicting the results of cartilage regeneration therapies [60]. Furthermore, Mankowsky et al. used computational and experimental methods to determine appropriate 3D bioprinted scaffold topologies for incubation in a perfusion bioreactor. They identified the ideal pore shape by investigating how fluid flow responded to various scaffold pore geometries and porosities. The findings demonstrated that the suggested comparison research may help with application-based material selection that is appropriate for perfusion bioreactor incubation. The long-term effects of perfusion on cell survival and proliferation, however, were not examined in this work; it might be a topic for future investigation [61]. Additionally, Rezk et al.'s objective was to create a unique 3D fibrous scaffold that resembled cotton and contained simvastatin in spatiotemporally specified patterns that were ideal for osteogenesis paired with angiogenesis. They used computer simulation to show the 3D fiber deposition method, functionalized the scaffolds with HA nanoparticles, and looked at how simvastatin affected angiogenesis and osteogenic differentiation. The results demonstrated that the 3D scaffold facilitated the production of osteogenic markers and encouraged endothelial cell migration and tube formation [62].

Additionally, Zhang et al. suggested using a voxel model as the basis for their modeling approach for porous bone scaffolds. Using CT scans and the moving cube technique, they initially rebuilt the scaffold's surface. They then created the voxel model and created a fast slicing technique to get the cross-sectional profile. The voxel model served as the basis for the design of the pore unit, and it was used to create the processing route as well. The experiment demonstrated that the properties of the scaffold, such as its porosity and pore size, could be controlled using the suggested technique. However, further studies might focus on the scaffolds' biocompatibility and mechanical properties [63]. In another study, Zhang et al. developed a 3D computer model to study adhesion and cell movement in a porous scaffold during perfusion seeding. Cell adhesion to scaffolds, cell dynamics, and cell-fluid interaction were all included in the model. The effects of initial cell concentration and perfusion pressure on scaffold permeability, adhesion rates, cell cluster formation, uniformity and efficiency of seeding, and seeding kinetics were examined using simulations. The findings provide insights for improving the seeding settings and emphasize the significance of adhesion dynamics and cell-fluid interaction in simulating the dynamic seeding process. The simplistic cell and scaffold geometries are a limitation, and more intricate and plausible situations may be explored in further studies [64]. Additionally, Foroutan et al. used freeze-drying to produce a unique porous bionanocomposite scaffold composed of wollastonite, graphene nanosheets, and sodium alginate. They assessed the biological and mechanical aspects and used the theories of Timoshenko and Euler-Bernoulli beams to perform buckling analysis. The sample containing 1 wt% graphene had encouraging mechanical and biological properties, indicating that it may be used as a lightweight alternative to bone. Limitations include the need for in vivo investigations; further work may include optimizing the amount of graphene and the architecture of the scaffold to enhance mechanical performance [65].

Table 2.2.6 presents the summary of the studies that used other numerical methods to model and analyze bone tissue scaffolds.

Table 2.2.6: Summary of the literature on the use of other numerical methods to model and analyze bone tissue scaffolds.

Objectives	Simulation or modeling method	Software used	Scaffold composition	Results	Limitations	Future recommendations	Reference
Computational model of cartilage regeneration by scaffold	A simulation tool to determine the mechanical conditions provided by the polymeric scaffold	ABAQUS	Polymeric scaffold	Scaffold provides favorable mechanical environment for cartilage regeneration	Model parameters not fully validated with experiments	Further experimental validation of computational model	[60]
Identify suitable 3D bioprinted scaffold architectures for perfusion bioreactor	Methodology was not directly specified	SolidWorks, NTOPOLOGY, and RHINO 6.0	8% Alginate, 4% alginate-4% CMC, and 2% alginate-6% CMC	Determined optimum pore geometry and scaffold material for perfusion bioreactor	–	Further optimization of scaffold architecture and bioreactor design	[61]
Develop a dual-functional 3D scaffold for angiogenesis and osteogenesis	Electric field simulation	COMSOL	Simvastatin-loaded poly(ε-caprolactone)-poly (glycerol-sebacate)-HA nanoparticles (3D-PGHS)	The 3D-PGHS scaffold enhanced the expression of osteogenic markers in hMSCs and promoted the migration and tube formation of HUVECs	–	Explore the potential of the 3D-PGHS scaffold as a point-of-care therapy for controlled tissue regeneration	[62]
Rapid modeling and manufacturing of porous bone scaffold	Voxel-based modeling	–	Porous structure based on voxel model	Porous scaffold constructed, parameters controlled, and processing path generated	–	–	[63]

(continued)

Table 2.2.6 (continued)

Objectives	Simulation or modeling method	Software used	Scaffold composition	Results	Limitations	Future recommendations	Reference
Investigate perfusion cell seeding process, cell transport, and adhesion	Immersed boundary-lattice Boltzmann method	FORTRAN	Porous scaffold and irregular geometry	Adhesion rates, cell cluster formation, seeding uniformity and efficiency, and scaffold permeability	Simplifications in cell and fluid modeling, and lack of experimental validation	Experimental validation, incorporating more detailed cell/scaffold interactions	[64]
Fabrication and analysis of a porous graphene scaffold	Euler-Bernoulli and Timoshenko beam theories, and Galerkin differential quadrature (GDQ) method	–	Sodium alginate, graphene nanosheets, and wollastonite powder	Scaffold with 1 wt% graphene showed good mechanical and biological properties	Mechanical properties of the scaffold still need improvement for load-bearing applications	Optimize the scaffold composition and structure to further enhance the mechanical properties	[65]

2.2.4 Discussion and future research directions

In regenerative medicine, bone tissue engineering plays a critical role in treating abnormalities in bones via the integration of biomaterials, cells, and bioactive substances. Structural scaffolds facilitate cell adhesion and proliferation by imitating the extracellular matrix. In order to improve bone integration, a variety of scaffold types, including those made of synthetic and natural polymers, are engineered for the best possible mechanical properties and degradability. For the purpose of enhancing nutrition delivery, anticipating mechanical behavior, and optimizing scaffold design, computational techniques such as CFD and FEA are essential. These methods save expenses and time, which makes it easier to develop orthopedic and dental therapies.

The findings revealed that new scaffold designs, such as the LC-1400 and the Hexanoid, improved cell proliferation and bone ingrowth, respectively. Graphene and HA nanoparticles were added, which enhanced the mechanical and biological characteristics of different kinds of scaffolds. While alginate incorporation had varying impacts on mechanical performance, infill patterns had a substantial influence on stiffness. Effective MSC-scaffold-bone interactions were verified using validated numerical models, and the gyroid scaffold had mechanical properties similar to those of bone, attaining a compressive modulus similar to cortical bone. For the distribution of oxygen and tissue regeneration, high porosity and biomimetic mechanics were essential. Compressive strength increased with increasing titanium concentration, and modeling projections closely matched experimental findings. Furthermore, the DES model performed better for 3D scaffolds than conventional models, and the best scaffold material and pore geometry promoted osteogenic marker expression in hMSCs and promoted cartilage regeneration. The results underscore the significance of scaffold design factors in stimulating cellular activity and enhancing healing outcomes during the regeneration of bone and cartilage.

There were several limitations, such as the fact that many research only evaluated one biomaterial, which limited how broadly the results could be applied. The lack of complete experimental validation for several model parameters has led to questions about their dependability. The examination was limited to brief time periods, and difficulties in maintaining uniform pore and layer widths further confounded the findings. Furthermore, other studies focused only on a specific scaffold design, producing results that varied greatly. Others only looked at a single type of scaffold and offered little information on how bone regeneration works in porous biomaterials. The robustness of the results of several other studies was restricted by simplifications in cell and fluid models and validation against a single in vivo dataset. Furthermore, since dynamic testing was not performed, certain evaluations were limited to static and low-frequency loads, which required a large amount of computational power. Moreover, the mechanical properties of the scaffolds still need to be improved for efficient load-bearing applications.

In order to improve osteogenic differentiation, future research should focus on biomimetic scaffolds made of a variety of biomaterials and experimenting with different compositions and geometries. Scaffold designs may be optimized for real-world applications by incorporating results with clinical rehabilitation. Conducting experimental validation and optimizing scaffold composition and structure is essential, as is investigating how design affects mechanical properties. In addition to taking into account various turbulence models for 3D scaffold design, researchers should look at the impact of various scaffold designs and stimulation settings on cell differentiation. In order to evaluate biocompatibility and degradation, long-term in vitro and in vivo investigations are essential, with a particular emphasis on the mechanical performance of scaffolds during bone regeneration. Furthermore, evaluating the scaffold's efficacy in clinical settings will require modifying the scaffold design and bioreactor growth conditions in addition to dynamic loading experiments. Ultimately, the area of bone tissue engineering will progress with a thorough strategy that includes different loading conditions and intricate cell-scaffold interactions.

2.2.5 Conclusion

Bone tissue engineering is essential to regenerative medicine because it combines biomaterials, bioactive substances, and cells for the treatment of bone defects. The development of innovative scaffold designs, including LC-1400 and Hexanoid, has shown substantial increases in bone ingrowth and cell proliferation, underscoring the significance of scaffold properties in fostering efficient tissue regeneration. Scaffold optimization has been made easier by developments in computational techniques, but there are still a number of drawbacks, such as the use of distinct biomaterials and the need for further experimental validation. The investigation of various biomaterials and scaffold geometries, together with the clinical application of discoveries, needs to be the primary focus of future studies. The evaluation of biocompatibility and degradation requires long-term research, and the mechanical performance of scaffold designs for load-bearing applications will be improved by dynamic testing and scaffold design optimization. Improved patient outcomes and more effective solutions for bone regeneration may be achieved in the area of bone tissue engineering by using a holistic approach that includes intricate interactions and varying loading conditions.

2.2.6 Declarations

2.2.6.1 Conflict of interests

The authors declare that there is no conflict of interest.

2.2.6.2 Data availability

No data was used or generated in this study.

2.2.6.3 Funding

No funding source is available for this study.

2.2.6.4 Declaration of generative AI and AI-assisted technologies in the writing process

During the preparation of this work, the authors used Assistant, QuillBot, and Grammarly to rephrase and edit the text grammatically and in terms of typos. After using these tools/services, the authors reviewed and edited the content as needed and take full responsibility for the content of the publication.

References

[1] Xue N, et al. Bone tissue engineering in the treatment of bone defects. Pharmaceuticals. 2022; 15(7): 879.
[2] Kanczler JM, Wells JA, Gibbs DMR, Marshall KM, Tang DKO, Oreffo Richard OC. Chapter 50 - Bone tissue engineering and bone regeneration. In: Robert Lanza, Robert Langer, Joseph P. Vacanti, Anthony Atala, editors. Principles of Tissue Engineering (Fifth Edition), Academic Press. 2020. pp. 917–935. ISBN 9780128184226, https://doi.org/10.1016/B978-0-12-818422-6.00052-6.
[3] Pedrero SG, Llamas-Sillero P, Serrano-López J. A multidisciplinary journey towards bone tissue engineering. Materials. 2021; 14(17): 4896.
[4] Mashhadi Keshtiban M, et al. Biological and mechanical response of graphene oxide surface-treated polylactic acid 3D-printed bone scaffolds: Experimental and numerical approaches. Advanced Engineering Materials. 2024; 26(3): 2301260.
[5] Gautam G, Kumar S, Kumar K. Processing of biomaterials for bone tissue engineering: State of the art. Materials Today: Proceedings. 2022; 50: 2206–2217.
[6] Ding Z, et al. Silk biomaterials for bone tissue engineering. Macromolecular Bioscience. 2021; 21(8): 2100153.

[7] Wang C, et al. 3D printing of bone tissue engineering scaffolds. Bioactive Materials. 2020; 5(1): 82–91.
[8] Bahraminasab M, et al. Bone scaffolds: An incorporation of biomaterials, cells, and biofactors. ACS Biomaterials Science and Engineering. 2021; 7(12): 5397–5431.
[9] Koushik TM, Miller CM, Antunes E. Bone tissue engineering scaffolds: Function of multi-material hierarchically structured scaffolds. Advanced Healthcare Materials. 2023; 12(9): 2202766.
[10] Noroozi R, et al. In vitro static and dynamic cell culture study of novel bone scaffolds based on 3D-printed PLA and cell-laden alginate hydrogel. Biomedical Materials. 2022; 17(4): 045024.
[11] Rico-Llanos GA, et al. Collagen type I biomaterials as scaffolds for bone tissue engineering. Polymers. 2021; 13(4): 599.
[12] Yadav LR, et al. Chitosan-based 3D–printed scaffolds for bone tissue engineering. International Journal of Biological Macromolecules. 2021; 183: 1925–1938.
[13] Ressler A. Chitosan-based biomaterials for bone tissue engineering applications: A short review. Polymers. 2022; 14(16): 3430.
[14] Li Z, et al. Bioinspired mineralized collagen scaffolds for bone tissue engineering. Bioactive Materials. 2021; 6(5): 1491–1511.
[15] Yang X, et al. The application of polycaprolactone in three-dimensional printing scaffolds for bone tissue engineering. Polymers. 2021; 13(16): 2754.
[16] Hou Y, Wang W, Bartolo P. Investigation of polycaprolactone for bone tissue engineering scaffolds: In vitro degradation and biological studies. Materials and Design. 2022; 216: 110582.
[17] Javid-Naderi MJ, et al. Synthetic polymers as bone engineering scaffold. Polymers for Advanced Technologies. 2023; 34(7): 2083–2096.
[18] Donate R, Monzón M, Alemán-Domínguez ME. Additive manufacturing of PLA-based scaffolds intended for bone regeneration and strategies to improve their biological properties. e-Polymers. 2020; 20(1): 571–599.
[19] Mohammadi Zerankeshi M, Bakhshi R, Alizadeh R. Polymer/metal composite 3D porous bone tissue engineering scaffolds fabricated by additive manufacturing techniques: A review. Bioprinting. 2022; 25: e00191.
[20] Janmohammadi M, et al. Cellulose–based composite scaffolds for bone tissue engineering and localized drug delivery. Bioactive Materials. 2023; 20: 137–163.
[21] Gritsch L, et al. Combining bioresorbable polyesters and bioactive glasses: Orthopedic applications of composite implants and bone tissue engineering scaffolds. Applied Materials Today. 2021; 22: 100923.
[22] Kumar A, et al. Load-bearing biodegradable PCL-PGA–beta TCP scaffolds for bone tissue regeneration. Journal of Biomedical Materials Research, Part B: Applied Biomaterials. 2021; 109(2): 193–200.
[23] Kalsi S, et al. Biomaterials for tissue engineered bone scaffolds: A review. Materials Today: Proceedings. 2023; 81: 888–893.
[24] Qi J, et al. Preparation and properties of a biodegradable poly(lactide-co-glycolide)/poly (trimethylene carbonate) porous composite scaffold for bone tissue engineering. New Journal of Chemistry. 2020; 44(34): 14632–14641.
[25] Girón J, et al. Biomaterials for bone regeneration: An orthopedic and dentistry overview. Brazilian Journal of Medical & Biological Research. 2021; 54: e11055.
[26] Kanwar S, Vijayavenkataraman S. Design of 3D printed scaffolds for bone tissue engineering: A review. Bioprinting. 2021; 24: e00167.
[27] Shirzad M, et al. Design, evaluation, and optimization of 3D printed truss scaffolds for bone tissue engineering. Journal of the Mechanical Behavior of Biomedical Materials. 2021; 120: 104594.
[28] Kennedy SM, et al. Experimental and finite element analysis on the effect of pores on bio-printed polycaprolactone bone scaffolds. Bioprinting. 2023; 34: e00301.

[29] Bin Noordin MA, et al. Finite element analysis of porosity effects on mechanical properties for tissue engineering scaffold. Biointerface Research in Applied Chemistry. 2021; 11(2): 8836–8843.
[30] Pires T, et al. Challenges in computational fluid dynamics applications for bone tissue engineering. Proceedings of the Royal Society A. 2022; 478(2257): 20210607.
[31] Saatchi AR, et al. Computational fluid dynamics in 3D-printed scaffolds with different strand–orientation in perfusion bioreactors. Iranian Journal of Chemistry and Chemical Engineering. 2020; 39(5): 307–320.
[32] Sahbafar H. Mechanical and Biological Properties of Polylactic Acid (PLA)-hardystonite-Fe3O4 Scaffolds Made by 3D Printing for Bone Tissue Engineering, Iran: University of Tehran; Iran, 2023.
[33] Fu M, Wang F, Lin G. Design and research of bone repair scaffold based on two-way fluid-structure interaction. Computer Methods and Programs in Biomedicine. 2021; 204: 106055.
[34] Gupta A, Rana M, Mondal N. Determination of optimum design parameters for gyroid scaffolds to mimic a real bone-like condition in vitro: A fluid structure interaction study. Journal of Engineering and Science in Medical Diagnostics and Therapy. ASME Journal of Medical Diagnostics. 2023; 6(3): 030909 (7 pages).
[35] Makowski P, Kuś W. Optimization of bone scaffold structures using experimental and numerical data. Acta Mechanica. 2016; 227(1): 139–149.
[36] Wieding J, Wolf A, Bader R. Numerical optimization of open–porous bone scaffold structures to match the elastic properties of human cortical bone. Journal of the Mechanical Behavior of Biomedical Materials. 2014; 37: 56–68.
[37] Wu J, et al. On the various numerical techniques for the optimization of bone scaffold. Materials. 2023; 16(3): 974.
[38] Wang Y, et al. Bone ingrowth simulation within the Hexanoid, a novel scaffold design. 3D Printing and Additive Manufacturing. 2023; 11(6).
[39] Tabatabaee S, et al. Modeling of the PHEMA–gelatin scaffold enriched with graphene oxide utilizing finite element method for bone tissue engineering. Computer Methods in Biomechanics and Biomedical Engineering. 2023; 26(5): 499–507.
[40] Scocozza F, et al. Prediction of the mechanical response of a 3D (bio)printed hybrid scaffold for improving bone tissue regeneration by structural finite element analysis. Journal of the Mechanical Behavior of Biomedical Materials. 2023; 142: 105822.
[41] Batoni E, et al. A computational model for the release of bioactive molecules by the hydrolytic degradation of a functionalized polyester-based scaffold. Pharmaceutics. 2023; 15: 10.3390/pharmaceutics15030815.
[42] Perier-Metz C, et al. An in silico model predicts the impact of scaffold design in large bone defect regeneration. Acta Biomaterialia. 2022; 145: 329–341.
[43] Uklejewski R, et al. Bone density micro-CT assessment during embedding of the innovative multi-spiked connecting scaffold in periarticular bone to elaborate a validated numerical model for designing biomimetic fixation of resurfacing endoprostheses. Materials. 2021; 14: 10.3390/ma14061384.
[44] Sampson K, et al. Cultivation of hierarchical 3D scaffolds inside a perfusion bioreactor: Scaffold design and finite-element analysis of fluid flow. SN Applied Sciences. 2021; 3(12): 884.
[45] Pais AI, Alves JL, Belinha J. Using a radial point interpolation meshless method and the finite element method for application of a bio-inspired remodelling algorithm in the design of optimized bone scaffold. Journal of the Brazilian Society of Mechanical Sciences and Engineering. 2021; 43(12): 557.
[46] Lowe B, et al. FEA evaluation of material stiffness changes for a polymer assisted 3D polycaprolactone/β-tricalcium phosphate scaffold in a mandibular defect reconstruction model. Ceramics International. 2021; 47(6): 8075–8081.
[47] Perier-Metz C, Duda GN, Checa S. Mechano-biological computer model of scaffold–supported bone regeneration: Effect of bone graft and scaffold structure on large bone defect tissue patterning. Frontiers in Bioengineering and Biotechnology. 2020; 8. https://doi.org/10.3389/fbioe.2020.585799.

[48] Maslov LB. Biomechanical model and numerical analysis of tissue regeneration within a porous scaffold. Mechanics of Solids. 2020; 55(7): 1115–1134.
[49] Esmaeili S, et al. A porous polymeric-hydroxyapatite scaffold used for femur fractures treatment: Fabrication, analysis, and simulation. European Journal of Orthopaedic Surgery and Traumatology. 2020; 30(1): 123–131.
[50] Baneshi N, et al. Investigation the mechanical properties of a novel multicomponent scaffold coated with a new bio-nanocomposite for bone tissue engineering: Fabrication, simulation and characterization. Journal of Materials Research and Technology. 2021; 15: 5526–5539.
[51] Channasanon S, et al. Scaffold geometry and computational fluid dynamics simulation supporting osteogenic differentiation in dynamic culture. Computer Methods in Biomechanics and Biomedical Engineering. 2024; 27(5): 587–598.
[52] Karaman D, Ghahramanzadeh Asl H. The effects of sheet and network solid structures of similar TPMS scaffold architectures on permeability, wall shear stress, and velocity: A CFD analysis. Medical Engineering & Physics. 2023; 118: 104024.
[53] Liu Z, et al. Oxygen and glucose transportation and distribution on 3D osteochondral scaffold in silico model. Journal of Bionic Engineering. 2022; 19(4): 1100–1108.
[54] Li J, et al. Integrated evaluation of biomechanical and biological properties of the biomimetic structural bone scaffold: Biomechanics, simulation analysis, and osteogenesis. Materials Today Bio. 2024; 24: 100934.
[55] Liang H, et al. Fabrication and characterization of wollastonite-titanium porous scaffold for pharmaceutical application: Representative volume element simulation. Materials Science and Engineering: B. 2022; 280: 115684.
[56] Li X, et al. Experimental measurement and simulation of mechanical strength and biological behavior of porous bony scaffold coated with alginate-hydroxyapatite for femoral applications. Composites Science and Technology. 2021; 214: 108973.
[57] Jiang L, et al. Fabrication of porous polymeric-based scaffold for dental tissue repair in fracture healing: RVE simulation and ANN optimization. Materials Science and Engineering: B. 2023; 297: 116770.
[58] Suffo M, López-Marín CJ. A comparative study of turbulence methods applied to the design of a 3D-printed scaffold and the selection of the appropriate numerical scheme to simulate the scaffold for tissue engineering. Applied Sciences. 2022; 12: 10.3390/app12010191.
[59] Azizi P, et al. Simulating the mechanical stimulation of cells on a porous hydrogel scaffold using an FSI model to predict cell differentiation. Frontiers in Bioengineering and Biotechnology. 2023; 11. https://doi.org/10.3389/fbioe.2023.1249867.
[60] Márquez-Flórez K, et al. Computational model of articular cartilage regeneration induced by scaffold implantation in vivo. Journal of Theoretical Biology. 2023; 561: 111393.
[61] Mankowsky J, et al. Identifying suitable three-dimensional bio-printed scaffold architectures to incubate in a perfusion bioreactor: Simulation and experimental approaches. Journal of Medical Devices. 2023; 17(2).
[62] Rezk AI, et al. De novo dual functional 3D scaffold using computational simulation with controlled drug release. Journal of Colloid and Interface Science. 2022; 625: 12–23. J. Med. Devices. Jun 2023, 17 (2): 020903 (14 pages).
[63] Zhang Z-y, et al. Study on rapid modeling and manufacturing method of porous bone scaffold based on voxel model. Journal of the Brazilian Society of Mechanical Sciences and Engineering. 2021; 43(12): 566.
[64] Zhang Z, et al. A 3D computational model of perfusion seeding for investigating cell transport and adhesion within a porous scaffold. Biomechanics and Modeling in Mechanobiology. 2020; 19(5): 1461–1475.
[65] Foroutan S, Hashemian M, Khandan A. A novel porous graphene scaffold prepared using freeze-drying technique for orthopedic approaches: Fabrication and buckling simulation using GDQ method. IUST. 2020; 17(4): 62–76.

Jayanta Kumar Biswas and Masud Rana*

Chapter 2.3 Fundamentals of computational modeling of biomechanics in the musculoskeletal system

Abstract: Computational modelling is a vital tool for analysing the complexities of the musculoskeletal (MSK) system, integrating mathematical formulations with computer simulations. Given the intricate nature of musculoskeletal structures, numerical approaches such as multi-body dynamics, finite element methods (FEM), and computational fluid dynamics (CFD) enable the study of motion, stress distributions, and fluid interactions within joint spaces. These techniques support applications in bone tissue engineering, fracture risk assessment, and scaffold optimization. This study explores computational modelling of bones and joints, emphasizing bone microstructure, patient-specific CT-based modelling, and fracture healing simulations. It also examines biomechanical modelling of key joints, including the shoulder, elbow, wrist, hip, knee, ankle, and spine, highlighting their structural and functional characteristics. Additionally, muscle, tendon, and ligament models provide insights into force transmission and joint stability. Despite advancements, challenges remain in validating computational models due to limited experimental data and the complexity of biological tissues. However, advances in imaging, material property modelling, and personalized simulations are enhancing model reliability. As computational power grows, these models will refine clinical interventions, optimize rehabilitation strategies, and improve implant and prosthetic designs, making them indispensable in musculoskeletal research and healthcare.

2.3.1 Introduction

Computational modeling is essentially a form of mathematical modeling that leverages computer simulations, requiring significant computational power to investigate the underlying mechanisms of intricate biomechanical systems. Given the complexity of the musculoskeletal (MSK) system, analytical solutions are often impractical. This

*__Corresponding author: Masud Rana__, Department of Aerospace Engineering and Applied Mechanics, Indian Institute of Engineering Science and Technology Shibpur, Howrah 711103, India, e-mail: masud.me50@gmail.com
Jayanta Kumar Biswas, Department of Mechatronics, National Institute of Technology Patna, Ashok Rajpath, Mahendru, Patna 800005, Bihar, India, e-mail: jayanta.mae@nitp.ac.in

https://doi.org/10.1515/9783111386027-008

modeling approach encompasses virtual modeling and computer simulations, and it has gained popularity, thanks to substantial improvements in software and hardware capabilities. By integrating computational modeling with experimental studies, researchers can create the necessary environments for input while also validating outputs quantitatively. This approach is particularly effective for tackling complex biological challenges. In the context of MSK systems, computational modeling primarily employs multibody dynamic analysis to assess motion and loading at the skeletal level, along with solid mechanics analysis to evaluate stresses and strains at the tissue level using the finite element method (FEM) [1,15,19]. Furthermore, computational fluid dynamics (CFD) is useful for studying fluid movement within joint spaces [2]. While experimental studies can be resource-intensive and time-consuming, computational modeling – coupled with limited experimental testing – provides valuable insights, enables parametric screening, facilitates optimization, and guides design for future experiments.

A generic computational approach is also a vast computational study that has validated different approaches with various applications. In this context, the physical problem is often simplified so that it can be described and analyzed mathematically. This may include material properties, the geometry of the MSK system, and physiological boundary conditions. The governing equations are established for numerical analysis, often using commercial software. It is ensured that this method is acceptable for verification and validation through physical means. Even though the accuracy of verifying and validating the computational model might raise some doubts, there are practical ways to establish the model's reliability. A simple way to check is by looking at factors like the number and quality of nodes and elements, mesh size, solver tolerance, sensitivity, and time steps for nonlinear problems [3].

2.3.2 Computational modeling of bone and joint

The study of computational modeling of bone is a rapidly growing area that uses sophisticated computer simulations to gain insights into how bone tissue behaves under different conditions. These models play a key role in designing and improving scaffolds for bone tissue engineering, which help support bone regeneration and healing. By applying techniques like FEMs and CFD, researchers can predict how mechanical forces and fluid movements behave within these scaffolds. Advanced models have made it possible to simulate the complex process of bone fracture healing, shedding light on the biological and mechanical factors that drive recovery. This understanding can lead to better treatment strategies for patients [4]. Constitutive modeling, which involves creating mathematical representations of bone tissue's properties, has also been a focus of research [17 1989; 5]. Recent developments include models that address microcracks in cortical bone and phase-field models for understanding trabecular bone behavior [6]. Research shows that fracture risk in bone is not determined by bone mineral density alone but also depends on other factors such as bone strength, shape, structure, and material properties [7, 8]. Modern imaging techniques can now produce detailed 3D models of the skeletal and MSK systems. These advances allow scientists to build more accurate computational models to predict bone strength and fracture risk [9].The internal structure of the bone provides mechanical and biomechanical support for the skeletal system, allowing for the transport of nutrients, providing sites for cellular activities [10]. The mechanical environment of bone microstructure regulates and controls cellular behavior, which can be influenced by altering the microstructure. Thus, in biomedical applications, the mechanisms of bone microstructure play a crucial role. Finite element modeling of bone microstructure is essential for various biomedical applications, though there is still no clear protocol for developing computational models of bone microstructure.

Patient-specific computed tomography (CT) scan-based three-dimensional modeling can predict the microstructure and mechanical properties of bone. This three-dimensional model is also capable of nonlinear finite element analysis. Researchers have been using image processing software such as Mimics (Materialise Inc.) to construct three-dimensional finite element models and determine the mechanical properties of bone structures.

The biomechanical computational models such as bones, joints, muscles, tendons, and ligaments are used in computational analysis [17] and are given in Table 2.3.1.

Table 2.3.1: Body parts used in computational models.

Body element	Computational model element
Bone	Deformable body or rigid link
Joint	Standard joint (frictionless or with friction pair)
Muscle and tendon	Actuators (responsible for moving or controlling a mechanism or system)
Ligament	Controller springs

2.3.3 Biomechanics of key joints

2.3.3.1 Shoulder joint

The shoulder joint consists of four joints: the sternoclavicular joint, acromioclavicular joint, scapulothoracic joint, and glenohumeral joint. However, the glenohumeral joint generally refers to the shoulder joint. Several muscles surround these joints, protecting them and controlling their movement. Common shoulder muscles include the deltoid, trapezius, pectoralis major, and sternocleidomastoid.

2.3.3.2 Elbow joint

The elbow joint is a fascinating and complex structure that plays a crucial role in the movement of the upper limb. It is a synovial hinge joint that connects the upper arm to the forearm, involving three key bones: the humerus, radius, and ulna. These bones come together to form two main joints. The humeroulnar joint is found between the trochlea of the humerus and the trochlear notch of the ulna, while the humeroradial joint sits between the capitulum of the humerus and the head of the radius. Several ligaments, including the ulnar collateral ligament, radial collateral ligament, and annular ligament, help keep the elbow joint stable. The elbow's main movements include bending and straightening (flexion and extension) as well as turning the forearm up and down (supination and pronation).

2.3.3.3 Wrist joint

The wrist joint, also called the radiocarpal joint, is a complex and crucial connection between the forearm and the hand. It offers a broad range of movements: you can bend and straighten your wrist, move it side to side, and even rotate it. The wrist joint is a condyloid-type synovial joint, formed where the end of the radius meets the first row of carpal bones (except the pisiform bone). The main bones involved here are the radius, scaphoid, lunate, and triquetrum. Stability comes from several ligaments, including the palmar and dorsal radiocarpal ligaments, and the ulnar and radial collateral ligaments.

2.3.3.4 Hip joint

The hip joint, or acetabulofemoral joint, is a strong ball-and-socket joint that connects the thigh bone (femur) to the hip bone (pelvis). It is essential for bearing the body's weight, whether you are standing still or moving around. The surfaces of the femoral head and the acetabulum are coated with articular cartilage, which reduces friction and helps absorb shock during movement. This joint gets its strength and stability from ligaments like the iliofemoral, pubofemoral, and ischiofemoral ligaments. Built for stability and weight-bearing, the hip plays a critical role in balance and in supporting the upper body, making it one of the most durable joints we have.

2.3.3.5 Knee joint

The knee is the body's largest joint and is essential for both movement and stability. It works primarily as a hinge joint, allowing you to bend (flex) and straighten (extend) your leg. The knee is made up of two main parts: the tibiofemoral joint (between the tibia and femur) and the patellofemoral joint (between the patella and femur). The knee's stability comes from various ligaments and two C-shaped pieces of cartilage called the medial and lateral menisci, which act as shock absorbers. Along with flexion and extension, the knee also allows limited medial and lateral rotation.

2.3.3.6 Ankle joint

The ankle joint, also known as the talocrural joint, is a complex synovial hinge joint that connects the bones of the leg (tibia and fibula) with the talus of the foot. The bones involved are: (a) the tibia, the shinbone, which forms the medial malleolus; (b) the fibula, the smaller bone of the lower leg, forming the lateral malleolus; and (c) the talus, the top bone of the foot that fits into the socket formed by the tibia and fibula. The ankle joint is stabilized by two main ligaments: (a) the medial (deltoid) ligament, a strong, triangular ligament that resists over-eversion of the foot, and (b) the lateral ligaments, which include the anterior talofibular, posterior talofibular, and calcaneofibular ligaments, resisting over-inversion. The ankle's movements include dorsiflexion (lifting the foot upward) and plantar flexion (pointing the foot downward). The ankle joint also allows for some degree of pronation and supination in conjunction with the subtalar and midtarsal joints. The primary function of the ankle joint is to enable the movements necessary for walking, running, and jumping. It acts as a shock absorber during the initial phase of gait when the heel strikes the ground.

2.3.3.7 Spinal column

The spine, also known as the vertebral column, is a crucial part of the human skeletal system. It consists of 33 vertebrae, which are small bones stacked on top of each other, separated by intervertebral discs. These vertebrae are categorized into five regions: (a) Cervical vertebrae (C1–C7) are located in the neck, and they support the head and allow for its movement; (b) thoracic vertebrae (T1–T12) are connected to the ribs and form the upper back; (c) lumbar vertebrae (L1–L5) are found in the lower back, and bear much of the body's weight; (d) sacrum (five fused) are fused together and connect the spine to the pelvis; and (e) coccyx (four fused), also known as the tailbone, are the fused vertebrae at the very bottom of the spine. It serves as an attachment point for muscles, tendons, and ligaments. Each vertebra has a similar structure, consisting of a vertebral body (the weight-bearing part) and a vertebral arch (which forms the vertebral foramen, a canal that houses the spinal cord). The vertebral arch has several bony projections, including spinous and transverse processes, which serve as attachment points for muscles and ligaments. The important spinal functions are: protection (it encloses and protects the spinal cord, which is a vital part of the nervous system); support (it supports the weight of the body above the pelvis); and movement (it allows for flexibility and movement, including bending and twisting).

2.3.4 Understanding computational models in musculoskeletal biomechanics

Computational modeling helps us understand how our muscles, tendons, and ligaments work together to support movement and maintain stability. Fundamental muscle models reveal the basic mechanics of how muscles contract and generate force. One of the most well-known theories, the sliding filament theory, introduced by Huxley and Hanson in 1954, explains how muscle fibers shorten and produce force by actin and myosin filaments sliding past each other. This forms the foundation of our understanding of muscle contraction. A key part of muscle modeling is the force-length relationship, which explains how muscle force varies with muscle length. Gordon et al. [11] showed that muscles generate their peak force at an optimal length, where the overlap of actin and myosin is greatest. Deviations from this optimal length reduce the force output because there is either too little overlap (if the muscle is too long) or excessive filament crowding (if too short).

2.3.4.1 Applied muscle models: real-world use

While fundamental models address the basics, applied models are used for real-world scenarios like sports performance and medical applications. One widely used model, Hill's phenomenological model from 1938, simplifies the muscle behavior to make it computationally accessible. It uses a "contractile element" to represent force production and additional elastic elements for muscle and tendon elasticity. Researchers have extended Hill-type models to simulate entire MSK systems, aiding in the study of movement biomechanics, surgical outcomes, and athletic performance.

2.3.4.2 Tendon and ligament mechanics

Tendons and ligaments play key roles in force transfer and joint stability [22]. Tendons link muscles to bones, enabling movement, while ligaments connect bones to each other, providing joint stability [21]. Both are primarily composed of collagen, a protein that gives them tensile strength. Their stress-strain behavior is nonlinear: tendons initially stretch easily, then become stiffer under increased tension – a feature critical in activities like running and jumping, where tendons store and release energy efficiently [14]. Ligaments, by contrast, are more elastic to stabilize joints against multidirectional forces. Computational models need to reflect these unique characteristics to accurately simulate joint behavior and injury risks.

2.3.4.3 Mechanical properties of tendons, ligaments, and aponeuroses

The mechanical properties of these structures depend on their collagen composition and fiber arrangement. Tendons are generally stiffer than ligaments due to a higher collagen concentration. Certain tendons, like the Achilles, store energy during muscle contraction and release it for propulsion, which is essential in high-intensity activities. Aponeuroses, broader and flatter structures than tendons, distribute muscle forces over larger areas and help stabilize joints, especially for muscles involved in posture [12]. They also stretch in multiple directions under tension, contributing to joint stability.

2.3.4.4 Challenges in model validation

Computational models have become incredibly useful for understanding the mechanics of the human body, but they still have hurdles to clear, especially when it comes to validation. One of the biggest challenges is the scarcity of high-quality experimental data. Consider muscle-tendon interactions, for example [13,18]. When muscles contract in non-steady-state situations – like when they produce extra force after being stretched or less force after being shortened – capturing these behaviors accurately in models is tricky.

A particularly fascinating area of research revolves around titin, a giant protein that plays a key role in muscle elasticity and generating passive force. When a muscle is activated, titin becomes stiffer and adds extra force during stretching movements. This feature is essential in activities such as running, jumping, or any dynamic motion, and adding titin dynamics to models could significantly enhance predictions of muscle behavior [16,20].

2.3.5 Conclusion

In short, computational modeling is now indispensable for exploring the MSK system. These models allow us to tackle complex biomechanical questions that are hard to address with traditional experiments alone. By simulating the interactions of bones, joints, muscles, tendons, and ligaments, researchers can better understand how our bodies work under various conditions, which in turn helps improve medical treatments, rehabilitation strategies, and biomechanics research.

Techniques such as multibody dynamics, finite element analysis, and CFD allow scientists study everything from motion and stress distribution to how fluids flow through MSK tissues. Although these models are efficient and cost-effective, they still have limitations. Biological systems are incredibly complex, and our experimental data is not always up to par. Even so, advances in imaging, better modeling of material properties, and personalized simulations are steadily boosting the accuracy of these tools. As computational power continues to grow, these models will be more essential than ever, helping us refine clinical interventions, prevent injuries, and design effective treatments for MSK conditions.

References

[1] Abdullah M, Hulleck AA, Katmah R, Khalaf K, El-Rich M. Multibody dynamics-based musculoskeletal modeling for gait analysis: A systematic review. Journal of NeuroEngineering and Rehabilitation. 2024; 21(1): 178.

[2] Huning AJ, Chandrasekaran S, Garimella S. A review of recent advances in HTGR CFD and thermal fluid analysis. Nuclear Engineering and Design. 2021; 373: 111013.

[3] Xing W, Song C, Tin-Loi F. A scaled boundary finite element based node-to-node scheme for 2D frictional contact problems. Computer Methods in Applied Mechanics and Engineering. 2018; 333: 114–146.

[4] Sabet FA, Raelsi Najafi A, Hamed E, Jasiuk I. Modelling of bone fracture and strength at different length scales: A review. Interface Focus. 2016; 6(1): 20150055.

[5] Zannoni C, Mantovani R, Viceconti M. Material properties assignment to finite element models of bone structures: A new method. Medical Engineering & Physics. 1999; 20(10): 735–740.

[6] Maghami E, Moore JP, Josephson TO, Najafi AR. Damage analysis of human cortical bone under compressive and tensile loadings. Computer Methods in Biomechanics and Biomedical Engineering. 2022; 25(3): 342–357.

[7] Mccreadie BR, Goldstein SA. Biomechanics of fracture: is bone mineral density sufficient to assess risk. Journal of Bone and Mineral Research. 2000; 15(12): 2305–2308.

[8] Van der Meulen MCH, Jepsen KJ, Mikić B. Understanding bone strength: Size isn't everything. Bone. 2001; 29(2): 101–104.

[9] Luo Y. On challenges in clinical assessment of hip fracture risk using image-based biomechanical modelling: A critical review. Journal of Bone and Mineral Metabolism. 2021; 39: 523–533.

[10] Burr DB, Allen MR Eds. Basic and Applied Bone Biology, USA: Academic Press; 2019.

[11] Gordon AM, Huxley AF, Julian FJ. The variation in isometric tension with sarcomere length in vertebrate muscle fibres. Journal of Physiology. 1966; 184: 170–192.

[12] Azizi E, Roberts TJ. Biaxial strain and variable stiffness in aponeuroses. Journal of Physiology. 2009; 587: 4309–4318.

[13] Biewener AA. Muscle–tendon stresses and elastic energy storage during locomotion in the horse. Comparative Biochemistry and Physiology, Part B: Biochemistry and Molecular Biology. 1998; 120(1): 73–87.

[14] Bobbert MF. Dependence of human squat jump performance on the series elastic compliance of the triceps surae: A simulation study. Journal of Experimental Biology. 2001; 204: 533–542.

[15] Delp SL, Loan JP. A computational framework for simulation and visualization of musculoskeletal models. IEEE Computing in Science and Engineering. 2000; 2(5): 46–55.

[16] Hill AV (1938). The heat of shortening and the dynamic constants of muscle. *Proceedings of the Royal Society B: Biological Sciences*, 126, 136–195.
[17] Natali AN, Meroi EA. A review of the biomechanical properties of bone as a material. Journal of Biomedical Engineering. 1989; 11(4): 266–276.
[18] Neptune RR, McGowan CP, Fiolkowski P. The influence of muscle-tendon mechanics on running economy. Journal of Biomechanics. 2009; 42(10): 1658–1664.
[19] Rajagopal A. et al., Full-body musculoskeletal model for muscle-driven simulation of human gait. IEEE Transactions on Biomedical Engineering. 2016; 63(10): 2068–2079.
[20] Rode C, Siebert T, Blickhan R. Titin-induced force enhancement and force depression: A 'sticky–spring' mechanism in muscle contractions?. Journal of Theoretical Biology. 2009; 259(2): 350–360.
[21] Thorpe CT. et al., Tendon structure–function relationships in health and disease. Bone Joint Research. 2016; 5(10): 372–380.
[22] Zitnay JL, Weiss JA. Load transfer and matrix damage in tendons and ligaments: Experimental and computational approaches. Journal of Orthopaedic Research: Official Publication of the Orthopaedic Research Society. 2018; 36(10): 2769–2779.

Pechimuthu Susai Manickam, S. Balamurugan, and Sandipan Roy*

Chapter 2.4
Computational modeling of bone, muscles, soft tissues, and ligaments

Abstract: The finite element method (FEM) is a computer simulation technique that was first applied in civil engineering in the early 1940s. Orthopedic surgery and other medical specialties have also made use of FEM applications. Over time, advancements in computing technology have made it possible to analyze increasingly complicated issues, such as those pertaining to the spine. Computational modeling of the spine is a multidisciplinary field that combines biomechanics, anatomy, computer science, and engineering to simulate and analyze the behavior of the spine. Understanding how the spine contributes to stability and mobility can be aided by these models. These computational models have advanced to include intricate relationships inside the spine, including the distribution of load among vertebral elements and the behavior of the intervertebral disc under different conditions. With the use of this modeling, improved surgical implants and methods can be created, enabling patient-specific treatments. Furthermore, the computer modeling assists in understanding how the spine will react to various physical therapies and activities. The FEM advancements help improve rehabilitation strategies.

Keywords: Bone model, FEM, FEA, ligament, cartilage

2.4.1 Introduction

The complex structure known as the vertebral column, or spine, serves three primary purposes: (i) transferring weights and the bending moments that result from lifting the head, trunk, and any other weights to the pelvis; (ii) allowing physiologic motion between the body parts; and (iii) shielding the spinal cord from harmful forces or mo-

***Corresponding authors: Sandipan Roy,** Department of Mechanical Engineering, College of Engineering and Technology, SRM Institute of Science and Technology, Potheri, Kattankulathur, Chengalpattu 603203, Tamil Nadu, India, e-mail: sandipag@srmist.edu.in
Pechimuthu Susai Manickam, Department of Mechanical Engineering, College of Engineering and Technology, SRM Institute of Science and Technology, Potheri, Kattankulathur, Chengalpattu, 603203 Tamil Nadu, India, e-mail: susaimap@srmist.edu.in
S. Balamurugan, Department of Mechanical Engineering, College of Engineering and Technology, SRM Institute of Science and Technology, Potheri, Kattankulathur, Chengalpattu 603203, Tamil Nadu, India, e-mail: balamurs3@srmist.edu.in

https://doi.org/10.1515/9783111386027-009

tions brought on by trauma or physiologic movements [1, 2]. The 33 bony components called vertebrae that make up the spine are normally separated into 5 zones, ranging from the head to the pelvis [2].

Over the years, research on spine biomechanics has changed [3]. Given the complexity and difficulty of doing in vivo research on human spinal columns, computational models offer an essential tool. There are numerous anatomical components involved in the spine's incredibly complex biomechanics. These elements are modeled independently in computational simulations, with each one being linked to particular modeling attributes based on its constitutive model relationships and biomechanical response. The modeling of bone components, intervertebral discs, facet joints, the seven major spinal ligaments, and the cartilaginous endplates is necessary for a thorough finite element (FE) study of the spine [4, 5].

2.4.2 Computational modeling of bone

The architecture inside the bone is a microstructure. The bones provide the mechanical support for the entire human skeleton. The main function of the bone is to transport the nutrients. The bone is classified into cortical bone and the cancellous bone. The vertebra is the part of the human body in the human spine. The bone is a composition of minerals, which have 65% inorganic matrix and 35% organic matrix. In the bone, 65% is made up of hydroxyapatite and the remaining is water cells and mixture of collagen [5]. The type of the bones in the human body is classified as cortical bone, which is a compact bone, and the cancellous bone is also known as trabecular bone.

The outer surface of the bone is called as cortical bone, which is a dense solid material. On the other hand, the cancellous bone is a porous material that is made up of platen structures known as trabeculae. The cancellous bone is filled with a fluid mixture of marrow and blood [6].

In FE modeling, the anatomical components are as follows: the vertebral body is made up of cortical bone and cancellous bone. The material is modeled as isotropic elastic, transversely elastic, and poroelastic elastic behavior [7].

2.4.3 Computational modeling of soft tissues

The intervertebral disc acts as a shock absorber. When a compressive load is applied, the hydrostatic pressure will be increased in the nucleus pulposus. In the bending loads, the nucleus pulposus acts as a pivot for vertebral body to rotate around [8, 9]. The pressure acting on the intervertebral disc will cause the radial force, which acts on the annulus fibrosus and the bulge in the disc. The bulging seen in the axial tension was carried out by the annulus fibrosus, which supports the tensile load because nucleus pulposus behaves like liquid [10].

In the human spine, the intervertebral disc is called soft tissues. The annulus fibrosus, nucleus pulposus, and cartilaginous endplates are the three separate parts that make up the intervertebral disc. The intervertebral disc is surrounded by nucleus pulposus and the outer component is called annulus fibrosus. The combination of the annulus fibrosus and the nucleus pulposus determines the physiological behavior of the intervertebral disc. The hydrostatic pressure rises in the nucleus of a disc when a compressive stress is applied. Cartilage, tendons, and intervertebral discs are examples of soft tissues that are essential for load absorption and smooth joint motion [11].

Complex constitutive models are frequently used in computational models of soft tissues to explain their time-dependent, nonlinear, and anisotropic behavior [12]. In FE modeling, the anatomical components are as follows: In the intervertebral disc, the components divided into annular ground substance which exhibits hyperelastic behavior, the nucleus pulposus that exhibits isotropic elastic, incompressible fluid, empty cavity, and hyperelastic behavior and the collagen fibers exhibit isotropic elastic nonlinear behavior [12, 13].

2.4.4 Computational modeling of ligaments

The fibrous fibers that resemble ribbons which unite the bone elements to form joints are called ligaments. The ligaments will not support the compressive loads. It is effective only for the transmission of uniaxial tensile loading [14]. The anterior longitudinal ligament resists the extension and translation motion. The posterior longitudinal ligament activates during flexion and prevents translation motion. The ligamenta flava prevent the flexion. The interspinous ligament is narrow and exhibits weak tensile properties. It will resist the flexion. The capsular ligament is found in the articular processes of the adjacent vertebrae and a joint is formed. The major role of the facet joint will resist the axial rotation and the lateral bending. The capsular ligament limits the motion in the flexion and rotation. In the FE model, the ligaments were modeled on isotropic elastic, isotropic varying on strain rate, nonlinear behavior, and stress-strain data curve [15].

2.4.5 Computational modeling of muscles

The muscles in the human body are classified into heart, skeletal, and smooth. The majority of the body muscles are made up of skeletal muscles, which are responsible for producing force and movement [16]. The intricate network of skeletal muscles in the human neck is necessary to provide stability in an afforded posture, movement generating during physiological endeavor, and safeguarding the spine during damage [9].

A main active purpose of muscles is to generate a contractile force that shortens the muscle's length. Since muscles cannot extend on their own, the activation of the additional muscles is required to stretch the muscle. Pairs of muscles that together to relax or contract are known as antagonistic pairs. Additionally, there is always at least one antagonistic muscle in the human body in every muscle.

The cross-sectional region of a muscle determines the maximal force produced during contraction [17]. A muscle with a wide cross-sectional area can contract more muscle fibers than one with a small cross section. The physiological cross-sectional area is its effective area, measured midway between the insertion and origin. The cervical spine's muscles are reflected in the medial plane, and each muscle's action relies on whether it contracts bilaterally or unilaterally. The muscle contracts either bilaterally or unilaterally based on its place within the neck region. For instance, when a muscle pair located posteriorly and laterally of the cervical spine is contracted bilaterally, this will result in cervical spine extension, and when a muscle pair is contracted unilaterally, this will result in cervical spine rotation.

In the human body, the muscle tissue is complicated because of the muscle mechanics. The properties of muscle tissue differ based on the contraction state. Therefore, there are two components to muscle response: an active part and a passive part. The muscle contraction in microstructural mechanics level involves microfibers; furthermore, sarcomeres are not covered in this section. There will be no discussion of the intricate electrochemical mechanism involving ions and axons to cause muscle contraction.

The typical Hill-type muscle model is used in the model to represent the cervical spine muscles as 1D discrete elements. A common phenomenological model for explaining active-passive muscle behavior is the Hill-type muscle model. Since many scenarios involving neck impacts entail active muscular behavior, this model is an excellent fit for the cervical spine muscles. Additionally, since the neck is not frequently directly impacted, bulk muscle modeling is not necessary. Thus, there is a lack of data in the literature concerning muscle properties that can be associated with the Hill-type muscle model [18–21]. A contractile element (CE) and a parallel elastic element consisted of the Hill muscle model. A series elastic element is frequently, but not always, included for tendon compliance. The CE generates the active force in the muscle and is dependent on the muscle length, velocity, and activation state, whereas the parallel element is based on passive muscle properties and depends only on the current length of the muscle.

2.4.6 Methods for validating the finite element bone models

The validation of the computational FE modeling is a very crucial step [21–24]. In FE models, the subject-specific models can be validated using the in vitro and in vivo testing. The major challenges are the age, gender, and body weight variations, which are barriers for validation. The validation of in vivo is also recommended for improving the accuracy of the results. In the vertebrae, the division of the cortical bone and the cancellous bone is a very important parameter [25–27]. The choice of the element is very important. In the mathematical modeling of the human spine, we took FE study for our considerations. The FE analysis (FEA) is a very powerful tool, which helps us to analyze the complex biological systems. The process of determining whether simulation results accurately represent real-world outcomes is known as validation in FE method (FEM). Validation is the process of determining the extent to which a model accurately predicts the real-world solution. The process of verifying that the FEA was carried out correctly is known as verification. The methods of validation and verification gather proof that an FE model is accurate and correct for a given situation.

2.4.7 Conclusion

In the field of orthopedics, FEM has grown to be a potent tool that gives surgeons a better understanding of bone biomechanics in both normal and abnormal conditions. FEM is a very promising technique in computational modeling of bones, muscles, soft tissues, and the ligaments. FEM has a lot of promise to improve musculoskeletal health outcomes and advance individualized treatments as it develops further. By creating realistic scenarios, this potent tool aids in clinical decision-making and promotes safer, more successful surgical and therapeutic procedures.

References

[1] DeSai C, Reddy V, Agarwal A. Anatomy, Back, Vertebral Column. In: StatPearls. Treasure Island (FL): StatPearls Publishing; 2018.

[2] Panzer M. Numerical Modelling of the Human Cervical Spine in Frontal Impact, Master's thesis, Waterloo: University of Waterloo; 2006, https://uwspace.uwaterloo.ca/handle/10012/2865.

[3] Panzer MB, Cronin DS. C4–C5 segment finite element model development, validation, and load-sharing investigation. Journal of Biomechanics. 2009; 42(4): 480–490.

[4] Goel VK, Monroe BT, Gilbertson LG, Brinckmann P. Interlaminar shear stresses and laminae separation in a finite element model of the lumbar disc. Spine. 1995; 20(6): 689–698.

[5] Cowin SC. Bone Mechanics Handbook, 2nd Ed., Boca Raton, Florida: CRC Press LLC; 2001.
[6] Carter DR, Hayes WC. The compressive behavior of bone as a two-phase porous structure. JBJS. 1977; 59(7): 954–962.
[7] Manickam PS, Roy S. The biomechanical study of cervical spine: A finite element analysis. The International Journal of Artificial Organs. 2022; 45(1): 89–95.
[8] Panjabi MM, White AA III. Basic biomechanics of the spine. Neurosurgery. 1980; 7(1): 76–93.
[9] White AA III and Panjabi MM. Clinical biomechanics of the spine. Clinical Biomechanics of the Spine. Philadelphia, New York: Lippincott-Raven publisher; 1990.
[10] Gilad I, Nissan M. A study of vertebra and disc geometric relations of the human cervical and lumbar spine. Spine. 1986; 11(2): 154–157.
[11] Sciortino V, Pasta S, Ingrassia T, Cerniglia D. On the finite element modeling of the lumbar spine: A schematic review. Applied Sciences. 2023; 13(2): 958.
[12] Iatridis JC, Setton LA, Foster RJ, Rawlins BA, Weidenbaum M. Degeneration affects the anisotropic and nonlinear behaviors of human annulus fibrosus in compression. Journal of Biomechanics. 1998; 31(6): 535–544.
[13] Urban JPG, Roberts S. Degeneration of the intervertebral disc. Arthritis Research and Therapy. 2003; 5(3): 120–130.
[14] Yoganandan N, Kumaresan S, Pintar FA. Biomechanics of the cervical spine Part 2. Cervical spine soft tissue responses and biomechanical modeling. Clinical Biomechanics. 2001; 16(1): 1–27.
[15] Belytschko T, Liu WK, Moran B, Elkhodary K. Nonlinear Finite Elements for Continua and Structures, Hoboken, NJ, USA: John Wiley & Sons; 2014.
[16] Fung YC. Biomechanics: Mechanical Properties of Living Tissue, 2nd Ed., New York, NY: Springer; 1993.
[17] Winters JM, Stark L. Estimated mechanical properties of synergistic muscles involved in movements of a variety of human joints. Journal of Biomechanics. 1988; 21(12): 1027–1041.
[18] Winters JM, Stark L. Analysis of fundamental human movement patterns through the use of in-depth antagonistic muscle models. IEEE Transactions on Biomedical Engineering. 1985; 10: 826–839.
[19] Winters JM, Stark L. Estimated mechanical properties of synergistic muscles involved in movements of a variety of human joints. Journal of Biomechanics. 1988; 21(12): 1027–1041.
[20] Belytschko T, Liu WK, Moran B & Elkhodary K. Nonlinear Finite Elements for Continua and Structures, John Wiley & Sons; Hoboken, NJ, USA, 2014.
[21] Winter DA. Human balance and posture control during standing and walking. Gait & Posture. 1995; 3(4): 193–214.
[22] Manickam PS, Roy S. The biomechanical effects of S-type dynamic cage using Ti and PEEK for ACDF surgery on cervical spine varying loads. The International Journal of Artificial Organs. 2021; 44(10): 748–755.
[23] Manickam PS, Ghosh G, Shetty GM, Chowdhury AR, Roy S. Biomechanical analysis of the novel S-type dynamic cage by implementation of teaching learning based optimization algorithm – an experimental and finite element study. Medical Engineering & Physics. 2023; 112: 103955.
[24] Manickam PS, Roy S, Shetty GM. Biomechanical evaluation of a novel S-type, dynamic zero-profile cage design for anterior cervical discectomy and fusion with variations in bone graft shape: A finite element analysis. World Neurosurgery. 2021; 154: e199–e214.
[25] Manickam PS, Ghosh G, Roy S. Optimization of bone graft shapes of s-type cervical cage through genetic algorithm. International Journal for Multiscale Computational Engineering. 2022; 20 (1). 55–68.

[26] Kulkarni AG, Kumar P, Shetty GM, Roy S, Manickam PS, Dhason R, Adbalwad YM. Finite element analysis comparing the biomechanical parameters in multilevel posterior cervical instrumentation model involving lateral mass screw versus transpedicular screw fixation at the C7 vertebra. Asian Spine Journal. 2024; 18(2): 163.
[27] Balamurugan S, Chhatoi RK, Dixit V, Manickam PS. Mechanical response of trapezoidal cage on cervical 6–cervical 7 level. Journal of Engineering and Science in Medical Diagnostics and Therapy. 2023; 6(3): 55–68.

Alireza Feizkhah

Chapter 2.5 Computational modeling of articular cartilage and cell mechanics

Abstract: A special connective tissue called articular cartilage plays a crucial role in synovial joints, including those in the knees, hips, and shoulders. Articular cartilage engineering as well as cell mechanics engineering has both been benefited greatly from computer modelling, which is discussed in this chapter. After that, the articular cartilage's mechanical and biological characteristics, as well as its effect on the joint's overall function. After that, several computational methods that have been useful in predicting how cartilage will respond, both in the lab and in the clinic have been discussed. Lastly, the remaining problems have been summarized which needs to be overcome. The future prospects for computational modelling in cartilage research are discussed. Computational models will drive the development of more effective therapeutics for the treatment of joint illnesses, leading to improved patient outcomes. In this chapter, the ideas from biomechanics, cell biology, and computational science come together in a unified front.

Keywords: Articular cartilage, Cell mechanics, Extracellular matrix, Mechanotransduction, Finite element analysis

2.5.1 Introduction

Articular cartilage is a special connective tissue that is important for synovial joints, such as knee, hip, and shoulder joints. It provides a low-friction, load-bearing surface that allows free motion of the joint and, at the same time, is able to resist repeatedly generated compressive loads during daily function. The unique biomechanical properties of articular cartilage emanate from its structural composition; for example, a well-structured extracellular matrix (ECM) and a low density of chondrocytes. These elements work in harmony to ensure the mechanical strength and resilience of the tissue [1].

Alireza Feizkhah, Department of Medical Physics and Biomedical Engineering, Tehran University of Medical Sciences, Tehran, Iran; Biomedical Engineering and Bioinspired Technologies Research Center (BTRC), Sina Institute for Bioengineering Science and Technology, Rasht, Iran,
e-mail: Alireza.feizkhah@gmail.com

https://doi.org/10.1515/9783111386027-010

Though the ability to resist wear is relatively high, articular cartilage tends to get injured and degenerated easily due to features that are avascularity, aneural, and alymphatic in nature. Diseases like osteoarthritis (OA) – the principal cause of disability – result from the progressive breakage of this tissue, leading to pain, limited movements, and, subsequently, reduction of quality of life. This area of research has become very important with the realization that it is not able to efficiently self-heal once it has been subjected to injury because of the mechanics and biology associated with it. This knowledge is important for the design of new therapies such as tissue engineering, regenerative medicine, and advanced surgical techniques [2].

One of the strongest arms in this area of effort is computational modeling. Computational models richly illustrate the cross talk between mechanical forces and the cellular response and matrix composition through simulation of the behavior of articular cartilage under various sets of physiological conditions. These models make possible the study of advancing degeneration of cartilage, the effectiveness of potential treatments, and the optimization of designs for engineered tissue of cartilage. Computational approaches, including FEA, multiscale modeling, and agent-based simulations, are increasingly becoming indispensable tools in basic research and clinical use [3].

This chapter details how computational modeling has played a leading role in the progress of articular cartilage and cell mechanics. This will be followed by a discussion of the biological and mechanical properties of the articular cartilage and how the latter contribute to the functions of the tissue in the joint. We will then describe computational techniques applied in the prediction of cartilage behavior and their applications to research and clinical practice. Finally, the remaining challenges that must be surmounted and future perspectives for computational modeling in cartilage research are outlined. This is the line in which biomechanics, cell biology, and computational science bring insights together into a unified front about how computational models will power development for more effective therapies in the treatment of joint diseases to provide improved patient outcomes.

2.5.2 Biological and mechanical properties of articular cartilage

Articular cartilage is a highly specialized tissue that plays a vital role in the normal functioning of synovial joints. Its unique composition and mechanical properties enable it to withstand repetitive mechanical loads and provide a smooth, frictionless surface for joint movement. Understanding these properties is crucial for developing effective treatments for cartilage-related diseases and injuries [4, 5].

2.5.2.1 Composition of articular cartilage

The mechanical properties of the articular cartilage are due to the sophisticated but well-organized structures in its ECM [6, 7]. Structurally, the ECM is made of three main constituents: water, collagen, and proteoglycans. The chondrocyte is the only type of cell in the cartilage, occupying less than 10% of tissue volume but actually being responsible for the synthesis and turnover of ECM [8].

- **Water:** Water is the most abundant component of articular cartilage content and makes up between 65% and 80% of its wet weight. It occupies the ECM. Water is critical to the load-bearing capacity of this tissue, through its interaction with the negatively charged GAG chains of proteoglycans, which generate osmotic pressure, maintaining the turgidity of the tissue and hence resistance to compressive loads [7].
- **Collagen:** Type II collagen within the tissue gives a dense layout of fibrils, thus creating tensile strength in the tissue and providing structural integrity to it. The arrangement of the fibrils is different for the superficial and deep zones of the cartilage:
  - **Superficial zone:** In this zone, fibrils of collagen are all oriented parallel to the articulating surface, helping resist shear and providing a smooth, lubricated surface for motion.
  - **Middle zone:** In the middle zone, the collagen fibrils are less oriented, contributing to its character of absorbing compressive forces.
  - **Deep zone:** Within the deep zone, collagen fibrils run perpendicular to the joint surface, attaching the cartilage to the underlying bone and providing structural support [9].
- **Proteoglycan:** Proteoglycan molecules, such as aggrecan, contain a core protein linked to GAG chains like chondroitin sulfate and keratan sulfate, which attract water and help maintain the elasticity of cartilage tissues by resisting compression and preserving structural strength [10].
- **Chondrocytes:** Although they represent only a small fraction of the total tissue volume, chondrocytes maintain and regenerate the ECM. They synthesize collagen and other matrix proteins in the cartilage in response to mechanical and biochemical signals from their surrounding environment [11].

2.5.2.2 Mechanical behavior

Mechanical behavior in articular cartilage is governed chiefly by the interplay between its ECM components and the tissue's fluid. The sophisticated balance of these elements allows cartilage to effectively bear loads and perform lubricating functions [7, 12].

- **Viscoelasticity:** Articular cartilage exhibits viscoelastic behavior, that is, its response to mechanical loading is both time-dependent and strain-dependent. This

property is crucial for the tissue's ability to absorb shocks and distribute loads across the joint surface. Under constant load, the cartilage initially deforms rapidly, then continues to deform at a slower rate. When the load is removed, the tissue gradually returns to its original shape [13].
- **Compression resistance:** The high water content and the proteoglycan–collagen network in the ECM provide most of the tissue's resistance to compressive forces. When compression is applied, water is squeezed out from the ECM, while collagen fibrils and proteoglycans resist further deformation. The osmotic pressure developed by GAG chains within proteoglycans helps restore the tissue's shape once the load is removed.
- **Shear resistance:** The superficial layer of articular cartilage is specifically adapted to resist shear forces, which are prevalent during joint movement. The parallel arrangement of collagen fibrils in this zone, combined with the lubricating properties of synovial fluid, reduces friction and prevents surface wear [14].
- **Load distribution:** The unique zonal architecture of articular cartilage efficiently distributes mechanical loads across the tissue. The superficial zone handles initial shear forces, while the middle and deep zones absorb compressive forces and anchor the tissue to the subchondral bone. This hierarchical structure allows cartilage to withstand complex loading patterns encountered during joint movement [15].
- **Fluid flow and poromechanics:** Fluid flow within articular cartilage plays a key role in its mechanical behavior. Under load, fluids are expressed from the ECM, helping the tissue resist compression. This fluid flow also facilitates nutrient transport as articular cartilage is avascular and relies on synovial fluid for nourishment. The study of fluid interaction with the solid matrix in cartilage, known as poromechanics, is essential for understanding the tissue's mechanical responses under various loading conditions [16].

2.5.3 Cell mechanics in articular cartilage

Articular cartilage is a specialized connective tissue that covers the ends of bones in synovial joints. It thus provides a smooth, lubricated surface for articulation and enables loads to be transmitted with low friction. Central to its functionality and health are the chondrocytes: the resident cells embedded within this ECM. Viewed from the cellular point of view, understanding mechanics is quite critical in the clarification of mechanobiological processes involved in the maintenance, degeneration, and repair of cartilages [17].

2.5.3.1 Chondrocyte function and mechanobiology

In articular cartilage, chondrocytes are the only cells found in this avascular, aneural tissue. These cells are responsible for synthesizing the ECM and turning it over, including collagen, proteoglycans, and other glycoproteins. Chondrocyte activity is tightly regulated through mechanical loading, where a mechanical signal is transmitted through the ECM and sensed by cells – a process termed mechanotransduction – to influence gene expression, protein synthesis, and cell behavior [18].

2.5.3.2 Mechanotransduction pathways

Mechanotransduction in chondrocytes is mediated by a variety of mechanosensitive structures and pathways:

- **Integrins and focal adhesions:** Integrins are transmembrane receptors that connect the ECM to the cytoskeleton. They play a key role in sensing mechanical forces and initiating signaling cascades that influence chondrocyte behavior. Focal adhesions composed of integrins and other associated proteins act as mechanical linkages between the ECM and the cell, translating mechanical stimuli into biochemical signals [17].
- **Ion channels:** Mechanosensitive ion channels, such as stretch-activated channels, respond to mechanical deformation by altering ion fluxes across the cell membrane. These changes trigger downstream signaling events that affect cellular processes.
- **Primary cilium:** This single, hair-like organelle produced from the cell surface acts as a mechanosensor involved in sensing fluid flow and mechanical loading, which is essential for maintaining cartilage homeostasis.
- **Cytoskeleton:** The cytoskeleton, composed of actin filaments, microtubules, and intermediate filaments, provides structural support to the cell and plays a crucial role in mechanotransduction. Changes in cytoskeletal organization in response to mechanical loading can influence chondrocyte shape, proliferation, and matrix synthesis [19].

2.5.3.3 Microscale mechanics of chondrocytes

At the microscale, mechanics dictate cellular function in response to contact between chondrocytes and their surrounding ECM. The mechanical environment within cartilage is highly heterogeneous due to the anisotropy and heterogeneity of ECM composition, leading to local differences in mechanical properties. Chondrocytes experience a complex mechanical environment characterized by compressive forces, shear stresses, and hydrostatic pressures [3, 18].

2.5.3.4 Deformation and stress distribution

The deformation of chondrocytes under mechanical load is governed by the viscoelastic properties of both the cell and the ECM. When articular cartilage is subjected to compression, the ECM deforms and transmits stress to the embedded chondrocytes. The extent and nature of chondrocyte deformation depend on factors such as the stiffness of the surrounding matrix, the density of the pericellular matrix (PCM, a specialized ECM region surrounding the chondrocytes), and the overall mechanical properties of the cartilage tissue [20]:

- **PCM:** The PCM plays a critical role in modulating the mechanical environment of chondrocytes. It acts as a mechanical buffer, shielding the cell from excessive deformation while still allowing the transmission of mechanical signals. The properties of the PCM, including its composition and thickness, significantly influence how chondrocytes perceive mechanical loads.
- **Stress concentration:** Due to the heterogeneous nature of cartilage, stress concentration can occur around chondrocytes, particularly in regions where the ECM is less dense or more compliant. This localized stress can influence cellular behavior, potentially leading to changes in matrix synthesis or cell apoptosis in response to abnormal mechanical environments [21].

2.5.3.5 Cellular responses to mechanical stimuli

Chondrocytes respond to mechanical stimuli through a variety of mechanisms that influence their activity and contribute to cartilage homeostasis or degeneration:

- **Matrix synthesis:** Mechanical loading can stimulate chondrocytes to synthesize ECM components, enhancing the structural integrity of cartilage. However, excessive or abnormal loading can lead to altered matrix synthesis, contributing to cartilage degradation.
- **Cell proliferation and apoptosis:** The mechanical environment can influence chondrocyte proliferation and survival. For example, moderate loading may promote cell proliferation, while excessive mechanical stress can induce apoptosis, leading to cartilage thinning and degeneration.
- **Gene expression:** Mechanotransduction pathways modulate gene expression related to matrix production, inflammation, and other cellular processes. Cyclic loading upregulates anabolic gene expression, while static or abnormal loading induces catabolic gene expression associated with cartilage breakdown.

2.5.3.6 Implications for cartilage health and disease

The mechanical environment of chondrocytes is a key determinant of cartilage health. Abnormal mechanical loading – due to injury, obesity, or joint misalignment – disrupts the balance between matrix synthesis and degradation, leading to the development of degenerative conditions such as OA. Understanding mechanics at the cellular level provides insights into the pathogenesis of cartilage-related diseases and informs therapeutic strategies aimed at restoring normal mechanical function and promoting cartilage repair [22].

2.5.4 Computational modeling techniques

Advancement in computational modeling has revolutionized our comprehension of the material, biological behaviors associated with articular cartilage and their mechanical roles. Such models allow researchers to model complex biological and mechanical processes, yielding information that would be difficult or impossible to obtain experimentally. Computational models that mimic these intricate interplays between cellular activity, ECM properties, and mechanical forces represent a useful approach for the study of cartilage physiology as well as disease.

2.5.4.1 Finite element analysis (FEA)

Finite element analysis (FEA) is one of the most popular computational methods to investigate the mechanical behavior of articular cartilage. FEA is responsible for breaking down a complex structure into smaller parts that can be easily analyzed and tested regarding stress, strain, and deformation under different loading conditions.

Applications in cartilage research: FEA has been instrumental in understanding how different layers of cartilage respond to mechanical loads. By simulating the material properties of the ECM and the anisotropic nature of collagen fibers, FEA models can predict how cartilage deforms under compression, shear, and tension. These models are particularly valuable for studying the effects of injuries, such as meniscal tears or cartilage lesions, on joint mechanics. Additionally, FEA is used to evaluate the biomechanical impact of surgical interventions and the design of cartilage implants.

Challenges and considerations: FEA can provide detailed information on cartilage mechanics, but its output is only as accurate as the input data regarding material properties and boundary conditions, both of which are inherently difficult to measure due to viscous behavior under load. In addition, cartilage behaves as a viscoelastic and poroelastic material (it is time-dependent), considering the fluid effect on its deformations under mechanical solicitations. These complex behaviors, however, place a requirement for advanced algorithms and immense computational power in FEA models [23].

2.5.4.2 Multiscale modeling

Multiscale modeling is a method that combines different levels of biological organization, ranging from the molecular scale to the tissue level for an integrated and detailed understanding of cartilage mechanics. This approach is necessary to connect microscopic cellular processes with macroscopic tissue behavior.

Applications in cartilage research: Multiscale models are especially important for exploring the coupling of cellular activities, such as chondrocyte mechanotransduction and matrix synthesis with overall tissue mechanics. These models may, for example, simulate global cartilage stiffness and viscoelasticity as a function of changes in the properties that govern PCM or total collagen cross-linking at the molecular level. Such an integration, which indeed is possible by linking molecular dynamics to cellular mechanics with tissue-level biomechanics in a multiscale sense, will provide the global framework for function and degeneration of cartilage.

Challenges and considerations: The main challenge of multiscale modeling lies in integrating data from different scales, each of which may require distinct modeling techniques and assumptions. Accurately capturing the interactions between scales is crucial for the model's predictive power, but it also adds complexity to the modeling process. Despite these challenges, multiscale models are invaluable for understanding the hierarchical structure of cartilage and its role in joint health [24].

2.5.4.3 Biomechanical modeling

Biomechanical modeling includes a set of computational techniques used to emulate the mechanical behavior of biological tissues under different physiological situations. In the context of articular cartilage, these models are employed to study how mechanical forces influence tissue architecture and function as well as pathology.

Applications in cartilage research: Biomechanical models of articular cartilage often incorporate detailed descriptions of the ECM, chondrocytes, and their mechanical environment. These models can simulate the effects of cyclic loading, joint alignment, and abnormal forces on cartilage health. For example, biomechanical models have been used to study the development of OA, where repetitive stress and joint misalignment lead to cartilage degradation. By predicting how mechanical forces influence cellular responses and matrix remodeling, these models inform the design of interventions aimed at preserving or restoring cartilage function.

Challenges and considerations: Biomechanical modeling necessitates tissue material property representations that need to be accurately reflected, depending on age, health status, and anatomical region. Cartilage mechanics are nonlinear and time-dependent, which complicates the model development and its evaluation. However, these challenges are but a few of the very real reasons that physical insights into biomechanics have to be modeled [25].

2.5.4.4 Agent-based models

Agent-based models (ABMs) are a computational technique used to simulate the interactions of individual entities, or "agents," within a system. In the context of articular cartilage, these agents often represent chondrocytes, ECM components, or even molecules involved in signaling pathways.

Applications in cartilage research: ABMs are particularly well-suited for studying the complex, dynamic interactions that occur within cartilage at the cellular level. For example, ABMs can simulate how chondrocytes respond to mechanical stimuli by altering their matrix synthesis or degradation activities. These models can also explore how local environmental changes, such as variations in nutrient availability or mechanical stress, influence chondrocyte behavior over time. ABMs have been used to investigate the progression of cartilage diseases like OA, where localized changes in cell behavior lead to macroscopic tissue degeneration.

Challenges and considerations: The main challenge of ABMs is to define the rules that describe agent interactions, which have to be a valid representation of the biological reality. Moreover, since ABMs usually deal with a few hundred or more interacting agents at the same time, it could be a computationally heavy task. While not without its own technical difficulties, ABMs provide a potentially unique method to interrogate emergent properties of cartilage tissue and formulate strategies around modifying these selections in disease states [24].

2.5.5 Applications of computational models

Computational models have become essential tools in biomechanics, offering deep insights into the behavior of articular cartilage under various conditions. These models allow researchers and clinicians to simulate complex biological processes, predict disease progression, optimize tissue engineering strategies, and enhance surgical outcomes. This section highlights the significant applications of computational models in disease modeling, tissue engineering, and surgical planning models, which have become indispensable tools in the field of biomechanics, offering unprecedented insights into the behavior of articular cartilage under various conditions. These models enable researchers and clinicians to simulate complex biological processes, predict disease progression, optimize tissue engineering strategies, and improve surgical outcomes. This section explores the key applications of computational models in disease modeling, tissue engineering, and surgical planning [26].

2.5.5.1 Disease modeling

One of the most impactful applications of computational modeling is in the study of cartilage-related diseases, particularly OA. OA is a degenerative joint disease characterized by the progressive breakdown of articular cartilage, leading to pain, stiffness, and reduced mobility. Despite its prevalence, the exact mechanisms driving OA progression are not fully understood, making it a challenging condition to manage effectively.

Applications in disease modeling: Computational models enable researchers to simulate the intricate interplay of mechanical forces, biochemical signals, and cellular responses that drive cartilage degeneration in OA. By integrating patient-specific data, such as joint loading patterns and cartilage composition, these models can forecast how various factors – age, weight, or injury history – affect the onset and progression of OA. For instance, finite element models can evaluate how abnormal joint mechanics after an injury might accelerate cartilage wear, triggering early-stage OA.

Moreover, disease models can simulate the effects of pharmacological interventions and lifestyle modifications on cartilage health, offering valuable insights into potential treatment strategies. By understanding how different therapeutic approaches impact cartilage integrity at the molecular, cellular, and tissue levels, computational models can guide the development of more effective and personalized treatment plans.

Challenges and considerations: OA is a complex condition involving a blend of mechanical, biological, and genetic factors, posing significant challenges for computational models. Accurately modeling OA progression requires synthesizing data from multiple scales, from molecular pathways to joint mechanics. However, despite these difficulties, disease modeling remains a critical aspect of computational biomechanics, offering immense potential to revolutionize our understanding and treatment of degenerative joint diseases [26].

2.5.5.2 Tissue engineering

Tissue engineering seeks to create biological substitutes that restore, maintain, or improve tissue function. In the context of articular cartilage, the focus is on developing engineered cartilage constructs that can replace damaged tissue, offering a viable solution for patients with severe joint damage.

Applications in tissue engineering: Computational models play a critical role in the design and optimization of tissue-engineered cartilage. These models can simulate how various scaffold materials, cell types, and growth factors interact within a three-dimensional construct, enabling researchers to predict the mechanical and biological properties of the engineered tissue. For instance, multiscale models can be used to simulate how chondrocytes within a scaffold respond to mechanical loading, guiding the development of constructs that mimic the mechanical behavior of native cartilage.

Additionally, computational models can help optimize the manufacturing processes for engineered cartilage. By simulating how cells populate a scaffold, how nutrients diffuse through the construct, and how mechanical stimuli influence tissue growth, these models provide valuable insights that can improve the quality and functionality of the final product. This is particularly important for ensuring that the engineered cartilage integrates well with the surrounding tissue and functions effectively under physiological loads.

Challenges and considerations: Successfully engineering cartilage that mimics the complex architecture and mechanical properties of native tissue remains a significant challenge. Computational models must account for the dynamic and heterogeneous nature of cartilage, as well as the interactions between cells and their microenvironment. Nonetheless, computational models hold great promise in advancing tissue engineering, bringing us closer to clinical applications of engineered cartilage [26].

2.5.5.3 Surgical planning and outcomes prediction

Surgical interventions, such as cartilage repair or joint replacement, are often necessary for patients with advanced cartilage damage. However, the success of these procedures depends on careful planning and the ability to predict postoperative outcomes. Computational models offer powerful tools for improving surgical planning and predicting patient-specific outcomes, leading to more successful interventions.

Applications in surgical planning: Preoperative planning through computational models allows surgeons to simulate different surgical scenarios and predict their effects on joint mechanics and function. For example, finite element models can be used to assess the mechanical stability of a repaired cartilage defect or to optimize the positioning and orientation of a joint implant. By simulating the joint environment after surgery, these models assist surgeons in selecting the optimal approach for each patient, reducing complications and improving long-term outcomes.

Computational models also enable the prediction of how surgical interventions will influence the progression of joint diseases like OA. For instance, a model could simulate how changes in joint alignment post-surgery affect the distribution of mechanical loads across the cartilage, potentially helping to prevent further degeneration. These predictive capabilities are especially valuable in personalized medicine, where treatment plans are tailored to individual patients.

Challenges and considerations: While computational models offer significant benefits for surgical planning, their accuracy depends on the quality of the input data and the assumptions made during model development. Patient-specific factors, such as variations in tissue properties and joint mechanics, must be accurately represented to ensure reliable predictions. Additionally, integrating these models into clinical practice requires user-friendly interfaces and efficient computational resources, which can be a barrier to widespread adoption. Despite these challenges, the application of computational models in surgical planning is a rapidly growing area with the potential to enhance surgical precision and patient outcomes [27].

2.5.6 Challenges and future directions

As computational modeling continues to evolve, it brings immense opportunities as well as significant challenges in the study of articular cartilage and cellular mechanics. While these models have dramatically improved our understanding of cartilage behavior, limitations remain that affect their accuracy and applicability. This section addresses the current limitations of computational models, explores emerging techniques, and outlines potential future research directions.

2.5.6.1 Limitations of current models

Despite the significant progress in computational modeling of articular cartilage, current models still face several limitations that impact their accuracy, applicability, and generalizability.

1. **Simplifications and assumptions:** To make simulations manageable, many computational models rely on simplifying assumptions. These include assuming homogeneous material properties or linear elastic behavior, and often overlooking the complex interactions across biological scales. Such simplifications can create discrepancies between model predictions and real-world outcomes, particularly given the heterogeneous and anisotropic nature of cartilage tissue [28].

2. **Incomplete representation of biological complexity:** Articular cartilage is a dynamic tissue, where mechanical forces, biochemical signals, and cellular responses interact in complex ways. Current models often struggle to fully capture this biological complexity, particularly in terms of accurately modeling the behavior of chondrocytes and the ECM over time. The lack of detailed, patient-specific data further limits the ability to create personalized models that accurately reflect individual variations in cartilage composition and mechanics.
3. **Computational constraints:** High-fidelity models that incorporate detailed biological and mechanical processes require significant computational resources. The trade-off between model complexity and computational efficiency often forces researchers to make compromises, limiting the ability to perform large-scale simulations or to explore a wide range of scenarios in a timely manner.
4. **Validation and translation to clinical practice:** Validation is essential for ensuring the reliability of computational models, but obtaining high-quality experimental data that accurately corresponds to simulated conditions can be challenging. Furthermore, integrating these models into clinical practice remains difficult, requiring user-friendly interfaces and seamless integration with medical imaging and diagnostic technologies [29].

2.5.6.2 Emerging techniques and technologies

To overcome the limitations of current models, several emerging techniques and technologies are under development, offering the potential to significantly enhance the accuracy and applicability of computational models in cartilage research.

1. **Multiscale and multiphysics modeling:** New approaches in multiscale and multiphysics modeling aim to integrate different biological and physical processes within a single framework. These models simultaneously simulate molecular, cellular, and tissue-level phenomena, providing a more holistic understanding of cartilage mechanics. For example, combining molecular dynamics simulations with FEA allows researchers to see how molecular changes influence the larger-scale behavior of cartilage.
2. **Machine learning (ML) and artificial intelligence (AI):** ML and AI are increasingly being integrated into computational modeling to enhance predictive capabilities and reduce the need for simplifications. By analyzing large datasets, ML algorithms can identify patterns and relationships that might be missed by traditional modeling approaches. These techniques can also be used to optimize model parameters, making simulations more accurate and less reliant on assumptions.
3. **High-performance computing (HPC) and cloud-based platforms:** Advances in HPC and cloud platforms are making it possible to run more complex and large-scale simulations than before. These technologies allow researchers to model the full complexity of cartilage tissue and explore a broader range of scenarios.

Cloud-based platforms also enable easier collaboration by allowing researchers to share models and data.

4. **Enhanced imaging and data acquisition technologies:** The development of advanced imaging technologies, such as high-resolution MRI and micro-CT, along with improved data acquisition methods, provides more detailed and accurate input for computational models. These advances help create more anatomically accurate and personalized models, increasing the predictive power and clinical relevance of simulations [30, 31].

2.5.6.3 Future research directions

Looking forward, several key areas of research have the potential to further advance the field of computational modeling in articular cartilage and cell mechanics.

1. **Integration of multiscale models:** Future research should focus on improving the integration of multiscale models that link molecular-, cellular-, and tissue-level phenomena. Achieving seamless transitions between different scales will be crucial for capturing the full complexity of cartilage behavior and enhancing predictive capabilities.
2. **Personalized and predictive medicine:** There is a growing emphasis on developing models that can be tailored to individual patients. These models would use patient-specific data – such as genetic information, lifestyle factors, and anatomical measurements – to predict disease risk, progression, and treatment outcomes. Such personalized models will be vital for optimizing treatment plans and improving surgical outcomes.
3. **Model validation and experimental collaboration:** To enhance the reliability of computational models, greater collaboration is needed between computational researchers, experimental biologists, and clinicians. This collaboration can ensure that models are validated with high-quality experimental data. Standardizing validation protocols will also help ensure consistency and credibility across studies.
4. **Ethical and regulatory considerations:** As computational models become more integrated into clinical decision-making, ethical and regulatory challenges must be addressed. Transparency in model development and validation will be essential, along with establishing clear guidelines for clinical applications. Close collaboration with regulatory bodies will be necessary to ensure safe and effective use of these models in healthcare.
5. **Expanding applications beyond cartilage:** While much of the focus has been on articular cartilage, the techniques developed for cartilage modeling have potential applications in other musculoskeletal tissues, such as tendons, ligaments, and intervertebral discs. Expanding these models could offer new insights into a range of orthopedic conditions and improve overall understanding of joint mechanics [32].

2.5.7 Conclusion

Understanding the mechanical properties of articular cartilage is essential for comprehending joint function and advancing treatments for cartilage-related diseases. Computational modeling has become an indispensable tool in this area, offering insights that are often beyond the reach of experimental methods. This chapter has discussed the biological and mechanical properties of articular cartilage, explored a range of computational modeling techniques, and highlighted their applications in disease modeling, tissue engineering, and surgical planning.

2.5.7.1 Summary of key findings

Throughout this chapter, we have examined the intricate structure and function of articular cartilage, highlighting its unique composition and mechanical behavior. Key findings include:

Complex composition and zonal architecture: Articular cartilage is a highly structured tissue with distinct zones, each contributing uniquely to its mechanical properties. The interaction between water, collagen, and proteoglycans within the ECM is crucial to the tissue's ability to endure compressive forces and maintain smooth joint movement.

Mechanical behavior: The viscoelastic and poroelastic nature of cartilage plays a critical role in its ability to absorb shocks, distribute loads, and resist wear. Computational models have been instrumental in simulating these behaviors, deepening our understanding of how cartilage responds to various mechanical stresses.

Computational modeling techniques: Various techniques, including FEA, multiscale modeling, biomechanical simulations, and ABMs, have provided valuable insights into cartilage mechanics and cellular behavior. These methods have applications in both fundamental research and clinical practice.

Applications in disease modeling, tissue engineering, and surgical planning: Computational models have been successfully used to simulate disease progression, such as in OA, to optimize tissue engineering strategies, and to improve surgical planning. These applications underscore the versatility of computational models in advancing cartilage research.

2.5.7.2 Importance of computational modeling in cartilage research

Computational modeling has proven to be a powerful tool in cartilage research, offering several significant advantages:

Enhanced understanding of cartilage mechanics: By simulating complex biological and mechanical processes, computational models provide a detailed understanding of how cartilage functions under normal and pathological conditions. This knowledge is critical for developing effective treatments and interventions.

Predictive power: Models can predict disease progression, evaluate the impact of surgical interventions, and optimize tissue engineering approaches, allowing for personalized medicine and more targeted therapies.

Bridging experimental and clinical research: Computational models serve as a bridge between experimental research and clinical practice. They allow researchers to test hypotheses, validate experimental results, and translate findings into clinical applications, ultimately improving patient outcomes.

2.5.7.3 Final thoughts

Looking ahead, the continued integration of computational modeling into cartilage research is essential for addressing existing challenges and advancing our understanding of joint mechanics. The future of this field lies in the development of increasingly sophisticated multiscale models capable of capturing the full complexity of cartilage behavior, from molecular-level interactions to whole-joint mechanics. Moreover, the evolution of personalized models, driven by advancements in data acquisition, machine learning, and HPC, will transform how we approach the treatment of cartilage-related diseases. As these models become more refined and accessible, they will play an ever-greater role in clinical decision-making, enabling more tailored and effective patient care.

In conclusion, the future of cartilage research is promising, with computational modeling at the forefront of innovation. Continued advancements in this field will lead to breakthroughs in the prevention, diagnosis, and treatment of joint diseases, ultimately improving health outcomes for patients worldwide.

References

[1] Márquez-Flórez K, Garzón-Alvarado DA, Carda C, Sancho-Tello M. Computational model of articular cartilage regeneration induced by scaffold implantation in vivo. Journal of Theoretical Biology [Internet]. 2023 Mar; 561: 111393. Available from: https://linkinghub.elsevier.com/retrieve/pii/S0022519322003848.

[2] Miramini S, Smith DW, Gardiner BS, Zhang L. Computational Modelling for Managing Pathways to Cartilage Failure BT – Electromechanobiology of Cartilage and Osteoarthritis: A Tribute to Alan Grodzinsky on His 75th Birthday. In: Connizzo BK, Han L, Sah RL, editors. Electromechanobiology of Cartilage and Osteoarthritis, Cham, Switzerland: Springer International Publishing; 2023. pp. 83–93. Available from: https://doi.org/10.1007/978-3-031-25588-5_6.

[3] Halloran JP, Sibole S, van Donkelaar CC, Van Turnhout MC, Oomens CWJ, Weiss JA, et al. Multiscale mechanics of articular cartilage: Potentials and challenges of coupling musculoskeletal, joint, and microscale computational models. Annals of Biomedical Engineering [Internet]. 2012 Nov; 40(11): 2456–2474. Available from: http://www.ncbi.nlm.nih.gov/pubmed/22648577.

[4] Manley MT, Capello WN, D'antonio JA, Edidin AA, Geesink RGT. Fixation of acetabular cups without cement in total hip arthroplasty. A comparison of three different implant surfaces at a minimum duration of follow-up of five years*. The Journal Bone and Joint Surgery [Internet]. 1998 Aug; 80(8): 1175–1185. Available from: http://journals.lww.com/00004623-199808000-00011.

[5] Krettek C, Mannß J, Miclau T, Schandelmaier P, Linnemann I, Tscherne H. Deformation of femoral nails with intramedullary insertion. Journal of Orthopaedic Research: Official Publication of the Orthopaedic Research Society [Internet]. 1998 Sep 18; 16(5): 572–575. Available from: https://onlinelibrary.wiley.com/doi/10.1002/jor.1100160508.

[6] Mow VC, Ratcliffe A, Robin Poole A. Cartilage and diarthrodial joints as paradigms for hierarchical materials and structures. Biomaterials [Internet]. 1992; 13(2): 67–97. Available from: https://linkinghub.elsevier.com/retrieve/pii/0142961292900015.

[7] Sophia Fox AJ, Bedi A, Rodeo SA. The basic science of articular cartilage: Structure, composition, and function. Sport Heal A Multidiscip Approach [Internet]. 2009 Nov 2; 1(6): 461–468. Available from: http://journals.sagepub.com/doi/10.1177/1941738109350438.

[8] Goldring MB, Goldring SR. Articular cartilage and subchondral bone in the pathogenesis of osteoarthritis. Annals of the New York Academy of Sciences [Internet]. 2010 Apr 5; 1192(1): 230–237. Available from: https://nyaspubs.onlinelibrary.wiley.com/doi/10.1111/j.1749-6632.2009.05240.x.

[9] Ricard-Blum S. The collagen family. Cold Spring Harbor Perspectives Biology [Internet]. 2011 Jan 1; 3(1): a004978–a004978. Available from: http://cshperspectives.cshlp.org/lookup/doi/10.1101/cshperspect.a004978.

[10] Roughley PJ, Mort JS. The role of aggrecan in normal and osteoarthritic cartilage. Journal of Experimental Orthopaedics [Internet]. 2014 Dec 16; 1(1): 8. Available from: http://www.jeo-esska.com/content/1/1/8.

[11] Basu BV. Comparative analysis of i Coronaviridae i nucleocapsid and surface glycoprotein sequences. Frontiers in Bioscience [Internet]. 2020; 25(10): 4883. Available from: https://imrpress.com/journal/FBL/25/10/10.2741/4883.

[12] Mazzei DR, Ademola A, Abbott JH, Sajobi T, Hildebrand K, Marshall DA. Are education, exercise and diet interventions a cost-effective treatment to manage hip and knee osteoarthritis? A systematic review. Osteoarthritis and Cartilage [Internet]. 2021 Apr; 29(4): 456–470. Available from: https://linkinghub.elsevier.com/retrieve/pii/S1063458420311651.

[13] Setton LA, Mow VC, Howell DS. Mechanical behavior of articular cartilage in shear is altered by transection of the anterior cruciate ligament. Journal of Orthopaedic Research [Internet]. 1995 Jul 18; 13(4): 473–482. Available from: https://onlinelibrary.wiley.com/doi/10.1002/jor.1100130402.

[14] Cheng AH-D. Introduction. In: Poroelasticity. Theory and Applications of Transport in Porous Media, Cham, Switzerland: Springer; 2016, Vol. 27, pp. 1–59. https://doi.org/10.1007/978-3-319-25202-5_1.

[15] Wang M, Wang L, Li P, Fu Y. A novel modelling and simulation method of hip joint surface contact stress. Bioengineered [Internet]. 2017 Jan 2; 8(1): 105–112. Available from: http://www.ncbi.nlm.nih.gov/pubmed/27696938.

[16] Scheiner S, Pivonka P, Hellmich C. Poromicromechanics reveals that physiological bone strains induce osteocyte-stimulating lacunar pressure. Biomechanics and Modeling in Mechanobiology [Internet]. 2016 Feb; 15(1): 9–28. Available from: http://link.springer.com/10.1007/s10237-015-0704-y.

[17] Benschop RJ, Collins EC, Darling RJ, Allan BW, Leung D, Conner EM, et al. Development of a novel antibody to calcitonin gene-related peptide for the treatment of osteoarthritis-related pain. Osteoarthritis and Cartilage [Internet]. 2014 Apr; 22(4): 578–585. Available from: https://linkinghub.elsevier.com/retrieve/pii/S1063458414000338.

[18] Halilaj E, Hastie TJ, Gold GE, Delp SL. Physical activity is associated with changes in knee cartilage microstructure. Osteoarthritis and Cartilage [Internet]. 2018 Jun; 26(6): 770–774. Available from: https://linkinghub.elsevier.com/retrieve/pii/S106345841831135X.

[19] Zhang M, Meng N, Wang X, Chen W, Zhang Q. TRPV4 and PIEZO Channels Mediate the Mechanosensing of Chondrocytes to the Biomechanical Microenvironment. Membranes (Basel) [Internet]. 2022; 12(2): Available from: https://www.mdpi.com/2077-0375/12/2/237.

[20] Khoshgoftar M, Torzilli PA, Maher SA. Influence of the pericellular and extracellular matrix structural properties on chondrocyte mechanics. Journal of Orthopaedic Research [Internet]. 2018 Feb 22; 36(2): 721–729. Available from: https://onlinelibrary.wiley.com/doi/10.1002/jor.23774.

[21] Wilusz RE, Zauscher S, Guilak F. Micromechanical mapping of early osteoarthritic changes in the pericellular matrix of human articular cartilage. Osteoarthritis and Cartilage [Internet]. 2013 Dec 1; 21(12): 1895–1903. Available from: https://doi.org/10.1016/j.joca.2013.08.026.

[22] Boos MA, Lamandé SR, Stok KS. Multiscale strain transfer in cartilage. Frontiers in Cell and Developmental Biology. 2022; 10: 1–12.

[23] Lei S, Zhao X, Yu X, Hu A, Vukelic S, Jun MBG, et al. Ultrafast laser applications in manufacturing processes: A state-of-the-art review. Journal of Manufacturing Science and Engineering [Internet]. 2020 Mar 1; 142(3): Available from: https://asmedigitalcollection.asme.org/manufacturingscience/article/doi/10.1115/1.4045969/1072541/Ultrafast-Laser-Applications-in-Manufacturing.

[24] Corti A, Colombo M, Migliavacca F, Rodriguez Matas JF, Casarin S, Chiastra C. Multiscale computational modeling of vascular adaptation: A systems biology approach using agent-based models. Frontiers in Bioengineering and Biotechnology [Internet]. 2021 Nov 2; 9: Available from: https://www.frontiersin.org/articles/10.3389/fbioe.2021.744560/full.

[25] Miramini S, Smith DW, Gardiner BS, Zhang L. Computational modelling for managing pathways to cartilage failure. In 2023; 83–93. Available from: https://link.springer.com/10.1007/978-3-031-25588-5_6.

[26] Pearce D, Fischer S, Huda F, Vahdati A. Applications of computer modeling and simulation in cartilage tissue engineering. Tissue Engineering and Regenerative Medicine [Internet]. 2020 Feb 5; 17(1): 1–13. Available from: http://link.springer.com/10.1007/s13770-019-00216-9.

[27] Borgiani E, Duda GN, Checa S. Multiscale modeling of bone healing: Toward a systems biology approach. Frontiers in Physiology [Internet]. 2017 May 8; 8: Available from: http://journal.frontiersin.org/article/10.3389/fphys.2017.00287/full.

[28] Pierce DM. Multi-phase, large-strain constitutive models of cartilage for finite element analyses in 3-D. Archive of Applied Mechanics [Internet]. 2022 Feb 24; 92(2): 513–528. Available from: https://link.springer.com/10.1007/s00419-021-01959-5.

[29] Kazemi M, Dabiri Y, Li LP. Recent advances in computational mechanics of the human knee joint. Computational and Mathematical Methods in Medicine [Internet]. 2013; 2013: 1–27. Available from: http://www.hindawi.com/journals/cmmm/2013/718423/.

[30] Halloran JP, Sibole S, Van Donkelaar CC, van Turnhout MC, Oomens CWJ, Weiss JA, et al. Multiscale mechanics of articular cartilage: Potentials and challenges of coupling musculoskeletal, joint, and microscale computational models. Annals of Biomedical Engineering [Internet]. 2012 Nov; 40(11): 2456–2474. Available from: http://link.springer.com/10.1007/s10439-012-0598-0.
[31] Uhlrich SD, Uchida TK, Lee MR, Delp SL. Ten steps to becoming a musculoskeletal simulation expert: A half-century of progress and outlook for the future. Journal of Biomechanics [Internet]. 2023 Jun; 154: 111623. Available from: https://linkinghub.elsevier.com/retrieve/pii/S0021929023001926.
[32] Pierce DM, Trobin W, Holzapfel GA. A computational framework for patient-specific analysis of articular cartilage incorporating structural information from DT-MRI. GAMM-Mitteilungen [Internet]. 2009 Dec 4; 32(2): 157–177. Available from: https://onlinelibrary.wiley.com/doi/10.1002/gamm.200910014.

Raja Dhason*, Sandipan Roy, and Shubhabrata Datta

Chapter 2.6
Experimental and computational analysis for osteoporotic fracture implant failure

Abstract: Implant failure in osteoporotic fractures is a complex issue affected by the material's structural integrity, surgical methods, and the diminished quality of bone at the implantation site. Although peripheral bone density tests provide little predictive utility, local bone quality evaluations and computational models demonstrate potential in elucidating implant fixation mechanisms. Comparative studies indicate that locking plates offer enhanced stability at lower stress levels but are susceptible to periprosthetic fractures with increased loading, especially in osteoporotic bone. Experimental studies highlight the significance of trabecular bone failure prior to implant fixation failure, with methods such as cement augmentation and advanced implant coatings improving stability. Computational methods, like as finite element models, present a cost-efficient substitute for experimental investigations, yielding significant insights into the mechanical interactions between implants and bone. Advancements in modeling bone-implant interactions and biomechanical characteristics are essential for enhanced predictions. Examining the relationship between bone quality, implant design, and loading conditions through experimental and computational techniques is essential for enhancing implant results in osteoporotic patients.

Keywords: Osteoporosis, fracture mechanics, implant stability, experimental methods, computational modeling

2.6.1 Introduction

The failure of an osteoporotic fracture implant is a complicated problem that is influenced by a number of different circumstances. Osteosynthesis implant failure is mostly caused by flaws in the material's structural integrity as well as errors associated with surgical methods [1]. Visual evaluation of the local bone quality at the implant site may be more instructive than tests of bone density in the periphery, which

***Corresponding author: Raja Dhason**, Department of Mechanical Engineering, SRM Institute of Science and Technology, Chengalpattu, Tamil Nadu, India, e-mail rajad@srmist.edu.in
Sandipan Roy, Department of Mechanical Engineering, SRM Institute of Science and Technology, Chengalpattu, India, e-mail: sandipag@srmist.edu.in
Shubhabrata Datta, Department of Mechanical Engineering, SRM Institute of Science and Technology, Chengalpattu, India, e-mail: shubhabp@srmist.edu.in

https://doi.org/10.1515/9783111386027-011

demonstrate only a limited connection with implant failure [2]. Based on the results of experimental comparisons between locking plates and intramedullary (IM) nails for the purpose of femoral fracture fixation in osteoporotic bone, it has been determined that locking plates provide higher stability under conditions of lower stress. On the other hand, they also exhibit a higher prevalence of abrupt periprosthetic fractures under ramping loading, which is especially difficult to salvage in osteoporotic bone due to the decrease in the trabecular matrix and the thinner cortex [3]. However, through the utilization of computational models, it is possible to acquire a more comprehensive understanding of the mechanisms that are responsible for implant fixation, particularly in bone that is of low quality. It is possible to get detailed information about the failure of osteoporotic fracture implants through the use of computational analysis, which offers valuable insights. In the computational technique, finite element (FE) models are used to evaluate the implant function in an effective manner in a number of circumstances, offering a complementary solution to the requirement for expensive experimental studies and maybe eliminating the necessity for such studies altogether. Understanding the failure mechanisms in low-quality bone can be aided by these models, which are able to assess the mechanical interaction between implants and the tissues surrounding them [4, 5]. Recent developments in imaging and computation have made it possible to create high-resolution FE models; however, there is still need for improvement in the modeling of bone–implant interfaces in order to more accurately simulate the nonlinear mechanical behavior observed in vitro [6]. Continuum of linear elastic properties with a significant link between cycles to failure and average compressive main strain around screws, FE models have demonstrated that they have the potential to accurately forecast fatigue failure in locking plate fixation of osteoporotic proximal humerus fractures [7]. When it comes to selecting implants for osteoporotic patients, it shows how important it is to take into consideration diverse loading scenarios. Additionally, these computational approaches have the potential to function as effective screening tools for the optimization and development of implants.

2.6.2 Experimentation of implant failure

For the treatment of osteoporosis conditions, a significant amount of research published in the literature focuses on experimental investigations that use traditional plating, conventional plating, light alternatives plate, and IM nailing. Salas et al. developed and analyzed the FE model and also probabilistic investigation of the probability of periprosthetic fracture in distal femoral fractures treated with either locking plates or IM nails. The findings of the investigation indicate that locking plates have a significantly higher probability of periprosthetic fracture (21.8%) compared to IM nails (0.019%) when subjected to ramped bending loads because during low load, locking plates were found to be more stable than IM nails in osteoporotic bone. However, locking plates had a greater incidence of sudden periprosthetic fractures compared to IM nails under higher, ramped loading conditions [3]. In another study, peri-implant trabecular bone failure occurs before implant fixation failure in osteoporotic bone. The experimental data obtained from this study are extremely important for expanding and understanding of the quality of the bone-implant interface as well as the trabecular bone failure that occurs during the process of implant fixation failure [8]. Blimel et al. used condylar screws and locking plate with cement augmentation for distal femoral fracture in osteoporosis bone showed significant decrease in plastic deformation and increases the load to failure, compared to augmented screws [9].

Barrios et al. examines the healing complications due to implant failures or loss of fracture reduction in patients with intertrochanteric hip fractures treated with either a sliding screw plate or Ender nailing. The majority of implant failures (76%) occurred in patients with osteoporosis [10]. Similarly, Dvorak et al. studies found no significant relationship between osteoporosis/osteopenia and the occurrence of peri-implantitis or late implant failures in postmenopausal women. This study divided participants into three groups based on their bone health status: osteoporosis, osteopenia, and healthy controls. Data from 203 women with a mean age of 63 ± 9 years and 967 dental implants were investigated [11]. Cho et al. showed that the bone-implant contact and the long-term durability of dental implants might be adversely affected by osteoporotic diseases. Peter et al. studied the concept for employing orthopedic implants as a drug delivery system. More specifically, the study demonstrated that bisphosphonates may be delivered to osteoporotic rats in order to boost the mechanical fixation and stability of the implant. Ghanem et al. studied osteoporotic bone, and showed that osteogenic coatings on implant surfaces can improve bone-implant contact. Peng et al. used three models of orchidectomy, ovariectomy, and immobilization in rats and predicted it will lead to decreased mechanical strength of the femoral neck, a sensitive indicator of bone loss in experimental osteoporosis. These studies were used on animal models with induced osteoporosis [12–15]. Linden et al. developed a new method of cement augmentation that improves implant fixation and lowers failure rates in osteoporotic proximal femoral fractures using human cadaver femur as experimental model [16]. In this work, authors examine the impact of sarcopenia and bone mineral density on implant failure after dorsal instrumentation in patients with osteoporotic vertebral fractures, and find that sarcopenia and osteoporosis are significant risk factors for implant failure [17].

2.6.3 Computational studies on implant failures

Osteoporosis is a disease that causes structural bone degradation and a reduction in bone mass density. Aged people are experiencing fractures in their remaining lifetime due to osteoporosis; in recent years, osteoporotic fractures have emerged as a significant challenge for public healthcare systems. Furthermore, as a result of an ageing population in developed countries and an inadequate prevention of osteoporosis, it is anticipated that the number of osteoporotic fractures will dramatically grow over the next few years, as will the expenditures that are incurred as a consequence of these fractures [18–21]. In the recent years, high stiffness implants showed reduced performance in the osteoporotic bone. The low-quality bones were treated with different approaches like using bone cement as a supportive material for implants used in the osteoporotic bones, providing multiple locking plates and screws in the fixation, and low stiffness/composite bone plate for low density bone fracture fixation [22, 23]. Additionally, mathematical models were used to predict the forces causing the titanium implant failure in the bone defect treatment [24]. Shim et al. developed a patient-specific computational model to predict implant failure in the pelvic ring fracture fixation. The developed model predictions matched well with the actual clinical outcomes in two case studies, demonstrating the potential of this technique for use in clinical practice for surgical planning and postsurgical management of pelvic ring fractures [25]. With respect to dental implant, an FE model was developed and analyzed to predict the relationship between bone resorption and stress induced failure. The results showed that the critical stress levels bone implant interface at a specific threshold of bone resorption, which ultimately resulted in the failure of the implant [26]. Furthermore, Liao et al. studied the biomechanical performance of different fixation constructs to treat thoracolumbar burst fracture for intact and osteoporosis bone conditions. A decrease in the stability of the spinal construct and an increase in the stress placed on the pedicle screws, both of which potentially result in the failure of the implant, were both caused by osteoporosis. The results showed the S-I-C construct offered more stability and observed very little stress on the vertebrae [27]. The risk of spinal implant failure is increased by osteoporosis, necessitating specific surgical planning and procedures to optimize implant stability. To increase implant stability and reduce problems, patients with osteoporosis need specific preoperative planning and surgical techniques [28]. Implant failure is caused due to inadequate bone quality, heavy loading, sudden impact, choice of implant for fixation. Still, failure mechanism and standards for implant selection were unknown. Additionally, detecting failed implants in clinical practice is difficult, making it difficult to assess the total failure rate [5].

2.6.4 Conclusion

The failure of implants for osteoporotic fractures is a complex issue caused by bone quality, choice of implants, and loading conditions. Experimental and computational assessments indicate the necessity for customized strategies to enhance implant durability and diminish failure rates in osteoporotic individuals. Innovations like cement augmentation, osteogenic coatings, and FE modeling have improved our comprehension of implant biomechanics and fixation processes. Nonetheless, considerable deficiencies persist in the standardization of implant selection and the precise modeling of bone-implant interactions. Advancing the integration of sophisticated computational tools with clinical practices can enhance surgical planning and patient outcomes, thereby decreasing the increasing prevalence of osteoporotic fractures in aging populations.

References

[1] Nica M, Cretu B, Ene D, Antoniac I, Gheorghita D, Ene R. Failure analysis of retrieved osteosynthesis implants. Materials. 2020; 13(5); 1201.
[2] Becker W, Hujoel PP, Becker BE, Willingham H. Osteoporosis and implant failure: An exploratory case-control study. Journal of Periodontology. 2000; 71(4); 625–631.
[3] Salas C, Mercer D, DeCoster TA, Reda Taha MM. Experimental and probabilistic analysis of distal femoral periprosthetic fracture: A comparison of locking plate and intramedullary nail fixation. Part A: Experimental investigation. Computer Methods in Biomechanics and Biomedical Engineering. 2011; 14(02); 157–164.
[4] Varga P, Inzana JA, Gueorguiev B, Südkamp NP, Windolf M. Validated computational framework for efficient systematic evaluation of osteoporotic fracture fixation in the proximal humerus. Medical Engineering & Physics. 2018; 57: 29–39.
[5] Wirth AJ, Müller R, van Lenthe GH. Computational analyses of small endosseous implants in osteoporotic bone. Eur Cells Mater. 2010; 20: 58–71.
[6] Steiner JA, Ferguson SJ, van Lenthe GH. Computational analysis of primary implant stability in trabecular bone. Journal of Biomechanics. 2015; 48(5); 807–815.
[7] Varga P, Grünwald L, Inzana JA, Windolf M. Fatigue failure of plated osteoporotic proximal humerus fractures is predicted by the strain around the proximal screws. Journal of the Mechanical Behavior of Biomedical Materials. 2017; 75: 68–74.
[8] Mueller TL, Basler SE, Müller R, Van Lenthe GH. Time–lapsed imaging of implant fixation failure in human femoral heads. Medical Engineering & Physics. 2013; 35(5); 636–643.
[9] Bliemel C, Oberkircher L, Bockmann B, Petzold E, Aigner R, Heyse TJ, . . . Buecking B. Impact of cement–augmented condylar screws in locking plate osteosynthesis for distal femoral fractures – A biomechanical analysis. Injury. 2016; 47(12); 2688–2693.
[10] Barrios C, Broström LÅ, Stark A, Walheim G. Healing complications after internal fixation of trochanteric hip fractures: The prognostic value of osteoporosis. Journal of Orthopaedic Trauma. 1993; 7(5); 438–442.
[11] Dvorak G, Arnhart C, Heuberer S, Huber CD, Watzek G, Gruber R. Peri-implantitis and late implant failures in postmenopausal women: A cross-sectional study. Journal of Clinical Periodontology. 2011; 38(10); 950–955.
[12] Cho P, Schneider GB, Krizan K, Keller JC. Examination of the bone–implant interface in experimentally induced osteoporotic bone. Implant Dentistry. 2004; 13(1); 79–87.
[13] Peter B, Gauthier O, Laïb S, Bujoli B, Guicheux J, Janvier P, . . . Pioletti DP. Local delivery of bisphosphonate from coated orthopedic implants increases implants mechanical stability in osteoporotic rats. Journal of Biomedical Materials Research Part A: An Official Journal of the Society for Biomaterials, the Japanese Society for Biomaterials, and the Australian Society for Biomaterials and the Korean Society for Biomaterials. 2006; 76(1); 133–143.

[14] Ghanem A, Kellesarian SV, Abduljabbar T, Al-Hamoudi N, Vohra F, Javed F. Role of osteogenic coatings on implant surfaces in promoting bone-to-implant contact in experimental osteoporosis: A systematic review and meta-analysis. Implant Dentistry. 2017; 26(5); 770–777.
[15] Peng Z, Tuukkanen J, Zhang H, Jämsä T, Väänänen HK. The mechanical strength of bone in different rat models of experimental osteoporosis. Bone. 1994; 15(5); 523–532.
[16] Von der Linden P, Gisep A, Boner V, Windolf M, Appelt A, Suhm N. Biomechanical evaluation of a new augmentation method for enhanced screw fixation in osteoporotic proximal femoral fractures. Journal of Orthopaedic Research. 2006; 24(12); 2230–2237.
[17] Krenzlin H, Schmidt L, Jankovic D, Schulze C, Brockmann MA, Ringel F, Keric N. Impact of sarcopenia and bone mineral density on implant failure after dorsal instrumentation in patients with osteoporotic vertebral fractures. Medicina. 2022; 58(6); 748.
[18] Chrischilles EA, Butler CD, Davis CS, Wallace RB. A model of lifetime osteoporosis impact. Archives of Internal Medicine. 1991; 151(10); 2026–2032.
[19] Ray NF, Chan JK, Thamer M, Melton LJ III. Medical expenditures for the treatment of osteoporotic fractures in the United States in 1995: Report from the National Osteoporosis Foundation. Journal of Bone and Mineral Research. 1997; 12(1); 24–35.
[20] Kanis JA, Johnell O. Requirements for DXA for the management of osteoporosis in Europe. Osteoporosis International. 2005; 16: 229–238.
[21] Burge R, Dawson-Hughes B, Solomon DH, Wong JB, King A, Tosteson A. Incidence and economic burden of osteoporosis-related fractures in the United States, 2005–2025. Journal of Bone and Mineral Research. 2007; 22(3); 465–475.
[22] Frankel BM, Jones T, Wang C. Segmental polymethylmethacrylate-augmented pedicle screw fixation in patients with bone softening caused by osteoporosis and metastatic tumor involvement: A clinical evaluation. Neurosurgery. 2007; 61(3); 531–538.
[23] Dhason R, Roy S, Datta S. Metal and composite bone plates for B1 periprosthetic femoral fracture in healthy and osteoporotic condition: A comparative biomechanical study. The International Journal of Artificial Organs. 2022; 45(8); 704–714.
[24] Nečas A, Urbanová L, Fürst T, Ženčák P, Tuček P. Mathematical modelling of crack fractography after implant failure of titanium 4.5 LCP used for flexible bridging osteosynthesis in a miniature pig. Acta Veterinaria Brno. 2011; 79(4); 621–626.
[25] Shim V, Höch A, Grunert R, Peldschus S, Böhme J. Development of a patient-specific finite element model for predicting implant failure in pelvic ring fracture fixation. Computational and Mathematical Methods in Medicine. 2017; 2017(1); 9403821.
[26] Michailidis N, Karabinas G, Tsouknidas A, Maliaris G, Tsipas D, Koidis P. A FEM based endosteal implant simulation to determine the effect of peri-implant bone resorption on stress induced implant failure. Bio-Medical Materials and Engineering. 2013; 23(5); 317–327.
[27] Liao JC. Impact of osteoporosis on different type of short-segment posterior instrumentation for thoracolumbar burst fracture – A finite element analysis. World Neurosurgery. 2020; 139: e643–e651.
[28] Schömig F, Becker L, Schönnagel L, Völker A, Disch AC, Schnake KJ, Pumberger M. Avoiding spinal implant failures in osteoporotic patients: A narrative review. Global Spine Journal. 2023; 13(1); 52S–58S.

Raja Dhason*, Vamsi Krishna Dommeti, and Sandipan Roy

Chapter 2.7 Computational modeling of fracture implants

Abstract: The rapid advancement of the field of computer modeling of fracture implants has been driven by the demand for improved orthopedic therapies and a more profound comprehension of bone healing processes. Fracture implants are indispensable in orthopedic surgery due to their capacity to stabilize fractured bones and facilitate their recovery. A robust framework for assessing and enhancing these implants has recently been built through the advancement of computational modeling techniques. This chapter explores the computational modeling of fracture implants that are widely used in the spine, pelvis, femur bone, knee joint, and tibia bone. The finite element models of these implants are also employed to evaluate the healing process of fractures. A comprehensive overview of the many computer modeling techniques used in the research and development of fracture implants, with a focus to improve patient outcomes in the clinical setting and expand our understanding of implant biomechanics is given in the chapter.

Keywords: Computational modeling, fracture implants, orthopedic therapies, finite element analysis, implant biomechanics

2.7.1 Introduction

Computational modeling has become a valuable instrument for analyzing implants used in fracture fixation, particularly in challenging scenarios in fractured and osteoporotic bones. Complex biomechanical interactions between implants and surrounding tissues can be simulated using finite element analysis (FEA) models, which can reveal failure mechanisms and the best fixation techniques. In particular, these models can include specific bone structure and density changes from computed tomography (CT) scans, which make it possible to identify osteoporosis conditions accurately [1, 2]. These models can be patient-specific, taking into account unique implant locations and fracture morphologies to reasonably anticipate potential failure sites. A cost-effective technique for preclinical trials and design optimization, FEA has been

*Corresponding author: Raja Dhason, Department of Mechanical Engineering, SRM Institute of Science and Technology, Chengalpattu, Tamil Nadu, India, e-mail: rajad@srmist.edu.in
Vamsi Krishna Dommeti, Department of Mechanical Engineering, SRM Institute of Science and Technology, Chengalpattu, India, e-mail: vamsikrd@srmist.edu.in
Sandipan Roy, Department of Mechanical Engineering, SRM Institute of Science and Technology, Chengalpattu, India, e-mail: sandipag@srmist.edu.in

https://doi.org/10.1515/9783111386027-012

used to treat a variety of fracture types, including proximal femur and pelvic ring fractures [3]. But a majority of models lack thorough validation, despite some agreement on modeling elements like geometries, meshing, and material properties. However, FEA has the ability to improve surgical guiding, optimize implant designs, and advance knowledge of fracture fixation failure with increased validation efforts and extensive systematic analysis [4]. Similarly, most of the studies have shown that stiff fixation may limit fracture mobility and hamper healing, whereas more flexible methods can promote better outcomes [5]. Additionally, mathematical modeling can forecast the forces that result in implant failure, which could guide clinical management to avoid difficulties [6]. Computational modeling of fracture implants is very important for the prediction of implant behavior during normal activities after surgery. In this chapter, different types of modeling techniques used for the development of FE models of fractured bone implants are provided to understand and improve the modeling approaches.

2.7.2 Computational modeling of spine implants

Several approaches have been employed to develop the computational spine model to replicate the spine anatomy. Computer-aided design (CAD) software was used to reproduce the simplified, unrealistic parameterized spine model with all its features [7–10]. Additionally, that can be easily modified to represent the patient-specific anatomies [11]. The design and assessment of spinal implants now heavily relies on computational modeling, especially FEA. It has become a key tool in designing and evaluating spinal implants [12]. During the surgery, the intervertebral disc is removed entirely or partially, and bone chips/metal/polymer blocks/artificial disc are typically augmented with a fixation device known as a cage/artificial disc. Several methods were used to implant cages into the intersomatic area. Many FEA models of cages and fixation devices were developed to study the range of motion and stress distribution in the surgical and adjacent level [13–17]. The cage/artificial disc and vertebral endplates have defined contact surfaces to prevent interpenetration while allowing for sliding and separation in specific instances [16, 18].

Computational models are used to study a variety of spinal devices. Large mesh densities are frequently used in implant modeling; nevertheless, mesh continuity with the remainder of the spine segment model would result in excessively high computational costs. As a result, mesh discontinuities are typically managed by numerical techniques [12]. One of the most recent developments is the creation of patient-specific models that can forecast the possibility of pedicle screw loosening, which is a major problem in lumbar spine fixation [19]. These models reflect the link between local bone strength and expected stress and include personalized joint loading, thus transforming risk assessment for implant fixation failure. Appropriate validation, a strong knowledge of morphology and material properties, and bioengineers' experience in converting clinical concepts into suitable finite element models define the correctness and dependability of computational models [20].

2.7.3 Computational modeling of femur bone implants

Realistic simulations of bone structures and implants under varied stress conditions are made possible by finite element models generated from CT scan data used to study the displacement in the fracture site and distribution of stress in intact and implanted femur models. These models are able to forecast probable fracture patterns and identify the regions with high concentrations of stress [21, 22]. Different materials were used for femur implant, especially medical-grade steel and titanium were used to examine the behavior of the implant under static loading; predicted titanium implant showed better movement and biocompatibility [23]. Additionally, a different type of fixation was modeled in FEA with plate and cable to compare the mechanical stability and to measure the axial stiffness [24, 25]. In some cases, metal plates were tried to be replaced by low stiffness material for better healing. Composite bone plates are being proposed for fracture fixation in periprosthetic femoral fractures. Metallic plates have a higher stiffness than bone, which causes stress shielding, reduces compression force at the fracture site, and affects healing. Reduced stiffness in the axial direction for the reasons stated above, without lowering stiffness in the transverse direction to avoid significant shear strain and thus avoiding instability at the fracture site, results in selective stress shielding. This is only possible with precisely developed fiber-reinforced composites. These composite plates were modeled layer by layer to increase the strength of the plate [26, 27].

Advanced modeling methods can also replicate periprosthetic fractures, increasing osteoporosis, and physiological motions like walking. A femur plate implant design for periprosthetic femoral fracture was proposed using a neural network method. Using a hybrid and variable fiber orientation, the genetic algorithm (GA) creates the ideal composite bone plate for fixing periprosthetic femur fractures. The artificial neural network surrogate models for the bone displacements in both axial and shear movement directions are created using the simulated data from the FEA. These models serve as the objective functions for the GA-driven optimization in a multiobjective manner. The solutions obtained during the design optimization process are further validated using FEA [28]. These computational techniques offer insightful information for surgical planning and implant design, hence possibly lowering the need for difficult biomechanical testing.

2.7.4 Computational modeling of knee joint implants

The knee joint model is typically rebuilt from a collection of CT scan images. Geometric modeling and image processing tools were used to reconstruct the 3D geometry from 2D images. Geometry reconstruction requires precise tissue boundary selection from images manually or automatically – named segmentation. Once the geometry is extracted from the images, there is need to improve the accuracy and smoothness of the extracted surface. Commercial software to improve the quality of the surface and also to develop the FE model of the knee joint [29] is available. The morphology of the human body is frequently analyzed using 3D models generated from CT scans to gather basic information for implant design.

Implant sizes and forms are also crucial since greater fixation and early stability are provided by prostheses that fit patient morphologies [30, 31]. There has been some dispute concerning how men's and women's knee joints differ anatomically, and if gender-specific prostheses are required [32, 33]. FE studies were widely used in post-total knee replacement (TKR) to predict the inner movement and behavior of the implant. KneeSIM was used to forecast the contact stress on patellar components during daily activities. Additionally, they examined the kinematic pattern of different TKR designs in comparison to a healthy knee and discovered that none of them replicated the typical kinematics [34]. In comparison to a square design, Watanabe et al. demonstrated that a rounded post-cam shape produced less stress concentration during flexion with rotation and during hyperextension. These findings are helpful for predicting the contact stress on the tibial post in posterior stabilized (PS), total knee arthroplasty (TKA) and for developing implant designs [35].

Using KneeSIM, Okamoto et al. investigated how tibial posterior slope affected TF joint kinematics, quadriceps force, and PF contact force. It was shown that a greater tibial slope enhanced the quadriceps muscles' exercise effectiveness and may lessen the PF joint's contact force. When the slope angle exceeded 5°, anterior translation was discovered for the implant alignment application [36]. When compared to the normal TKR design, Kang et al. discovered that a patient-specific TKR design could considerably lessen ligament tension during deep squatting and the gait cycle [37]. Another use of computational modeling is rehabilitation planning following TKR, albeit there are not many studies on this topic in the literature. MSM was used by Li et al. to comprehend the body-level distinction between TKR and healthy knee joints. According to their findings, TKR patients exhibited a "quadriceps avoidance" gait pattern, with the vasti muscle contributing minimally to the extension moment throughout the gait cycle's stance phase [38].

In order to compare the predicted kinematics, weight loss due to wear, and wear depth contours with the results of a force-controlled experimental knee simulator, Knight et al. further developed this dynamic simulation and investigated an adaptive

FE method that is capable of simulating the wear of a PE tibial insert [39]. Furthermore, population-level simulations have been made possible by the advent of design of experiments, probabilistic, and neural network approaches. However, the absence of customized material qualities and boundary constraints makes it difficult to create subject-specific models [40]. Notwithstanding these drawbacks, it is anticipated that computer modeling will become more significant in preclinical testing and TKR design, especially when paired with multi-scale analysis and supported by experimental evidence.

2.7.5 Computational modeling of tibia bone implants

Computational modeling of tibia bone implants Provides important information for surgical planning and implant design. Patient-specific simulations of bone–implant systems are made possible by automated procedures that use CT data and FEA to assess mechanical stability, stress distribution and also to predict the healing process [41, 42]. In the tibial fracture area, edge of the screw and hole, and bone plate edges were fine meshed. To replicate bone plate fixation, screws were attached to the hole surface of the bone plate and bone nail, and the tibial fracture surfaces' friction coefficient was set to 0.3 to mimic the real fixation condition [41]. In tibia, fracture healing getting delayed due to lack of required movement in the fracture site may be due to high stiffness of bone plate [42]. When it comes to the initial stages of the healing process, this highlights the significance of receiving biomechanical treatments that are both rigorous and appropriate.

The majority of the time, bone plates used in osteosynthesis are constructed out of metals like stainless steel, which have a Young's modulus that is significantly higher than that of bone. A stress-shielding effect and excessive contact stresses at bone surfaces are thus brought about as a result of their presence. Reoperation might be required to address the biological issues that are brought about by metal bone plates; many composite bone plates have been examined in terms of their mechanical appropriateness, biocompatibility, and biomechanical performance [43–46]. In addition, computational models can mimic bone remodeling processes after implantation and evaluate the safety of removing an external fixator in cases where the tibial defect is of critical magnitude. When it comes to determining the load-bearing capabilities of bone-implant systems, these studies offer insight on the significance of certain aspects such as the volume of the callus, the characteristics of the scaffold, and the geometry of the implant [47].

2.7.6 Conclusions

Computational modeling has significantly improved the design, assessment, and optimization of fracture implants. FEA and similar methodologies have improved understanding of biomechanical interactions between implants and bone, leading to patient-specific solutions and improved clinical results. Models for spinal, femoral, knee, and tibial implants can forecast stress distribution, detect failure processes, and enhance fixation procedures. Novel methodologies, including composite materials and individualized designs, address issues like stress shielding and protracted healing in osteoporotic and critical fractures. However, limitations persist, such as the need for comprehensive model validation and the lack of tailored material qualities in some models. Future advancements should focus on multi-scale modeling, rigorous experimental validation, and sophisticated simulation tools.

References

[1] Noor S, Pridham C, Fawcett T, Barclay M, Feng YT, Hassan O, Pallister I. Finite element analysis modelling of proximal femoral fractures, including post-fixation periprosthetic fractures. Injury. 2013; 44(6): 791–795.

[2] Wirth AJ, Müller R, Van Lenthe GH. Computational analyses of small endosseous implants in osteoporotic bone. European Cells & Materials Journal. 2010; 20: 58–71.

[3] Shim V, Höch A, Grunert R, Peldschus S, Böhme J. Development of a patient-specific finite element model for predicting implant failure in pelvic ring fracture fixation. Computational and Mathematical Methods in Medicine. 2017; 2017(1): 9403821.

[4] Lewis GS, Mischler D, Wee H, Reid JS, Varga P. Finite element analysis of fracture fixation. Current Osteoporosis Reports. 2021; 19(4): 403–416.

[5] Moazen M, Jones AC, Leonidou A, Jin Z, Wilcox RK, Tsiridis E. Rigid versus flexible plate fixation for periprosthetic femoral fracture – computer modelling of a clinical case. Medical Engineering & Physics. 2012; 34(8): 1041–1048.

[6] Nečas A, Urbanová L, Fürst T, Ženčák P, Tuček P. Mathematical modelling of crack fractography after implant failure of titanium 4.5 LCP used for flexible bridging osteosynthesis in a miniature pig. Acta Veterinaria Brno. 2011; 79(4): 621–626.

[7] Shirazi-Adl SA, Shrivastava SC, Ahmed AM. Stress analysis of the lumbar disc-body unit in compression a three-dimensional nonlinear finite element study. Spine. 1984; 9(2): 120–134.

[8] Galbusera F, Bellini CM, Aziz HN, Raimondi MT, Brayda-Bruno M, Fornari M. Parametric FE mesh generation: Application to the cervical spine. Journal of Applied Biomaterials and Biomechanics. 2008; 6(2): 95–103.

[9] Niemeyer F, Wilke HJ, Schmidt H. Geometry strongly influences the response of numerical models of the lumbar spine – A probabilistic finite element analysis. Journal of Biomechanics. 2012; 45(8): 1414–1423.

[10] Manickam PS, Roy S. The biomechanical study of cervical spine: A finite element analysis. The International Journal of Artificial Organs. 2022; 45(1): 89–95.

[11] Laville A, Laporte S, Skalli W. Parametric and subject-specific finite element modelling of the lower cervical spine. Influence of geometrical parameters on the motion patterns. Journal of Biomechanics. 2009; 42(10): 1409–1415.

[12] Noailly J, Malandrino A, Galbusera F. Computational Modelling of Spinal Implants. In: Zhongmin Jin, editors. Computational Modelling of Biomechanics and Biotribology in the Musculoskeletal System. Woodhead Publishing; 2014, 447–484. ISBN: 9780857096616.
[13] Kim YOUNG, Vanderby RAY. Finite element analysis of interbody cages in a human lumbar spine. Computer Methods in Biomechanics and Biomedical Engineering. 2000; 3(4): 257–272.
[14] Adam C, Pearcy M, McCombe P. Stress analysis of interbody fusion – finite element modelling of intervertebral implant and vertebral body. Clinical Biomechanics. 2003; 18(4): 265–272.
[15] Manickam PS, Roy S, Shetty GM. Biomechanical evaluation of a novel S-type, dynamic zero-profile cage design for anterior cervical discectomy and fusion with variations in bone graft shape: A finite element analysis. World Neurosurgery. 2021; 154: e199–e214.
[16] Daniel Glad Stephen JAM, Prakash M. The influence of the viscoelastic property of polycarbonate urethane as an artificial disc core material under various physiological motions at the L4–L5 level. The International Journal of Artificial Organs. 2022; 45(11): 957–965.
[17] Fantigrossi A, Galbusera F, Raimondi MT, Sassi M, Fornari M. Biomechanical analysis of cages for posterior lumbar interbody fusion. Medical Engineering & Physics. 2007; 29(1): 101–109.
[18] Galbusera F, Schmidt H, Wilke HJ. Lumbar interbody fusion: A parametric investigation of a novel cage design with and without posterior instrumentation. European Spine Journal. 2012; 21: 455–462.
[19] Fasser MR, Gerber G, Passaplan C, Cornaz F, Snedeker JG, Farshad M, Widmer J. Computational model predicts risk of spinal screw loosening in patients. European Spine Journal. 2022; 31(10): 2639–2649.
[20] Goel VK, Nyman E. Computational modeling and finite element analysis. Spine. 2016; 41: S6–S7.
[21] Noor S, Pridham C, Fawcett T, Barclay M, Feng YT, Hassan O, Pallister I. Finite element analysis modelling of proximal femoral fractures, including post–fixation periprosthetic fractures. Injury. 2013; 44(6): 791–795.
[22] Basafa E, Armiger RS, Kutzer MD, Belkoff SM, Mears SC, Armand M. Patient-specific finite element modeling for femoral bone augmentation. Medical Engineering & Physics. 2013; 35(6): 860–865.
[23] Dhason R, Roy S, Datta S. Metal and composite bone plates for B1 periprosthetic femoral fracture in healthy and osteoporotic condition: A comparative biomechanical study. The International Journal of Artificial Organs. 2022; 45(8): 704–714.
[24] Dubov A, Kim SYR, Shah S, Schemitsch EH, Zdero R, Bougherara H. The biomechanics of plate repair of periprosthetic femur fractures near the tip of a total hip implant: The effect of cable–screw position. Proceedings of the Institution of Mechanical Engineers, Part H: Journal of Engineering in Medicine. 2011; 225(9): 857–865.
[25] Shah S, Kim SYR, Dubov A, Schemitsch EH, Bougherara H, Zdero R. The biomechanics of plate fixation of periprosthetic femoral fractures near the tip of a total hip implant: Cables, screws, or both?. Proceedings of the Institution of Mechanical Engineers, Part H: Journal of Engineering in Medicine. 2011; 225(9): 845–856.
[26] Dhason R, Roy S, Datta S. A biomechanical study on the laminate stacking sequence in composite bone plates for Vancouver femur B1 fracture fixation. Computer Methods and Programs in Biomedicine. 2020; 196: 105680.
[27] Samiezadeh S, Avval PT, Fawaz Z, Bougherara H. On optimization of a composite bone plate using the selective stress shielding approach. Journal of the Mechanical Behavior of Biomedical Materials. 2015; 42: 138–153.
[28] Dhason R, Roy S, Datta S. Surrogate model assisted design optimization of composite bone plates to achieve selective stress shielding. Materials and Manufacturing Processes. 2023; 38(16): 2034–2043.
[29] Weiss JA, Gardiner JC, Ellis BJ, Lujan TJ, Phatak NS. Three–dimensional finite element modeling of ligaments: Technical aspects. Medical Engineering & Physics. 2005; 27(10): 845–861.
[30] Incavo SJ, Ronchetti PJ, Howe JG, Tranowski JP. Tibial plateau coverage in total knee arthroplasty. Clinical Orthopaedics and Related Research (1976–2007). 1994; 299: 81–85.

[31] Westrich GH, Haas SB, Insall JN, Frachie A. Resection specimen analysis of proximal tibial anatomy based on 100 total knee arthroplasty specimens. The Journal of Arthroplasty. 1995; 10(1): 47–51.
[32] Barrett WP. The need for gender-specific prostheses in TKA: Does size make a difference?. Orthopedics. 2006; 29(9): S53.
[33] Greene KA. Gender-specific design in total knee arthroplasty. The Journal of Arthroplasty. 2007; 22(7): 27–31.
[34] Morra EA, Rosca M, Greenwald JF, Greenwald AS. The influence of contemporary knee design on high flexion: A kinematic comparison with the normal knee. JBJS. 2008; 90(4): 195–201.
[35] Watanabe T, Koga H, Horie M, Katagiri H, Sekiya I, Muneta T. Post-cam design and contact stress on tibial posts in posterior–stabilized total knee prostheses: Comparison between a rounded and a squared design. The Journal of Arthroplasty. 2017; 32(12): 3757–3762.
[36] Okamoto S, Mizu-uchi H, Okazaki K, Hamai S, Nakahara H, Iwamoto Y. Effect of tibial posterior slope on knee kinematics, quadriceps force, and patellofemoral contact force after posterior–stabilized total knee arthroplasty. The Journal of Arthroplasty. 2015; 30(8): 1439–1443.
[37] Kang KT, Koh YG, Jung M, Nam JH, Son J, Lee YH, Kim SH. The effects of posterior cruciate ligament deficiency on posterolateral corner structures under gait-and squat-loading conditions: A computational knee model. Bone & Joint Research. 2017; 6(1): 31–42.
[38] Li K, Ackland DC, McClelland JA, Webster KE, Feller JA, De Steiger R, Pandy MG. Trunk muscle action compensates for reduced quadriceps force during walking after total knee arthroplasty. Gait & Posture. 2013; 38(1): 79–85.
[39] Knight LA, Pal S, Coleman JC, Bronson F, Haider H, Levine DL, Rullkoetter PJ. Comparison of long–term numerical and experimental total knee replacement wear during simulated gait loading. Journal of Biomechanics. 2007; 40(7): 1550–1558.
[40] Shu L, Li S, Sugita N. Systematic review of computational modelling for biomechanics analysis of total knee replacement. Biosurface and Biotribology. 2020; 6(1): 3–11.
[41] Liao B, Sun J, Xu C. A mechanical study of personalised Ti6Al4V tibial fracture fixation plates with grooved surface by finite element analysis. Biosurf Biotribol. 2921; 7: 142–153.
[42] Son DS, Mehboob H, Chang SH. Simulation of the bone healing process of fractured long bones applied with a composite bone plate with consideration of the blood vessel growth. Composites Part B: Engineering. 2014; 58: 443–450.
[43] Spivak JM, Zuckerman JD, Kummer FJ, Frankel VH. Fatigue failure of the sliding screw in hip fracture fixation: A report of three cases. Journal of Orthopaedic Trauma. 1991; 5(3): 325–331.
[44] Kim JH, Kim SH, Chang SH. Estimation of the movement of the inter–fragmentary gap of a fractured human femur in the presence of a composite bone plate. Journal of Composite Materials. 2011; 45(14): 1491–1498.
[45] Kim SH, Chang SH, Jung HJ. The finite element analysis of a fractured tibia applied by composite bone plates considering contact conditions and time–varying properties of curing tissues. Composite Structures. 2010; 92(9): 2109–2118.
[46] Roy D, Das MC, Dhason R, Roy S, Datta S. Effect of matrix material property on the composite tibia fracture plate: A biomechanical study. Biomedical Physics & Engineering Express. 2024; 10(3): 035026.
[47] Doyle H, Lohfeld S, Dürselen L, McHugh P. Computational modelling of ovine critical–sized tibial defects with implanted scaffolds and prediction of the safety of fixator removal. Journal of the Mechanical Behavior of Biomedical Materials. 2015; 44: 133–146.

3 Manufacturing

Mohammadjavad (Matin) Einafshar*, Mohadese Rajaeirad,
and Michael Skipper Andersen

Chapter 3.1
Patient-specific implant (PSI) by additive manufacturing

Abstract: Additive Manufacturing (AM), commonly known as 3D printing, has significantly advanced medical applications, particularly through the production of Patient-Specific Implants (PSIs). This chapter provides a detailed overview of AM techniques, emphasizing their customization capabilities, precision, and suitability for medical applications. Various AM methods such as Fused Deposition Modeling (FDM), Stereolithography (SLA), Selective Laser Sintering (SLS), Direct Metal Laser Sintering (DMLS), and Electron Beam Melting (EBM) are introduced, outlining their processes, suitable materials, and relevant medical applications. The chapter highlights the transformative role of AM in creating highly customized implants tailored to individual patient anatomy, thereby enhancing surgical outcomes. It further discusses existing challenges, including material limitations, manufacturing constraints, regulatory issues, and cost considerations, along with promising future directions that include material innovations, technological advancements, increased clinical adoption, and interdisciplinary research.

Keywords: Additive Manufacturing, Patient-Specific Implants, 3D Printing, Customization, Orthopedic Implants, Dental Implants, Craniofacial Implants, Electron Beam Melting, Direct Metal Laser Sintering, Surgical Applications

3.1.1 Introduction

3.1.1.1 Overview of additive manufacturing (AM)

Additive manufacturing (AM), commonly referred to as 3D printing, is a process for creating three-dimensional objects by adding material layer by layer, guided by a digital model. Unlike traditional manufacturing methods that often involve removing material from a larger block (subtractive manufacturing), AM focuses on building objects

*Corresponding author: **Mohammadjavad (Matin) Einafshar**, Department of Material and Production, Aalborg University, Aalborg, Denmark, e-mail: moei@mp.aau.dk
Mohadese Rajaeirad, Department of Biomedical Engineering, Faculty of Engineering, University of Isfahan, Iran, e-mail: m.rajaeirad@eng.ui.ac.ir
Michael Skipper Andersen, Department of Material and Production, Aalborg University, Denmark, e-mail: msa@mp.aau.dk

https://doi.org/10.1515/9783111386027-013

incrementally. This approach opens up possibilities for crafting intricate designs and complex geometries that are often challenging – or even impossible – to achieve using conventional manufacturing techniques.

The process begins with a 3D computer-aided design (CAD) model, which is digitally sliced into thin, horizontal layers. These slices serve as a blueprint for the printer, dictating how material is deposited step by step to form the final object [1].

3.1.1.2 Principles and key technologies of additive manufacturing

At the heart of AM lies its distinctive layer-by-layer construction method. This technique allows for unparalleled customization and the production of geometrically complex structures. One of the most significant advantages of AM is its material efficiency – unlike traditional methods that often generate substantial waste, AM uses only the material needed for the object.

Moreover, AM's inherent design flexibility enables the creation of structures that were previously considered unfeasible with traditional manufacturing techniques. These benefits have made AM an indispensable tool across various industries.

Several key technologies form the backbone of AM, each tailored to specific applications:

- **Fused deposition modeling (FDM):** This widely adopted method involves extruding a thermoplastic filament through a heated nozzle to build objects layer by layer. FDM is popular due to its simplicity, affordability, and compatibility with a broad range of materials (Fig. 3.1.1. A).
- **Selective laser sintering (SLS):** In this process, a high-powered laser fuses powdered materials to form solid structures. SLS is particularly advantageous for creating durable components with intricate geometries, making it ideal for industrial applications.
- **Stereolithography (SLA):** SLA uses a laser to cure liquid resin into solid parts with exceptional precision and detail. It is favored for applications requiring high-resolution outputs (Fig. 3.1.1. C).
- **Direct metal laser sintering (DMLS):** DMLS employs a laser to sinter metal powders, producing robust, highly detailed metal parts. It is especially suited for industries that demand high-strength components, such as aerospace and automotive manufacturing.

These technologies collectively showcase the versatility of AM, supporting a wide range of applications across diverse sectors. The specific features, advantages, and limitations of these methods will be explored in further detail.

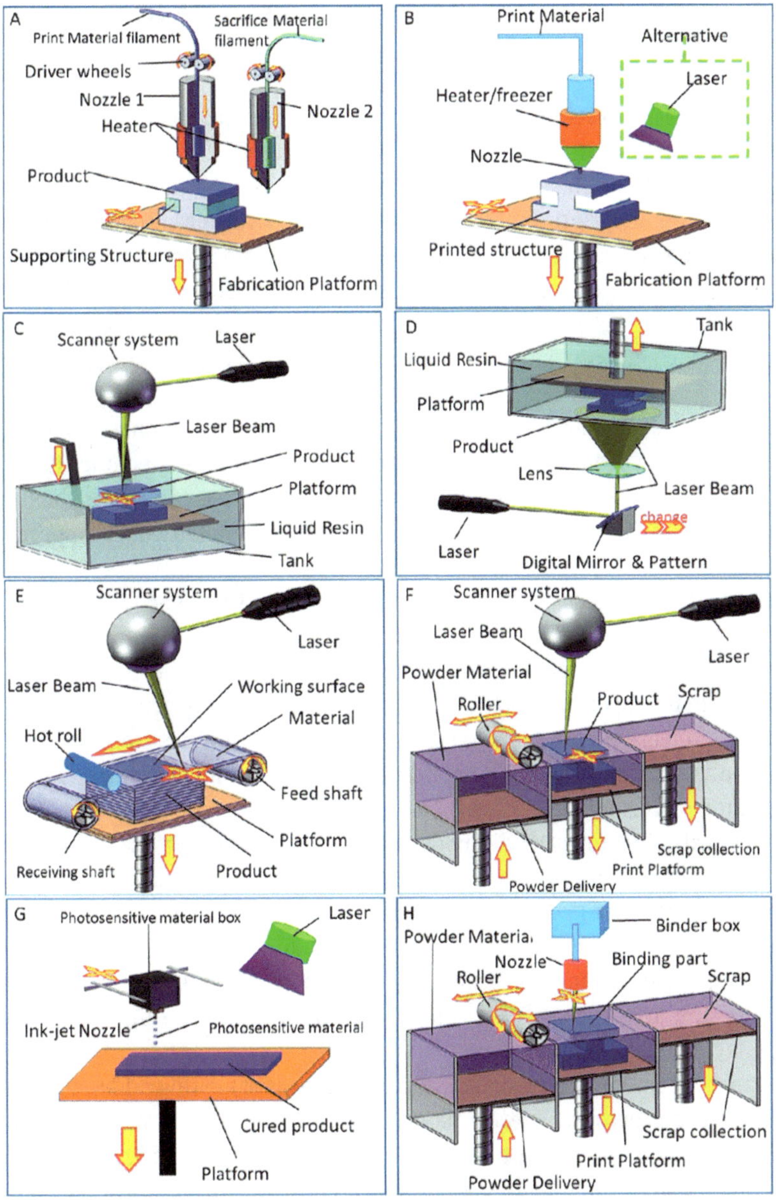

Figure 3.1.1: 3D printing methods include: (A) fused deposition modeling (FDM); (B) direct ink writing (DIW); (C) stereolithography (SLA); (D) digital light processing (DLP); (E) lamination (LOM); (F) selective laser sintering (SLS) and selective laser melting (SLM); (G) photopolymer jetting (PolyJet); and (H) binder jetting (3DP). This image is open access and distributed under the terms of the Creative Commons CC BY 4.0 license, which permits unrestricted use, distribution, and reproduction in any medium, provided the original work is properly cited. Permission not required (License: CC BY 4.0) [2].

3.1.1.3 Historical development and significance in medical applications

3D printing, the process of creating objects by layering material, has undergone remarkable evolution since its inception in the 1980s. To simplify the concept, imagine using a glue gun to stack thin layers of material, one atop the other, until a complete object emerges. While 3D printing has recently captured widespread attention, its roots trace back to the groundbreaking work of Chuck Hull, who invented the first 3D printer and patented the SLA process in 1986 https://en.wikipedia.org/w/index.php?title=Chuck_Hull&oldid=1231244558 [53]. Hull's SLA method involved curing layers of photosensitive resin to construct a three-dimensional object, laying the foundation for modern AM.

Interestingly, Hull's was not the only mind exploring this revolutionary concept. Around the same time, a team of French inventors led by Alain Le Mehaute filed a similar patent application. However, their work was ultimately abandoned, as the potential applications of 3D printing were not yet widely understood or appreciated (2023).

The 1980s saw the emergence of other AM techniques that broadened the scope of 3D printing. Carl Deckard developed SLS, while EOS GmbH introduced DMLS. Another key figure, S. Scott Crump, invented FDM and co-founded Stratasys, a company that played a pivotal role in advancing FDM technology. During this early period, 3D printers were costly and primarily used for industrial prototyping. However, technological advancements have since expanded the applications of 3D printing to industries such as aerospace, automotive, and healthcare [1, 3].

Among the various 3D printing methods, FDM remains the most accessible for consumer-grade printers. The process works by melting a thermoplastic filament, which is extruded through a heated nozzle and layered to build the desired object. SLA, another widely used method, employs a UV laser to cure liquid resin into detailed objects, though additional post-processing is often required. Digital light processing functions similarly but uses micromirrors to precisely direct light onto the resin, curing it layer by layer.

Other advanced techniques include SLS and selective laser melting (SLM), both of which utilize lasers to process powdered materials such as plastics, ceramics, or metals. The key distinction between the two is that SLS sinters the material, fusing particles together, while SLM fully melts the powder, creating a denser, more homogeneous structure. Another metal-based method, electron beam melting (EBM), uses an electron beam instead of a laser to melt metal powders.

The 2000s marked a turning point for 3D printing, as numerous foundational patents expired between 2002 and 2014. This shift paved the way for the development of affordable consumer-grade printers. One of the most notable early models was the RepRap, a self-replicating machine designed to produce many of its own components using 3D printing. The open-source philosophy behind the RepRap project ignited a wave of innovation, making 3D printing more accessible and fueling its rapid adoption across various fields [4].

3.1.1.4 Significance in medical applications

AM has profoundly impacted healthcare, paving the way for further advancements. One of its earliest applications was the creation of anatomical models for surgical planning. These 3D-printed models enable surgeons to hold and examine highly accurate replicas of complex anatomical structures. These models help surgeons plan surgeries more precisely by giving them a concrete, hands-on understanding of each patient's unique anatomy. Therefore, this lowers the risk of complications and improves surgical outcomes.

As AM technologies advanced, their potential to improve personalized medicine became increasingly evident. One notable breakthrough was the production of customized implants and prosthetics. Traditional manufacturing methods typically produced generic, "one-size-fits-all" solutions that often failed to meet the specific needs of individual patients. With AM, however, implants can now be tailored to fit each patient's unique anatomy. This personalized approach enhances both the functionality of the implants and the comfort of the patient, leading to better recovery outcomes and higher levels of satisfaction.

Perhaps one of the most transformative developments in healthcare is bioprinting – the application of AM techniques to create tissue and organ structures. Bioprinting holds immense promise for regenerative medicine and organ transplantation (Fig. 3.1.2). Researchers are looking into creative solutions to the ongoing organ donor shortage by creating tissues from living cells. Because this technology makes it possible to create substitute tissues or even entire organs in a lab, it has the potential to completely change therapeutic approaches [5].

The use of AM in medical implants has been very innovative. Unlike traditional manufacturing, which frequently generates implants in uniform shapes and sizes, AM enables the manufacture of complicated and porous structures that are uniquely adapted to the patient's body. For example, AM enables the production of bone implants with porous architectures designed to promote osseointegration – the natural bonding of the implant with surrounding bone tissue. This feature gives patients a quicker return to normalcy by speeding up recovery timeframes in addition to improving the implant's stability and functionality [5, 6].

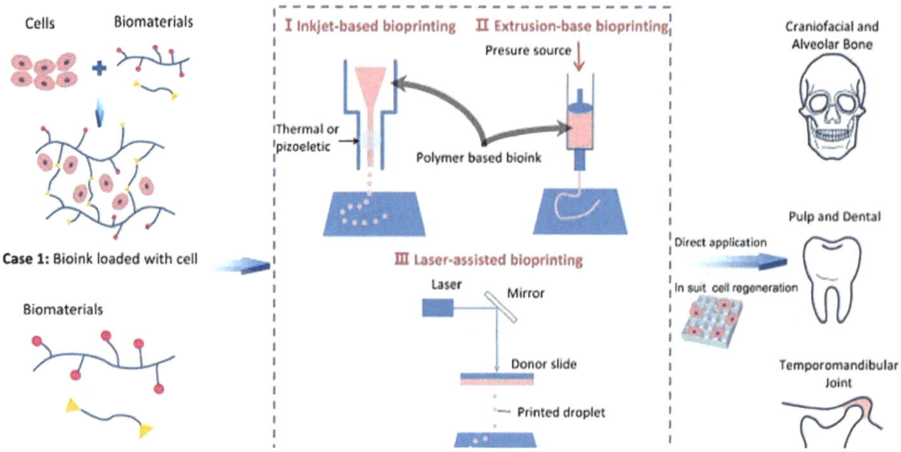

Figure 3.1.2: Diagrammatic overview outlining the definition, principles, processes, classifications, and uses of 3D bioprinting in craniofacial and dental tissue engineering. This image is distributed under the Creative Commons CC BY license, allowing unrestricted use, distribution, and reproduction in any format, provided proper credit is given to the original source. Permission not required (License: CC BY) [7].

3.1.2 Types of additive manufacturing techniques

AM encompasses a range of techniques, each offering distinct advantages depending on the materials used and the specific application. In the medical field, selecting the right AM technique is critical for creating patient-specific devices, implants, and anatomical models that meet the highest standards of precision and biocompatibility. Below, we explore some of the most widely employed AM techniques, detailing their processes, materials, and medical applications.

3.1.2.1 Fused deposition modeling (FDM)

FDM is one of the most popular 3D printing methods due to its simplicity and affordability. This technique works by melting thermoplastic filament and extruding it layer by layer to construct an object (Fig. 3.1.3). Critical parameters in the FDM process include layer height, which influences surface smoothness and detail, and extrusion temperature, which ensures proper adhesion between layers. Infill density also plays a vital role, affecting the strength and weight of the final part [8, 9].

Common materials used in FDM include:

- **Polylactic acid (PLA):** Known for its ease of use and smooth surface finishes.

- **Acrylonitrile butadiene styrene:** A durable and impact-resistant option.
- **Polyhydroxyalkanoates:** Biocompatible polymers ideal for medical applications [10].

FDM has been successfully employed in producing patient-specific implants (PSIs), such as cranial and dental implants, as well as biodegradable bone screws made from PLA. These applications showcase the potential of FDM for creating cost-effective and customizable medical devices [11].

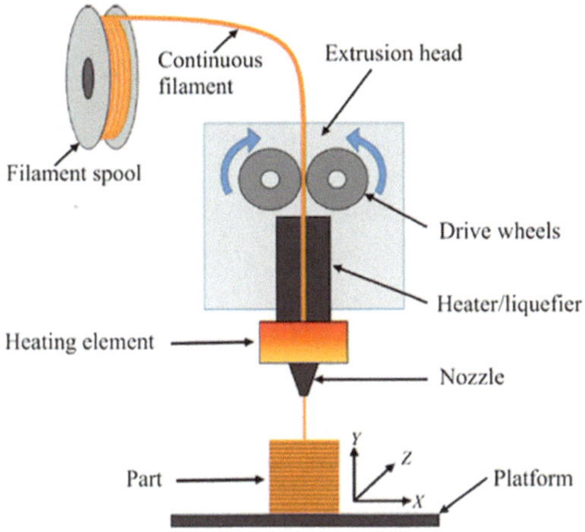

Figure 3.1.3: Schematic of FDM printer system. Reprinted from (Subhedar et al 2023) with permission from Elsevier [12].

3.1.2.2 Stereolithography (SLA)

SLA is a high-precision 3D printing method that uses a laser to selectively cure liquid resin, layer by layer, into solid objects (Fig. 3.1.4). Renowned for its ability to produce high-resolution parts with exceptionally smooth surfaces, SLA is widely used in healthcare for its accuracy and detail [13].

Biocompatible resins used in SLA include:

- Dental resins: For creating custom aligners and dental prosthetics.
- Medical-grade resins: For surgical guides and implants.

SLA's ability to create detailed anatomical models and custom-fit medical devices has significantly improved surgical precision and patient outcomes [14].

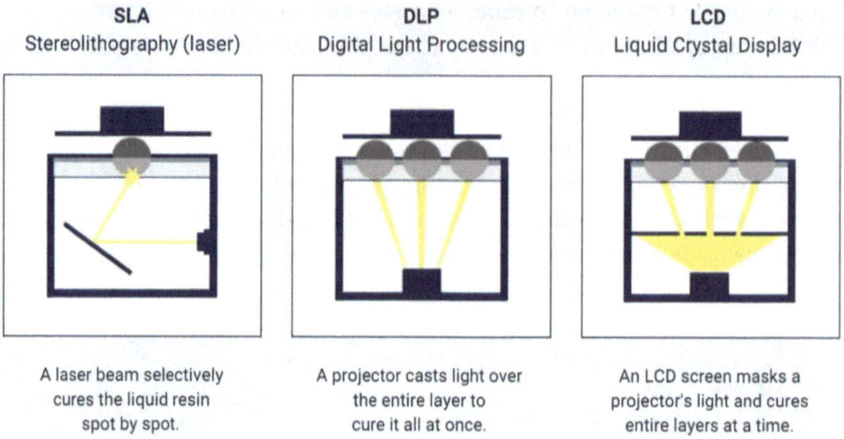

Figure 3.1.4: Main resin 3D printing methods (source: zongheng3d.com).

3.1.2.3 Selective laser sintering (SLS)

SLS involves using a laser to sinter powdered materials, layer by layer, into solid objects. This process does not require support structures, enabling the creation of complex geometries. The final product's resolution influences surface finish and detail.

Materials commonly used in SLS for medical applications include:

- Nylon: Known for its strength and flexibility.
- Thermoplastic elastomers: Soft and stretchable materials ideal for medical devices.

SLS is particularly valuable for producing orthopedic implants and bone scaffolds, which benefit from intricate structures designed to enhance tissue integration [15].

3.1.2.4 Direct metal laser sintering (DMLS) and selective laser melting (SLM)

DMLS and SLM are metal-based AM techniques that utilize lasers to transform metal powders into solid objects. While DMLS sinters the powder, SLM fully melts it, resulting in a denser and more homogeneous structure. Both techniques often require post-print heat treatments to enhance the mechanical properties of the final part.

Materials commonly used in DMLS and SLM include:

- Titanium: Valued for its strength and biocompatibility.
- Cobalt-Chrome: Highly wear-resistant, making it ideal for load-bearing implants.

These methods are instrumental in producing precise and durable implants, such as hip and knee replacements, where strength and accuracy are paramount [16].

3.1.2.5 Electron beam melting (EBM)

EBM is similar to DMLS and SLM but employs an electron beam instead of a laser to melt metal powder in a vacuum, reducing the risk of oxidation. This technique is particularly suitable for reactive metals like titanium, making it ideal for manufacturing load-bearing implants such as spinal and cranial devices [16].

3.1.2.6 Binder jetting

Binder jetting creates solid parts by depositing a liquid binding agent onto a powder bed (Fig. 3.1.5). While less common in medical applications, it shows potential for fabricating porous structures, essential for implants requiring tissue integration, such as bone scaffolds and cartilage replacements [16] (Fig. 3.1.6).

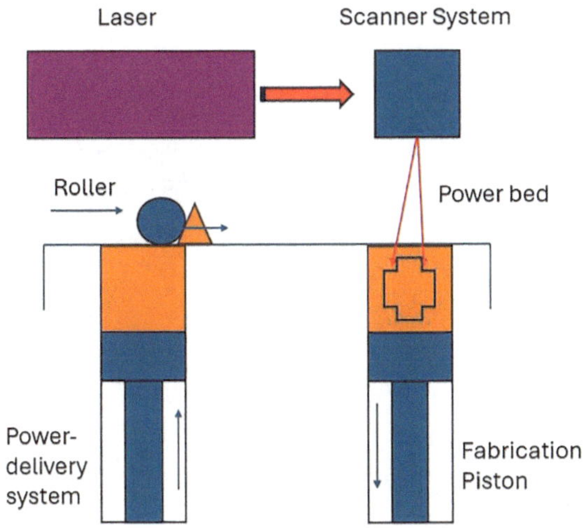

Figure 3.1.5: Schematic of a binder jetting system [17].

Each AM technique offers unique advantages, enabling tailored solutions for a wide range of medical applications. With advancements in technology, the integration of these methods into healthcare promises to deliver even greater precision, customization, and accessibility in the future (Fig. 3.1.7).

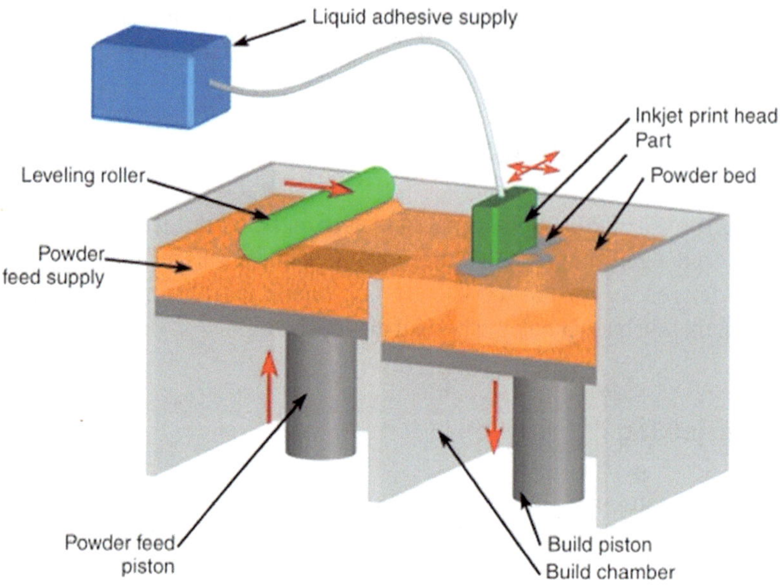

Figure 3.1.6: Schematic of binder jetting system (source: CustomPartNet. http://www.custompartnet.com/wu/3d-printing).

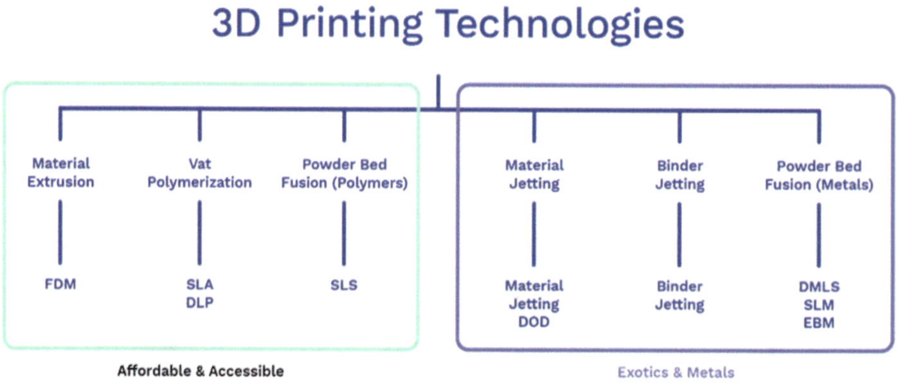

Figure 3.1.7: Summary of 3D printing technologies.

3.1.3 Patient-specific implants (PSIs) manufactured by AM

AM has completely changed the way PSIs are made, allowing for an unprecedented level of customization and precision. These personalized implants are a game-changer in medicine, where no two patients are the same, and standard solutions often fall short of what is needed.

3.1.3.1 Customization and precision of PSIs using AM

One of the most exciting things about AM is its ability to create implants tailored to fit a patient's unique anatomy. For example, researchers have used fused filament fabrication (FFF) to make 316 L stainless steel implants for animals, precisely crafted to match their anatomy [18]. Imagine the possibilities for human patients! AM also allows for intricate designs, which are essential for implants that need to integrate seamlessly into complex body structures [19].

In one heartwarming case, doctors used AM to create a lightweight prosthesis for a child with bone cancer. The customized design helped the implant integrate better with the child's bone and improved their mobility, demonstrating how personalized implants can significantly improve quality of life [20].

3.1.3.2 Advantages over traditional manufacturing methods

Compared to traditional methods like casting or milling, AM offers some major advantages:

- **Flexibility:** AM can create implants with incredibly intricate designs, which is critical for complex surgeries, like reconstructing facial bones [21].
- **Speed:** In emergencies, time matters. AM allows doctors go from design to production much faster than conventional methods [19].
- **Less waste:** Traditional methods often leave a lot of leftover material, but AM uses only what is needed. This is especially important when working with costly materials like titanium [19].
- **Perfect fit:** Since AM customizes implants for each patient, the fit and functionality are often much better, as seen in dental applications where precision is everything [22].

3.1.3.3 Orthopedic implants

In orthopedic surgery, AM has been successfully employed to create custom knee and hip implants with porous structures that promote osseointegration while reducing wear (Fig. 3.1.8). These implants are designed to match the patient's anatomy, providing superior stress distribution and encouraging bone growth (Fig. 3.1.9). Materials such as Ti-6Al-4V and Co-29Cr-6Mo alloys are frequently used due to their excellent mechanical properties and biocompatibility, making them ideal for both solid and porous implant designs [23]. Studies have shown that custom AM-produced implants improve recovery times and overall patient outcomes [24].

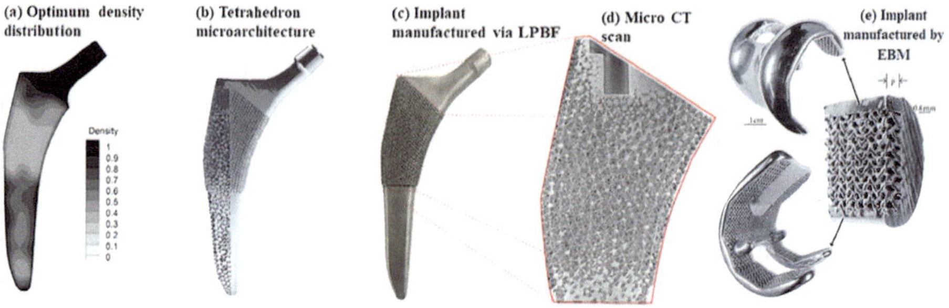

Figure 3.1.8: (a) Optimal relative density distribution for a fully porous implant; (b) creation of lattice microarchitecture from this density distribution using a strong tetrahedral structure; (c) implant manufacturing through laser powder bed fusion (LPBF); (d) micro-CT evaluation of the implant lattice in the proximal area [25]; (e) prototype of a Co-29Cr-6Mo alloy femoral knee implant made by electron beam melting (EBM) and hot isostatic pressing (HIP) per ASTM-F75 standards, highlighting the porous inner surface zone (p) of the monolith [23] (part (e) adapted from [23], licensed under CC BY 3.0 (https://creativecommons.org/licenses/by/3.0/)).

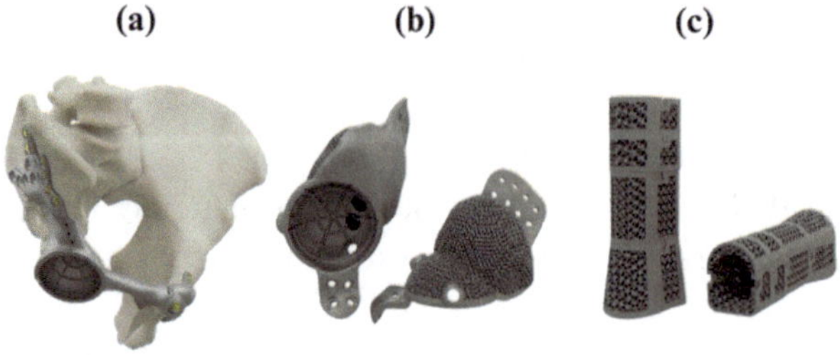

Figure 3.1.9: Additive manufacturing in orthopedic reconstruction: (a) custom pelvic implants for bone tumors, (b) custom triflange hip prosthesis, and (c) customized distal tibial cage. All implants are designed and produced by Fanavaran Jarahyar Sharif (FJSharif), Tehran, Iran.

3.1.3.4 Dental implants

AM has had a significant impact on the production of dental implants, including crowns, bridges, and orthodontic devices (Fig 3.1.10). Its ability to precisely customize implants to fit the patient's dental anatomy enhances treatment accuracy and patient comfort. AM processes frequently use biocompatible polymers and composites, which are suitable for the intricate structures required in dental applications. The orientation and thickness of the printed layers are important factors that can be optimized to enhance the mechanical properties of these dental implants [26].

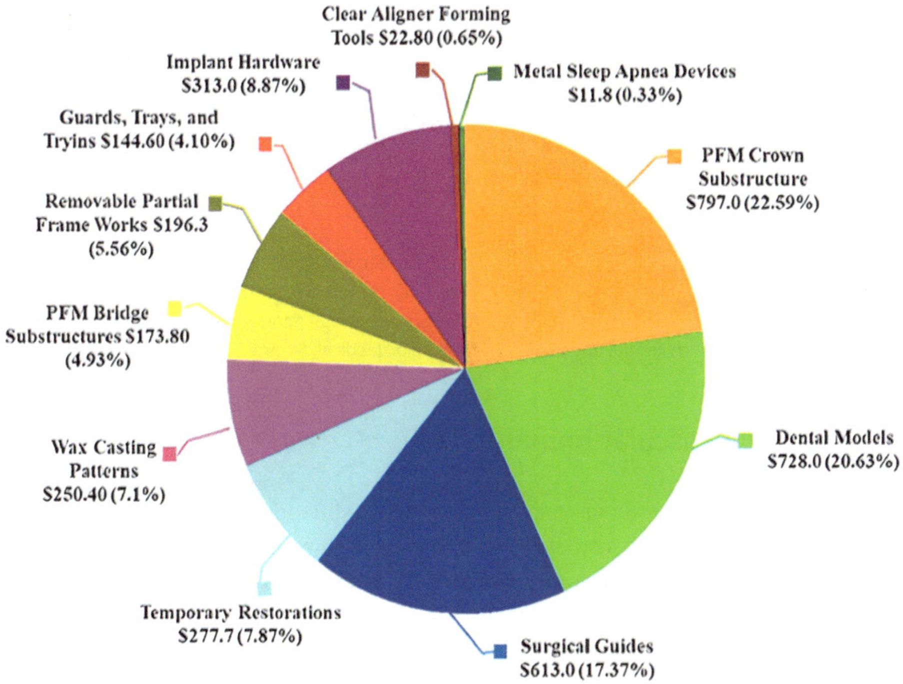

Figure 3.1.10: Projected market value of the dental 3D printing application sector by type for 2021 (in millions of USD). Reproduced with permission from [27]. Copyright 2024 American Chemical Society.

3.1.3.5 Craniofacial implants

In craniofacial surgery, AM enables the production of customized implants crucial for reconstructing complex anatomical structures. Techniques such as FFF and DMLS allow for the precise creation of implants based on a patient's specific craniofacial features. For instance, 3D-printed titanium implants have demonstrated excellent anatomical conformity in craniomaxillofacial reconstruction, resulting in optimal fit and aesthetic outcomes [28] (Fig. 3.1.11). The use of CAD/CAM technology and detailed CT data further enhances the precision and efficacy of these implants, leading to reduced surgical time and costs [29].

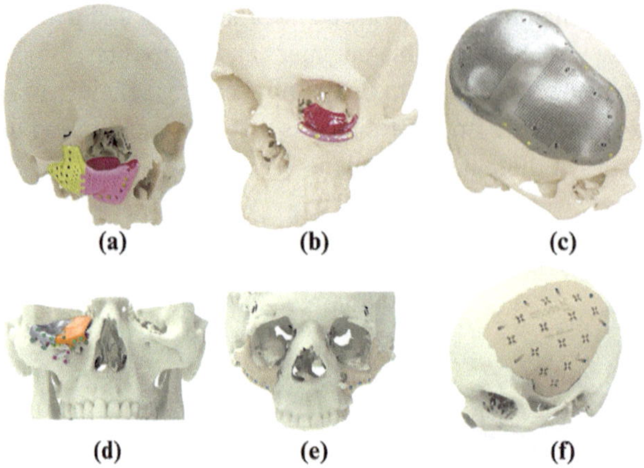

Figure 3.1.11: Patient-specific implants created using additive manufacturing for craniofacial and cranial reconstruction: (a) facial contouring implant; (b) orbital floor reconstruction; (c) cranial plate from Fanavaran Jarahyar Sharif (FJSharif), Tehran, Iran; (d) two-piece orbital floor; (e) bilateral zygoma augmentation PEEK; and (f) cranial implants PEEK from Xilloc Medical, Sittard-Geleen, the Netherlands.

3.1.3.6 Cardiovascular implants

Although AM's use in cardiovascular applications is still emerging, its potential is evident in the production of custom stents and heart valves (Fig. 3.1.13). AM's precision allows the creation of devices tailored to a patient's vascular anatomy, improving safety and efficacy (Fig. 3.1.12). Moreover, AM has been used to simulate cardiovascular procedures, such as coronary angioplasty, enhancing stent design and treatment outcomes [30].

3.1.3.7 Spinal implants

AM enables the development of spinal implants tailored to the contours of bony defects, facilitating less invasive surgeries and faster recovery (Fig. 3.1.14). These implants promote osseointegration and provide stability in complex spinal procedures, making them indispensable in addressing unique patient needs [33, 34].

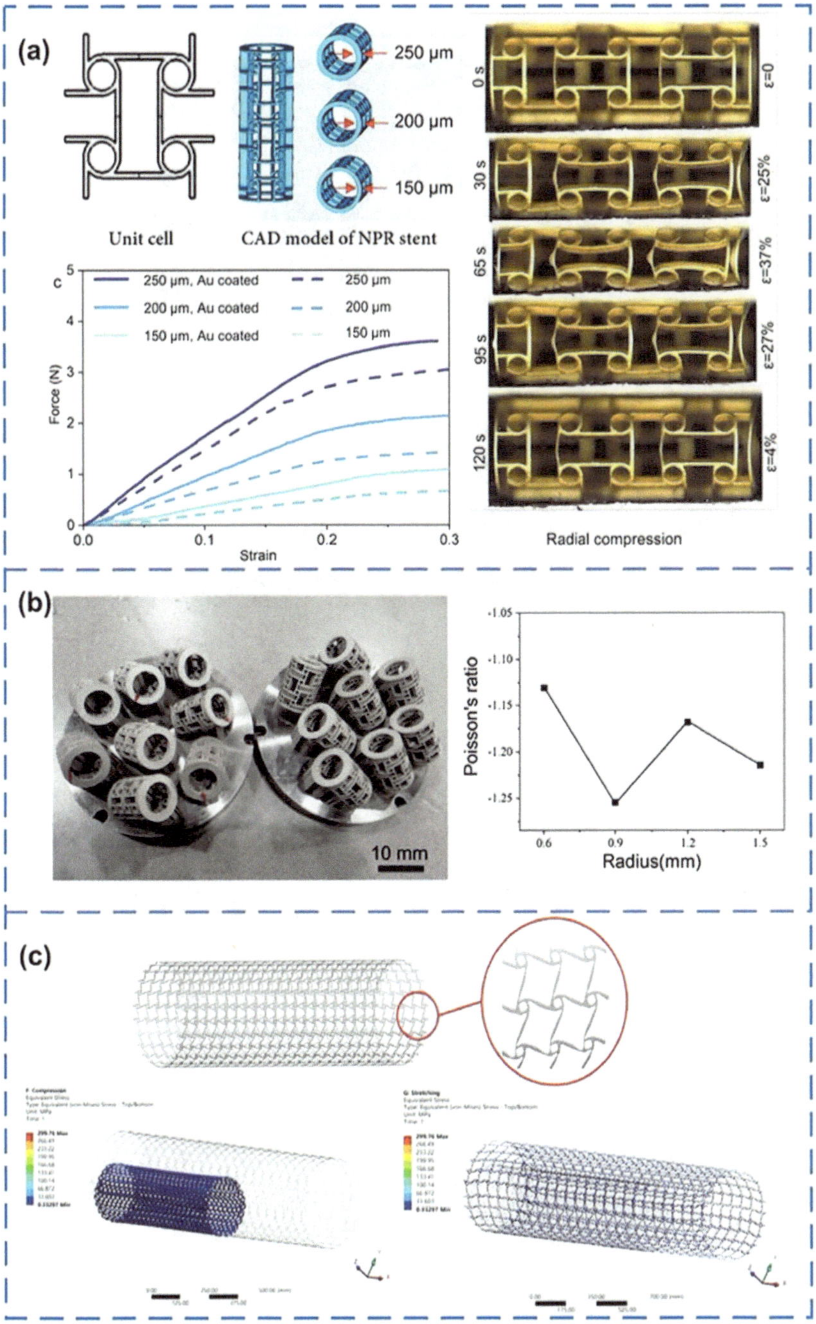

Figure 3.1.12: Additive manufacturing of vascular stents, emphasizing design precision and material optimization. Reproduced under the terms of the Creative Commons CC-BY license, which permits unrestricted use, distribution, and reproduction in any medium, provided the original work is properly cited [31].

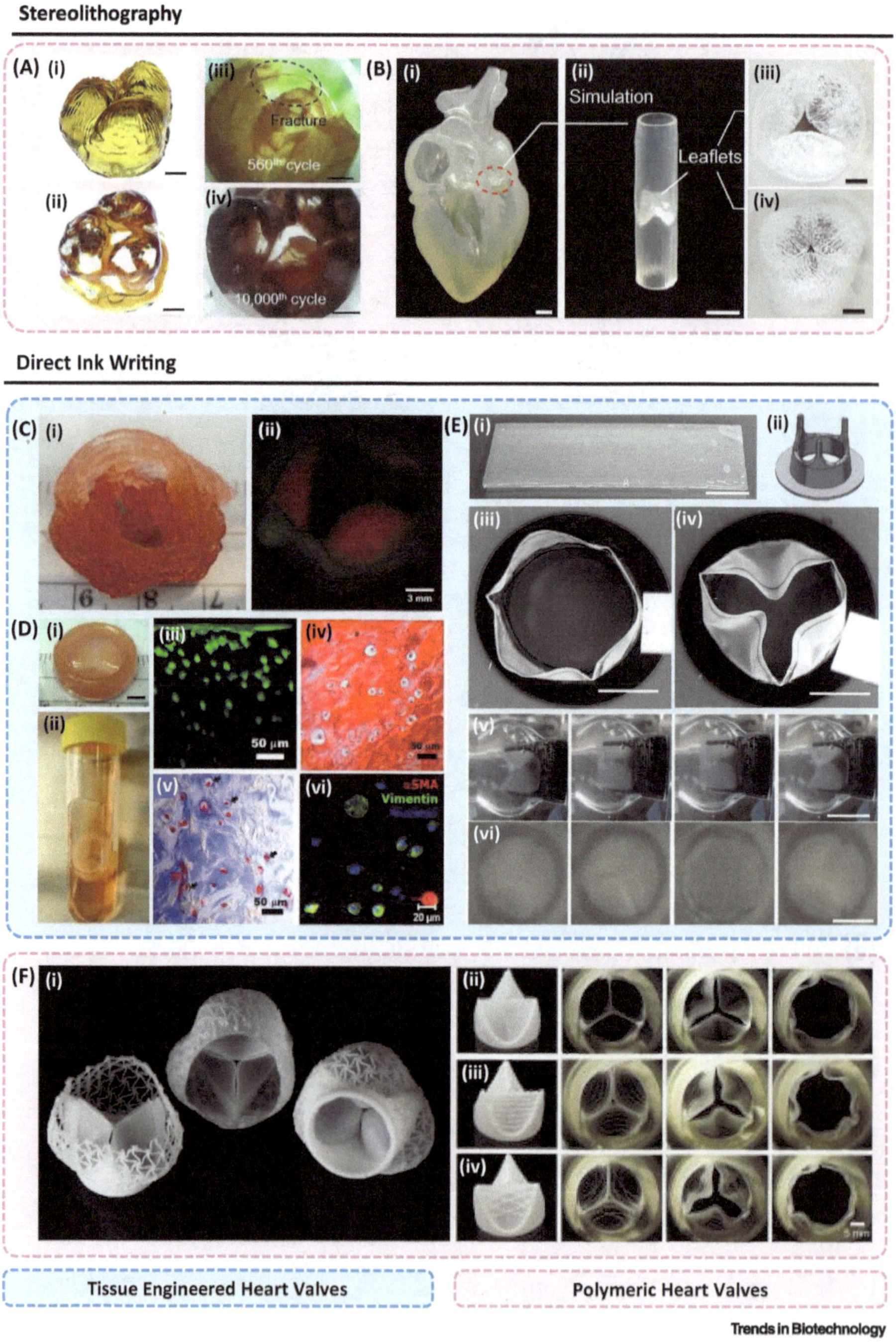

Figure 3.1.13: Heart valves fabricated using selective laser melting (SLM) and direct ink writing (DIW) technologies. Reprinted from (Vernon et al 2024) with permission from Elsevier [32].

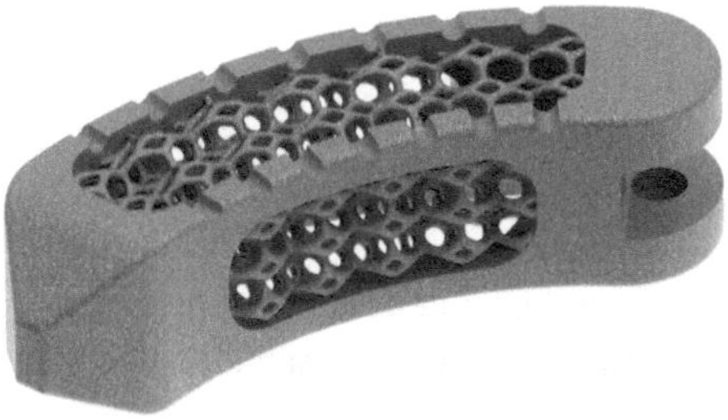

Figure 3.1.14: TLIF spinal cage produced by AM, featuring a porous design to enhance integration and stability (Xilloc Medical, Netherlands).

3.1.4 Complications and challenges

AM in the medical field, while very successful, is not without its challenges. The intricacies of creating customized implants for specific patient needs bring several complications that span material selection, manufacturing processes, regulatory compliance, and cost implications. These challenges are necessary to be aware of the limitations of current AM technologies and the obstacles that need to be addressed for broader adoption in clinical practice. A detailed exploration of these limitations will help clarify the scope of the challenges faced by both manufacturers and healthcare providers.

3.1.4.1 Material limitations

Material selection AM of medical implants is crucial due to the stringent requirements for biocompatibility and mechanical strength. Commonly used materials, like titanium and its alloys (such as Ti-6Al-4V), are favored for their excellent biocompatibility and mechanical properties. However, ensuring consistent material properties across different production batches and maintaining the mechanical integrity of printed parts remains a challenge [35, 36]. Residual materials or contaminants introduced during the printing process can pose biocompatibility risks, underscoring the importance of stringent quality control [37].

The long-term durability and wear of AM-produced implants also present concerns. While materials like titanium are resistant to corrosion and wear, the surface finish and microstructure of AM parts significantly affect their long-term performance [38]. Extensive research on the degradation of AM implants highlights the importance of thorough testing and evaluation to ensure implants remain functional over time [54].

3.1.4.2 Manufacturing challenges

AM technologies face limitations in terms of precision and resolution, which can impact the fit and function of medical implants. The layer-by-layer process of AM can result in surface roughness and geometric inaccuracies, often necessitating post-processing to meet clinical standards [54]. For example, while SLA offers high resolution, it is restricted by resin viscosity and the need for support structures during printing [39].

Post-processing is essential to enhance surface finish and improve the mechanical properties of AM implants. Procedures such as polishing, cleaning, and sterilization help eliminate surface roughness and contaminants, ensuring the biocompatibility and overall performance of the implant [37]. However, the need for extensive post-processing increases production time and costs, which poses a challenge for widespread AM adoption in clinical settings [40].

3.1.4.3 Regulatory and compliance issues

Regulatory approval for AM-produced medical devices is complex, with stringent requirements from agencies like the FDA and the European Union. These regulations ensure implants meet safety and efficacy standards, but the unique nature of AM processes presents challenges for compliance. Material variability and the high degree of customization in AM devices complicate regulatory pathways [37, 40].

Obtaining approval for AM implants requires comprehensive documentation that includes design validation, material characterization, and post-processing protocols. This documentation proves that the implants meet regulatory standards and are safe for clinical use. However, the iterative nature of AM design can make the documentation process more complex, necessitating clear communication between manufacturers and regulators [41].

3.1.4.4 Cost and accessibility

The costs associated with AM for medical implants are driven by several factors, such as raw material prices, machine operation, and post-processing needs. While AM minimizes material waste and allows for rapid prototyping, the initial investment in equipment and biocompatible materials can be substantial [42]. Studies suggest that AM becomes cost-effective in scenarios requiring customization or small batch production [42].

The scalability and accessibility of AM in smaller clinics or hospitals are limited by high costs and the need for technical expertise. Despite the advantages of customization and rapid production, economic and logistical challenges hinder its widespread clinical use. Lowering costs and simplifying processes are key to increasing accessibility and fostering broader adoption in clinical environments [43].

3.1.5 Potentials and future directions

Despite the challenges associated with AM, ongoing research and technological innovations continue to push the boundaries of what is possible in this field. The potential for AM to transform patient care by offering customized, precise medical implants and devices is enormous. Advancements in materials, manufacturing processes, and clinical adoption are paving the way for a future where AM is a cornerstone of personalized medicine. This section explores these future directions, focusing on how material innovations, improved technologies, and collaborative research are opening new opportunities in the medical landscape.

3.1.5.1 Advancements in materials

Research into new biocompatible materials for AM in medical applications is advancing rapidly. For instance, tantalum has gained attention for its excellent biocompatibility, showing promise in orthopedic and biomedical implants. The laser powder bed fusion (LPBF-LB/M) process is being explored to adapt tantalum's mechanical and biomedical properties for innovative implant designs [44]. Zirconia-based bio-ceramics are also being optimized for use in dental and orthopedic applications due to their mechanical stability and biocompatibility [45].

Improvements in materials are enhancing implant performance by boosting mechanical properties and functionality. Advancements in magnesium alloy AM, known for biodegradability and biocompatibility, are addressing challenges such as high corrosion rates and suboptimal mechanical properties. Techniques like LPBF enable the customization of implant size and shape to suit individual patients [46]. Additionally, diamond-like carbon coatings on Ti-6Al-4V Eli implants have been shown to enhance bioactivity and promote bone cell proliferation, improving the performance of AM implants [47].

3.1.5.2 Technological innovations

Advances in AM technology continue to improve the precision and speed of the manufacturing process. Laser-based techniques like LPBF are known for their ability to produce intricate 3D forms with high accuracy and reproducibility. These methods are being refined to reduce porosity and prevent defects like hot ripping and cracking, which are crucial for producing high-quality medical implants [48].

The integration of data-driven technologies, such as machine learning, is transforming the design of PSIs. By analyzing large datasets, these technologies optimize implant designs for better fit, function, and patient outcomes. AI-assisted design tools enable rapid iteration and testing of different configurations, resulting in more effective and personalized implant solutions.

3.1.5.3 Clinical adoption and customization

The use of AM in clinical settings is steadily growing, with hospitals and clinics successfully employing AM to create patient-specific anatomical models for preoperative planning, surgical guides, and customized implants which improve surgical accuracy and enhances patient outcomes [49]. The ability to produce on-demand, customized prosthetics and orthotic devices further boosts the clinical utility of AM [50].

AM is also supporting personalized medicine, as it allows for the production of medical implants and devices, which are created based on a patient's unique anatomy. This personalized approach to treatment improves the precision and patient satisfaction, thus transforming the delivery of healthcare.

3.1.5.4 Research and development

Ongoing research focuses on exploring new materials, refining AM processes and techniques, and developing novel applications in the field of PSIs. Studies are investigating biocompatible polymers, metals, and ceramics to expand the range of implant applications [44, 51]. A growing trend is the use of biodegradable and bioactive materials designed to blend naturally with the body's tissues, promoting better integration and healing. Collaboration between universities, companies, and hospitals is essential in order to improve AM techniques and applications. These partnerships facilitate knowledge exchange, generating new ideas and accelerating the translation of research findings into clinical practice. Joint efforts are also vital for overcoming regulatory challenges and ensuring that AM-produced implants meet necessary safety and efficacy standards.

References

[1] Pawar, R. P., Nikumbh, R. S., & Mali, S. K. (2021). Overview of additive manufacturing technologies. International Research Journal of Engineering and Technology (IRJET), 2021; 8(5): 1322–1329.

[2] Yuanyuan X, Xiaoyue W, Guo X, Kong B, Zhang M, Qian X, Shengli M, Sun W. The boom in 3D-printed sensor technology. Sensors. 2017; 17; 1166. doi: 10.3390/s17051166.

[3] Muralidhara HB, Banerjee S, Guna V, Ilangovan M, Prajwal B. Historical Development and Modern-Day Trends in Additive Manufacturing, Boca Raton: Apple Academic Press; 2021, pp. 1–13. 10.1201/9781003145349-1.

[4] Kennedy E. A brief history of 3D printing. 2021. https://openpress.universityofgalway.ie/designingthedigitalworld/chapter/brief-history-3d-printing/.

[5] Buj-Corral I, Tejo-Otero A, Fenollosa-Artés F. Development of AM technologies for metals in the sector of medical implants. Metals. 2020; 10; 686. doi: 10.3390/met10050686.

[6] Peng W, Liu Y, Wang C. Definition, measurement, and function of pore structure dimensions of bioengineered porous bone tissue materials based on additive manufacturing: A review. Frontiers in Bioengineering and Biotechnology. 2023; 10; 1081548. doi: 10.3389/fbioe.2022.1081548.

[7] Hailong X, Zhang Y, Zhang Y, Zhao Z, Xue T, Wang J, Mengmeng L, Zhao S, Zhang H, Ding Y. 3D bioprinting advanced biomaterials for craniofacial and dental tissue engineering – A review. Materials & Design. 2024; 241; 112886. doi: 10.1016/j.matdes.2024.112886.

[8] Wicaksono LH, Farid A. Influence of annealing on warping angle on polylactic acid in fused deposition modeling 3D printer. Journal of Science and Applied Engineering. 2023; 6(1): 58. doi: 10.31328/jsae.v6i1.4711.

[9] Faik Y, Baki ŞahİN İ, Koç F, UrtekİN L. The effects of different process parameters of PLA + on tensile strengths in 3D printer produced by fused deposition modeling. El-Cezeri Fen Ve Mühendislik Dergisi. 2023; doi: 10.31202/ecjse.1179492.

[10] Gregory DA, Fricker ATR, Mitrev P, Ray M, Asare E, Sim D, Larpnimitchai S. et al., Additive manufacturing of polyhydroxyalkanoate-based blends using fused deposition modelling for the development of biomedical devices. Journal of Functional Biomaterials. 2023; 14(1): 40. doi: 10.3390/jfb14010040.

[11] Ismail R, Fajar Fitriyana D, Priharyoto Bayuseno A, Munanda R, Chandra Muhamadin R, Wisda Nugraha F, Rusiyanto. et al., Design, manufacturing and characterization of biodegradable bone screw from PLA prepared by Fused Deposition Modelling (FDM) 3D printing technique. Journal of Advanced Research in Fluid Mechanics and Thermal Sciences. 2023; 103(2): 205–215. doi: 10.37934/arfmts.103.2.205215.

[12] Subhedar P, Padmanabhan D, Agrawal R, Singh G. A comprehensive review summarising the reinforcement of UHMWPE composite scaffolds as an artificial implant biomaterials. Materials Today: Proceedings. 2023; doi: 10.1016/j.matpr.2023.06.435.

[13] Pagac M, Hajnys J, Quoc-Phu M, Jancar L, Jansa J, Stefek P, Mesicek J. A review of vat photopolymerization technology: Materials, applications, challenges, and future trends of 3D printing. Polymers. 2021; 13(4): 598. doi: 10.3390/polym13040598.

[14] Milovanović A, Sedmak A, Golubović Z, Zelić Mihajlović K, Žurkić A, Trajković I, Milošević M. The effect of time on mechanical properties of biocompatible photopolymer resins used for fabrication of clear dental aligners. Journal of the Mechanical Behavior of Biomedical Materials. 2021; 119; 104494. doi: 10.1016/j.jmbbm.2021.104494.

[15] George, M. J., & Crawford, R. H. (2010). The effects of dry heat sterilization on parts using selective laser sintering. In 2010 International Solid Freeform Fabrication Symposium. University of Texas at Austin.

[16] Kafle A, Luis E, Silwal R, Matthew Pan H, Lal Shrestha P, Kumar Bastola A. 3D/4D printing of polymers: Fused Deposition Modelling (FDM), Selective Laser Sintering (SLS), and Stereolithography (SLA). Polymers. 2021; 13(18): 3101. doi: 10.3390/polym13183101.

[17] Peltola SM, Melchels FPW, Grijpma DW, Kellomäki M. A review of rapid prototyping techniques for tissue engineering purposes. Annals of Medicine. Taylor & Francis. 2008; 40(4): 268–280. doi: 10.1080/07853890701881788.

[18] Chacón JM, Núñez PJ, Caminero MA, García-Plaza E, Vallejo J, Blanco M. 3D printing of patient-specific 316 L–stainless–steel medical implants using fused filament fabrication technology: Two veterinary case studies. Bio-Design and Manufacturing. 2022; 5(4): 808–815. doi: 10.1007/s42242-022-00200-8.

[19] Zhai Y, Zhang H, Wang J, Zhao D. Research progress of metal-based additive manufacturing in medical implants. Reviews on Advanced Materials Science. 2023; 62(1): 20230148. doi: 10.1515/rams-2023-0148.

[20] Tilton M, Lewis GS, Hast MW, Fox E, Manogharan G. Additively manufactured patient-specific prosthesis for tumor reconstruction: Design, process, and properties. PLoS ONE. 2021; 16(7): e0253786. doi: 10.1371/journal.pone.0253786.

[21] Bawa R, Verma N, Girdhar P, Hashmi Z. Patient specific implants (PSI) in maxillofacial rehabilitation-a systematic review. International Journal of Advanced Research. 2023; 11(06): 88–97. doi: 10.21474/IJAR01/17053.

[22] Ligang H, Jiang G. 3D printing techniques in environmental science and engineering will bring new innovation. Environmental Science & Technology. 2017; 51(7): 3597–3599. doi: 10.1021/acs.est.7b00302.

[23] Murr LE, Gaytan SM, Martinez E, Medina F, Wicker RB. Next generation orthopaedic implants by additive manufacturing using electron beam melting. International Journal of Biomaterials. 2012; 245727. doi: 10.1155/2012/245727.

[24] Neamah, Z. H., Al-Kindi, L. A., & Al-Kindi, G. (2022, May). Additive Manufacturing of Custom Orthopedic Implants: A Review. In 2022 International Conference for Natural and Applied Sciences (ICNAS) (pp. 11–16). IEEE.Z.

[25] Dias JM, Correia Pereira da Silva FS, Gasik M, Georgina Macedo Miranda M, Jorge Fernandes Bartolomeu F. Unveiling additively manufactured cellular structures in hip implants: A comprehensive review. The International Journal of Advanced Manufacturing Technology. 2024; 130(9): 4073–4122. doi: 10.1007/s00170-023-12769-0.

[26] Farkas AZ, Galatanu S-V, Nagib R. The influence of printing layer thickness and orientation on the mechanical properties of DLP 3D-printed dental resin. Polymers. 2023; 15(5): 1113. doi: 10.3390/polym15051113.

[27] Chaudhary S, Kumar Avinashi S, Rao J, Gautam C. Recent advances in additive manufacturing, applications and challenges for dentistry: A review. ACS Biomaterials Science & Engineering. American Chemical Society. 2023; 9(7): 3987–4019. doi: 10.1021/acsbiomaterials.2c01561.

[28] Jardini AL, Sócrates Najar Lopes É, Pellizzer Gabriel L, Calderoni D, Maciel Filho R, Kharmandayan P. Bone reconstruction surgery of complex craniomaxillofacial deformities using additive manufacturing customized implants – A case report. Rapid Prototyping Journal. 2021; 27(5): 872–878. doi: 10.1108/RPJ-07-2020-0169.

[29] Memon AR, Jianning L, Egger J, Chen X. A review on patient-specific facial and cranial implant design using Artificial Intelligence (AI) techniques. Expert Review of Medical Devices. 2021; 18(10): 985–994. doi: 10.1080/17434440.2021.1969914.

[30] Datz, Janina C., Ivo Steinbrecher, Christoph Meier, Nora Hagmeyer, Leif-Christopher Engel, Alexander Popp, Martin R. Pfaller, Heribert Schunkert, and Wolfgang A. Wall. "Patient-specific coronary angioplasty simulations – A mixed-dimensional finite element modeling approach." Computers in Biology and Medicine. 2025; 189: 109914.

[31] Yageng L, Shi Y, Yuchen L, Li X, Zhou J, Zadpoor A, Wang L. Additive manufacturing of vascular stents. Acta Biomaterialia. 2023; 167: doi: 10.1016/j.actbio.2023.06.014.
[32] Vernon MJ, Mela P, Dilley RJ, Jansen S, Doyle BJ, Ihdayhid AR, De-Juan-Pardo EM. 3D printing of heart valves. Trends in Biotechnology. Elsevier. 2024; 42(5): 612–630. doi: 10.1016/j.tibtech.2023.11.001.
[33] Martinez-Marquez D, Jokymaityte M, Mirnajafizadeh A, Carty CP, Lloyd D, Stewart RA. Development of 18 quality control gates for additive manufacturing of error free patient-specific implants. Materials. 2019; 12(19): 3110. doi: 10.3390/ma12193110.
[34] Tredan DAM, Mobbs RJ, Maharaj M, Parr WCH. Combining virtual surgical planning and patient-specific 3D-printing as a solution to complex spinal revision surgery. Journal of Personalized Medicine. 2022; 13(1): 19. doi: 10.3390/jpm13010019.
[35] Jun L, Cui X, Lindberg GCJ, Alcala-Orozco CR, Hooper GJ, Lim KS, Woodfield TBF. Hybrid fabrication of photo-clickable vascular hydrogels with additive manufactured titanium implants for enhanced osseointegration and vascularized bone formation. Biofabrication. 2022; 14(3): doi: 10.1088/1758-5090/ac6051.
[36] Simchi A, Petzoldt F, Hartwig T, Boris Hein S, Barthel B, Reineke L. Microstructural development during additive manufacturing of biomedical grade Ti-6Al-4 V alloy by three-dimensional binder jetting: Material aspects and mechanical properties. The International Journal of Advanced Manufacturing Technology. 2023; 127(3–4): 1541–1558. doi: 10.1007/s00170-023-11661-1.
[37] Morrison RJ, Kashlan KN, Flanangan CL, Wright JK, Green GE, Hollister SJ, Weatherwax KJ. Regulatory considerations in the design and manufacturing of implantable 3D-Printed medical devices. Clinical and Translational Science. 2015; 8(5): 594–600. doi: 10.1111/cts.12315.
[38] Gayathry, G. (2017). Long term Osseointegration study of Laser Additive Manufactured Commercially Pure Titanium (LAM-Cp-Ti) Implants in Rabbit femur model by Histopathological and Molecular Analysis" (Doctoral dissertation, SCTIMST).
[39] Bao Y, Paunović N, Leroux J. Challenges and opportunities in 3D printing of biodegradable medical devices by emerging photopolymerization techniques. Advanced Functional Materials. 2022; 32(15): 2109864. doi: 10.1002/adfm.202109864.
[40] Adamo JE, Grayson WL, Hatcher H, Swanton Brown J, Thomas A, Hollister S, Steele SJ. Regulatory interfaces surrounding the growing field of additive manufacturing of medical devices and biologic products. Journal of Clinical and Translational Science. 2018; 2(5): 301–304. doi: 10.1017/cts.2018.331.
[41] Min Y, Zhang J, Liu B, Zhong L. Discussion on technical review guidance for registration of personalized additive manufacturing medical devices of passive implantable bone, joint and oral hard tissues. Zhongguo Yi Liao Qi Xie Za Zhi = Chinese Journal of Medical Instrumentation. 2021; 45(2): 200–204. doi: 10.3969/j.issn.1671-7104.2021.02.016.
[42] Zahan, N., Jony, M. F. I., & Nahar, K. (2018). Proposition of additive manufacturing technique for artificial bone implantations and its feasibility assessment via cost analysis. In Proceedings of the International Conference on Mechanical, Industrial and Energy Engineering (ICMIEE 2018) (ICMIEE18-156). Khulna, Bangladesh.
[43] Schmitt P, Zorn S, Gericke K. Additive manufacturing research landscape: A literature review. Proceedings of the Design Society. 2021; 1; 333–344. doi: 10.1017/pds.2021.34.
[44] Mohsan AUH, Wei D. Advancements in additive manufacturing of tantalum via the laser powder bed fusion (PBF-LB/M): A comprehensive review. Materials. 2023; 16(19): 6419. doi: 10.3390/ma16196419.
[45] Kumaresan, S., Vaiyapuri, S., Kang, J.-H., Dubey, N., Manivasagam, G., Yun, K.-D., & Park, S.-W. (2022). Perspective Chapter: Additive Manufactured Zirconia-Based Bio-Ceramics for Biomedical Applications. IntechOpen. doi: 10.5772/intechopen.101979
[46] Liu C, Ling C, Chen C, Wang D, Yang Y, Xie D, Shuai C. Laser additive manufacturing of magnesium alloys and its biomedical applications. Materials Science in Additive Manufacturing. 2022; 1(4): doi: 10.18063/msam.v1i4.24.

[47] Yick S, Reneman J, Martin PJ, Evans MDM, Bean PA, Söhnel T, Tse NMK, Bendavid A. Enhancing the biocompatibility of additively manufactured Ti-6al-4V Eli with diamond-like carbon coating. Advanced Materials Interfaces. 2023; 10(29): 2300225. doi: 10.1002/admi.202300225.

[48] Sharma SK, Singh Grewal H, Kumar Saxena K, Mohammed KA, Chander Prakash JPD, Buddhi D, Raju R, Mohan DG, Tomków J. Advancements in the additive manufacturing of magnesium and aluminum alloys through laser-based approach. Materials. 2022; 15(22): 8122. doi: 10.3390/ma15228122.

[49] Singh A. Recent advancements in 3-D printing in medical applications. International Journal of Scientific Research in Engineering and Management. 2023; 07(07): doi: 10.55041/IJSREM24845.

[50] Sathies T, Senthil P, Anoop MS. A review on advancements in applications of fused deposition modelling process. Rapid Prototyping Journal. 2020; 26(4): 669–687. doi: 10.1108/RPJ-08-2018-0199.

[51] Xiaobo L, Wanping L, Xiayi X, Wang Y, Chen S-C. Advanced optical methods and materials for fabricating 3D tissue scaffolds. Light: Advanced Manufacturing. 2022; 3(3): 1. doi: 10.37188/lam.2022.026.

Sharayu Jadhav, Vaibhav Girase, and Pankaj Dhatrak*

Chapter 3.2
Development of artificial skin using composites

Abstract: The development of artificial skin has long been a goal in biomedical engineering, with possible uses in wound healing, prostheses, and tissue engineering. The skin is an intricate organ that plays a critical role in sustaining vital host physiological functions. Researchers have investigated the use of numerous materials and methods to develop skin substitutes that can replicate the intricate structure and functions of human skin. This chapter reviews the applications of artificial skin and the techniques employed in its development. The current level of research into composites used to build artificial skin is also discussed. While synthetic polymers offer greater control over chemical composition and mechanical qualities, natural biopolymers exhibit superior cell responsiveness. The continued exploration of new biomaterial combinations and advanced fabrication techniques will undoubtedly lead to the development of even more effective and versatile artificial skin substitutes that can address the complex challenges of skin regeneration and replacement.

Keywords: Skin substitutes, wound healing, tissue engineering

3.2.1 Introduction

It is essential to comprehend the structure of the skin and the functions it serves in order to build artificial skins that may be a substitute for natural skin. This provides us a framework to direct us in the creation of artificial skins with numerous applications [1] The types of artificial skin is depicted in Figure 3.2.1.

The skin is an intricate organ that plays a critical role in sustaining vital host physiological functions like limiting fluid loss, regulating body temperature, and transmitting sensory information. Furthermore, it has a highly specialized immune niche that is essential for defense, repair, and tissue homeostasis [2].

Skin plays a crucial role in wound healing; however, following severe skin injuries (burns) or in some dermatological diseases (chronic wounds), this vital mechanism can become lost or dysregulated, leading to serious consequences. Research has concentrated

***Corresponding author: Pankaj Dhatrak**, Dr. Vishwanath Karad, MIT World Peace University, Pune 411038, Maharashtra, India e-mail: pankaj.dhatrak@mitwpu.edu.in
Sharayu Jadhav, Vaibhav Girase, Dr. Vishwanath Karad, MIT World Peace University, Pune 411038, India e-mail: sharayu1612@gmail.com, vaibhavgirase17@gmail.com

https://doi.org/10.1515/9783111386027-014

on creating tissue-engineered skin substitutes, which aim to replace and regenerate the damaged skin, in order to prevent this [3]. Skin was the first tissue-engineered organ to be used in patient care after being developed in a lab. In recent times, a range of synthetic and bioengineered alternatives have been created. These are often placed into the injury and serve as a barrier, protecting against pathogens, easing pain from wounds, and encouraging tissue regeneration to promote wound healing [4, 5].

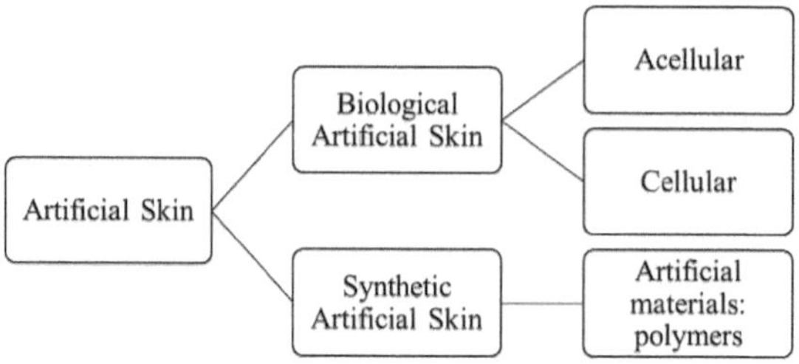

Figure 3.2.1: Types of artificial skin [6, 7].

3.2.2 Properties of artificial skin

Modern science has produced amazing artificial skins that mimic the intricate structure and functions of real skin. Their qualities are essential to their efficacy in treating wounds and for other uses. The key properties that are considered while developing artificial skins include their thermal properties, ability to heal, and their mechanical strength.

3.2.2.1 Thermal properties

3.2.2.1.1 Moisture retention

Moisture retention is an artificial skin's most important characteristic. Skin is special, in that it can be somewhat elastic, but it's also very resilient and keeps a moisture balance that helps it stay elastic without being soggy. To preserve the artificial skin's sensitivity and softness over an extended period of time, the composite substance that has been created must have the ability to moisturize. Chemical-physical cross-linked hydrogel skin possesses outstanding self-healing capabilities, high sensitivity, and exceptional moisture retention for the creation of the next generation of soft artificial skin [8].

3.2.2.1.2 Flexibility

The flexibility that comes with having natural skin is very beneficial to human bodies and should also be accurately replicated in synthetic skins. Skins that are more flexible have greater movement for the underlying structure they cover (i.e., human bones and joints, robotic systems, etc.), greater resistance and durability to punctures, and protection from the environment, without compromising its function [9, 10].

3.2.2.1.3 Self-healing properties

Human skin is a unique type of material that serves a variety of purposes, including touch detection, self-healing mechanical damage, and for perfectly covering delicately formed body parts [11]. The unresolved problem of fracture repair has greatly hampered the practical application of artificial skin. Thus, the ability of artificial skin to promptly heal damage experienced during use is crucial to its practical application. This technology has several advantages, including low cost, long aging, easy preparation, and offering a range of material alternatives. The mechanisms of microvascular self-repairing, microcapsule self-healing, and reversible bond self-healing have all been gradually studied in an effort to increase the longevity of artificial skin [12].

3.2.2.1.4 Mechanical strength

In both dry and wet settings, the mechanical strength of artificial skin must satisfy the following requirements:
- Maximum flexibility to conform to the shape of the wound without leaving any healing scars
- Sufficient strength to adhere to the wound's surface and remain there
- Maximum adaptability to fit the wound surface's correspondingly applied force
- Enough structural integrity to maintain the required porous behavior and simplicity of removal after healing, particularly in conditions that are wet [13]

3.2.3 Materials used for the development of artificial skin

A variety of materials have been researched and developed in order to produce artificial skin. They can be broadly divided into two categories: natural and synthetic. Composites are created to maximize the benefits of each of these two.

3.2.3.1 Natural biomaterials

Numerous studies have been conducted on the biocompatibility, biodegradability, and capacity to promote cell adhesion and proliferation of natural biomaterials, like collagen, chitosan, and alginate [14]. Skin tissue engineering has found great success with collagen, the main structural protein in the extracellular matrix (ECM) of skin, because of its superior mechanical qualities and capacity to attach cells [15]. The potential of chitosan, a polysaccharide produced from chitin, to heal wounds and have antibacterial action has also been investigated [14].

3.2.3.1.1 Chitosan

Studies show that chitosan's porous structure, which allows for enhanced water absorption, gas permeability, cell interaction, and the integration of drugs, growth factors, stem cells, or DNA molecules, makes it an excellent material for skin tissue engineering. These characteristics make it a suitable replacement for wound healing since it prevents wounds from drying out and causing bacterial infections [13]. Developments in chitosan-based scaffolds have resulted in the creation of active substances containing functional electrospun nanofibers [16]. Chitosan-blended alginate has been reported as a polyelectrolyte complex (PEC) membrane. These biodegradable chitosan-alginate PEC membranes demonstrated improved resilience to pH fluctuations and were more efficient as controlled-release membranes than either chitosan or alginate alone [17].

3.2.3.1.2 Chitosan-based hydrogel composites

Hydrogels are soluble materials that can absorb a significant volume of liquid without getting dissolved. While the solid behavior of gel offers a physical support and mimics the nature of soft tissue, their high water content makes them soft and biocompatible with living tissues. Hydrogels' delicate, flexible nature guarantees that the nearby living tissues suffer as little damage as possible. Depending on how much the gel structure has been cross-linked, the porosity of hydrogels can be readily changed. Hydrogels' porous structure also makes it possible to incorporate different macro- and micromolecules that improve tissue healing. When utilized as a tissue substitute, hydrogels also serve as a reservoir and offer a localized drug concentration [18]. Due to all of these benefits, chitosan-based hydrogels are becoming more and more popular for usage as skin substitutes and in a variety of biomedical applications, including drug administration, cellular immobilization, and supportive materials for cell adhesion [19] Some applications of hydrogel in biomedical industries are depicted in Figure 3.2.2.

Effective dressing materials are necessary for the treatment of burns in order to heal the dermal and epidermal tissues. However, preventing the spread of bacterial infections is a crucial requirement in burn treatment situations. To avoid harm during dressing changes, natural polymer hydrogel dressings with antibacterial properties are recommended. According to a study, chitosan-alginate hydrogel has sufficient mechanical qualities that are needed for a durable dressing material. In this case, microspheres containing the antimicrobial medication tetracycline hydrochloride were placed onto gelatin. These microspheres had a strong antibacterial activity against Staphylococcus aureus and Escherichia coli, when added to chitosan-alginate (4:1) hydrogels. When compared to the hydrogel without medicated microspheres, the tetracycline-loaded microspheres implanted in the hydrogel demonstrated a superior sustained release of the medicine over a 4-day period. The drug's consistent release lasted for up to 2 weeks, making this hydrogel-based composite an efficient antibacterial biocomposite for use as wound dressing. A dual barrier is created for the release of the water-soluble medication by the hydrogel's use of microspheres. Hydrogel-based solutions work well for wound dressings because they can readily reorganize to conform to the contours of the wound, giving them an advantage over other types of skin substitutes.

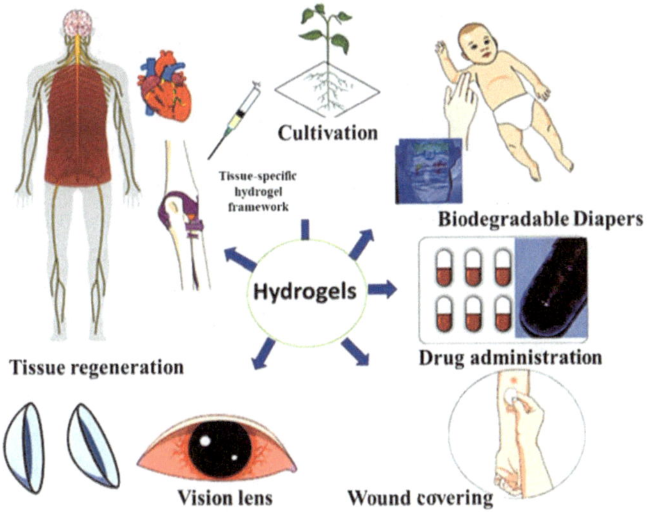

Figure 3.2.2: Hydrogel applications across a range of industries [19]

3.2.3.1.3 Collagen

Compared to other proteins, collagen is naturally present in tissues in greater quantities and serves as a substrate for a variety of animal cell types. Animal tissues can readily yield and purify collagen [20]. Its exceptional mechanical qualities and capacity to bind cells have made it a preferred option for skin tissue engineering [15]. Collagen is classified into three types such as I, II, and III, which are commonly employed in clinical applications. Fish skin is the most abundant source of collagen, making it ideal for bioprocessing and biomedical engineering applications. Collagen-derived composites have shown effectiveness in clinical applications such as porous sponges, gels, sheets, prosthetic implants, tissue engineering, and medication delivery [21]. Collagen can be used to create artificial skin, bone graft alternatives (collagen composites, including inorganic particles), dental implants, artificial tendons and blood arteries, corneal implants, nerve and cartilage regeneration, and a variety of other tissues [22].

3.2.3.1.4 Cellulose

The most prevalent biopolymer present in plant cell walls is cellulose, a polysaccharide made up of β-D-glucose repeating units joined by β-1,4 linkages [23].

Because of its complex porous three-dimensional (3D) structure that resembles the ECM of the skin and promotes tissue regeneration, cellulose finds use in a variety of applications. Potential applications for bacterial cellulose include medicine delivery, bioprinting, implants, and artificial organs [23, 24]. In order to efficiently combat infections at wound sites, cellulose can be combined with additional materials, such as antimicrobials [25]. Cellulose-based biocompounds are intriguing for wound-healing/dressing materials because of their capacity to be processed into a variety of forms, including nano/microfibers, hydrogels, films, and colloids. Their hydrophilic properties have been used to improve the effectiveness of many natural and synthetic wound-healing/dressing products. Several bioactive compounds, including antibiotics, vitamins, growth hormones, peptides, and natural chemicals, have been transported to the wound site via cellulose-based structures. The natural breakdown of cellulose-based biomaterials in the human body is difficult, which is not a problem for wound-healing/dressing purposes. They are simply removed from the wound bed after achieving the desired effect [26].

3.2.3.1.5 Alginate

Alginate is a naturally occurring polysaccharide found in the bacterial capsules of *Pseudomonas* sp. and *Azotobacter* sp. as well as the cell walls of algae. The presence of alginate in the cell walls of brown algae, also known as brown seaweeds, gives the algae flexibility and a robust structure while protecting them from potential harm when exposed to powerful seawater waves [27]. Alginate has been combined with other materials to create alginate composite, which improves the physical properties of alginate. Natural polymers like collagen, chitosan, and gelatin; synthetic polymers like polypyrrole and polylacticide; and inorganic substances like hydroxyapatite and tetraethylorthosilicate are used to create alginate composites [28, 29]. Alginates have also been identified as natural polymers used in hydrogel-based nanocomposites. Chemical modification of the polymer is a common strategy used in the development of alginate-based nanocomposites to strengthen the interaction between the polymer matrix and the nanoparticles, resulting in a hydrogel with stable mechanical properties. However, chemical alteration of natural polymers can impair their biocompatibility, which is why unmodified biopolymers should be used [30].

3.2.3.1.6 Hyaluronic acid

Hyaluronic acid (HA) is a highly versatile substance with distinct features. It is a glycosaminoglycan that occurs naturally in the body and has no sulfate linkages [31]. HA is a significant ECM component. HA is a crucial biological material for healing wounds since it has a variety of biological functions [32]. The role is usually determined by the molecular weight of the HA. For example, low-molecular-weight HA is required for recovery and the development of scars, whereas high-molecular-weight HA could promote tissue integrity. Due to water absorption, HA gives flexibility to animal tissue along with lubrication in muscle connective tissues [33]. HA has remarkable water retention capabilities at both high and low relative humidity levels. HA can also preserve the turgidity, hydration, and flexibility of the skin [34]. As a significant ingredient of all vertebrates' connective tissue (ECM), HA promotes the formation of a fibrin clot as well as the generation of interleukins and proinflammatory cytokines [35, 36].

3.2.3.2 Synthetic polymers

Creating scaffolds that replicate the natural physical fibrillar structure and nanoscale architecture of ECM is a popular approach used in the development of scaffolds for tissue regeneration and repair, including skin. Scaffolds composed of micro- or nano-polymeric fibers that are structurally comparable to the ECM could be very advantageous [37]. Furthermore, cellularized skin substitutes that are designed to hasten the healing process are created by seeding and cultivating cells on synthetic polymer scaffolds and delivering them into wounds [38]. These polymer-based wound dressings can promote cell proliferation and skin tissue repair by acting as transporters for cells and growth factors, in addition to supporting and shielding the wound sites when applied to the skin [39].

3.2.3.2.1 Polyvinylpyrrolidone

As it does not irritate the skin, polyvinylpyrrolidone (PVP) is a biocompatible, water-soluble, and biodegradable polymer that is frequently utilized as a hydrogel membrane for skin replacement goods. It is safe, has strong water vapor transfer, and when PVP membranes are cross-linked via radiation polymerization, they become impervious to germs and show increased elasticity [26]. Furthermore, in a humid, tropical climate, PVP membranes may be appropriate for use as skin dressings due to their high water retention capacity [40]. PVP-based hydrogels are widely employed in biomedical applications. However, they are often combined with other polymers to enhance undesired characteristics [41].

3.2.3.2.2 Polylactic acid

A biodegradable polyester made from renewable resources like sugar cane, corn, or starch is polylactic acid (PLA) [42]. The outstanding mechanical integrity and biocompatibility of PLA and its copolymers have made them commonly utilized in healthcare applications [43]. They are nontoxic to humans or the environment, and can be made from natural sources [44]. Creating PLA-based composites is a key solution to addressing issues with PLA scaffolds. Using composite techniques expands the range of biomaterial components and applications. Combining different materials with PLA may result in optimum physical and biological qualities [45].

3.2.3.2.3 Poly-3-caprolactone

Poly-3-caprolactone (PCL) is a semicrystalline, hydrophobic polymer that has garnered significant attention in the biomedical industry due to its unique properties. PCL's relatively low melting point, coupled with its excellent biocompatibility and high solubility, make it a highly versatile material with promising applications in various medical fields. One of the areas where PCL has been extensively studied is skin tissue engineering, where its properties have proven particularly advantageous.

In the context of skin regeneration, PCL is most commonly utilized through the process of electrospinning, a technique that has become the method of choice for fabricating skin scaffolds. PCL electrospinning is favored because it offers precise control over critical parameters such as pore size, pore distribution, fiber diameter, and fiber orientation within the scaffold. These parameters are crucial in skin tissue engineering because they influence how cells interact with the scaffold, how nutrients and waste products are exchanged, and how the tissue ultimately regenerates.

The ability to finely tune the characteristics of electrospun PCL scaffolds makes them highly effective in promoting cell adhesion, proliferation, and differentiation, which are essential processes in skin regeneration. By controlling pore size and distribution, researchers can ensure that the scaffold provides an optimal environment for cellular activity while also allowing for the diffusion of nutrients and oxygen necessary for tissue growth [46, 47]. Additionally, by adjusting the fiber diameter and orientation, the mechanical properties of the scaffold can be tailored to more closely mimic the natural ECM of the skin, further enhancing the regenerative process.

PCL's versatility and effectiveness in skin tissue engineering extend beyond just the creation of scaffolds. Its biocompatibility means that it can be combined with various bioactive molecules, growth factors, or even other polymers to create more complex, multifunctional scaffolds that can further enhance the healing process. For instance, PCL can be engineered to release growth factors in a controlled manner, promoting faster and more efficient tissue regeneration [48].

In summary, PCL stands out as a highly promising material in the biomedical industry, particularly in the field of skin tissue engineering. Its semicrystalline and hydrophobic nature, combined with its low melting point, biocompatibility, and high solubility, make it an excellent choice for long-term applications. The slow degradation rate of PCL compared to PLA and PLGA allows for sustained support in tissue engineering applications, while the use of PCL electrospinning offers unparalleled control over scaffold properties such as pore size, fiber diameter, and orientation. These attributes make PCL an invaluable tool in the development of effective skin regeneration strategies, with the potential to significantly improve outcomes for patients with skin injuries or conditions requiring tissue engineering solutions.

3.2.3.3 Composite biological systems

Composite biomaterial systems, comprising synthetic as well as natural polymers, can be used to construct biomaterials with desired properties to satisfy physical, chemical, and mechanical requirements. These biomaterial composites are intended for long-term medication release, and are biocompatible and biodegradable [49]. Collagen-chitosan composites, for instance, have been demonstrated to have better mechanical strength, better cell adhesion, and faster wound healing. Comparably, polyurethane-collagen composites are proven to have increased elasticity and tensile strength, which qualifies them for use in skin tissue engineering applications [50].

Research into the properties of collagen and HA blends has already commenced, and a number of intriguing biomaterials derived from these blends have been suggested. Numerous research teams have looked into the interactions between HA and collagen as well as the potential for exploiting this combination to create novel materials [51–53]. New materials based on a combination of HA with collagen, chitosan, and other biopolymers may fill the gap between the demand and supply of biomaterials for tissue regeneration [54]. The collagen-based composites and the fabrication techniques employed for the development of artificial skin are illustrated in Table 3.2.1.

Table 3.2.1: Collagen-based composites used for the development of artificial skin.

Reference	Biomaterials used	Fabrication techniques	Key points
Ma et al. [55]	Collagen-chitosan	Freeze drying + GA cross-linking	Chitosan inhibited the scaffold's biodegradation
Romanova et al. [56]	Chitosan-collagen	Freeze-drying	Reduced contraction and increased HDF proliferation
Bonvallet et al. [57]	Collagen-PCL	Electrospinning (non-cross-linked) + micropores	Quicker wound healing and skin regeneration, along with increased production of blood vessels and hair follicles
Huang et al. [58]	PCL-collagen	Electrospinning	Aligned fibers encouraged the shape of longer cells, sped up migration, and led to myofibroblast differentiation
Yang et al. [59]	PLGA-collagen	Electrospinning	Enhanced extracellular matrix deposition, HDF adhesion, and proliferation
Huang et al. [60]	Gelatin and collagen	Freeze thawing and gelation	Encourage the healing of wounds and re-epithelialization

3.2.4 Biofabrication techniques

3.2.4.1 Electrospinning

Artificial skin needs to match the high elasticity and breathability of natural human skin to be effective [61]. When designing artificial skin, it's crucial to select materials that replicate the flexibility and resilience of human skin. This includes not only choosing skin-like materials but also consistently evaluating how the artificial skin performs in real-world use. Important factors to consider are the comfort of wearing the artificial skin, ensuring it fits well without causing irritation, and closely replicating the physical and chemical properties of natural skin.

Electrospinning technology offers a promising method for creating such flexible, skin-like materials. This technique is highly versatile and scalable, making it popular in tissue engineering for producing fibers with diameters that range from nanometers to micrometers. The process involves applying an electric field to a polymer solution or melt, which generates a charged jet. As this jet is expelled, it stretches and solidifies, forming fine fibers [62, 63]. These fibers can be collected into a mesh or mat, serving as a foundation for artificial skin.

The fibers created through electrospinning are particularly useful for artificial skin because they can be engineered to have properties similar to natural skin, such as flexibility, strength, and breathability. The fine structure of electrospun fibers also contributes to the artificial skin's comfort and functionality, especially when worn for extended periods [64].

Beyond artificial skin, electrospinning has numerous applications, including filtration, tissue engineering, and drug delivery. In tissue engineering, electrospun fibers can create scaffolds that support cell growth and tissue regeneration, while in drug delivery, they can be used to produce fibers that encapsulate medications for controlled release. The ability to fine-tune the properties of electrospun fibers makes electrospinning an invaluable tool in developing artificial skin and in other biomedical applications [65]. By carefully selecting and processing polymers, researchers can create artificial skin that not only mimics the mechanical properties of human skin but also supports biological processes vital for skin healing and regeneration [66]. Consequently, electrospinning-based flexible equipment represents a practical and promising approach for fabricating artificial skin, with the potential to greatly enhance the lives of patients who need skin replacements or improvements.

3.2.4.2 3D printing

Using state-of-the-art technology, additive manufacturing, often known as 3D printing, it is possible to build intricate 3D objects layer by layer. Recently, 3D printing has been applied to a wide range of fields, most notably biomedical engineering, where tissue engineering and regenerative medicine have showed remarkable promise [67, 68]. In addition to its potential use in medicine, 3D bioprinting can be used to simulate disorders for pharmaceutical testing and to investigate tissue development and homeostasis [69]. The 3D-printed scaffolds can be used to generate a 3D microenvironment that promotes cell adhesion, proliferation, and differentiation [70]. Bioactive materials such as growth factors and peptides can also be functionalized to enhance the behavior of cells. High-throughput drug testing and tailored treatment are made possible by the additional benefit of bioprinting, which can be scaled by expanding the printing area or using several printing machines at once, while assuring consistent quality [71]. Bioprinted skin is one of the most promising manufacturing technologies because it can do things that conventional skin engineering cannot, like high throughput manufacturing, high repeatability, and short production cycles. Additionally, 3D bioprinting can supply the right ECM for skin replacements, which is advantageous for cell-extracellular milieu interaction [72]. There are several advantages of this technology. Some of these are mentioned below:

- Neurosurgeons employ 3D-printed neuroanatomical models to design surgical procedures that show the physical representation of some challenging human structures. As a result, the surgeon can see and comprehend issues pertaining to delicate areas, such as connections between cerebral structures, cranial nerves, arteries, and the form of the skull, which is not achievable with 2D technology.

Therefore, by improving the medical technique after viewing the image, surgical errors can be decreased.
- The primary distinction between 2D and 3D printing technologies is that the latter does not adhere to the conventional production method. It is also a highly effective method for rapidly inducing, at a cost that is profitable, patient-specific, high-constancy medical hallucinations.
- When compared to alternative methods, the 3D bioprinting methodology makes it possible to identify the differences between naturally occurring tissues and artificially manufactured tissue constructs.
- Exceptional flexibility and the ability to distribute cells and biomaterials with clear control over spatial circulation are offered by 3D organ printing.

3.2.4.3 Self-assembly

The use of self-assembly techniques has shown promise in producing artificial skin to mimic the structure and function of real skin while lowering its immunogenicity. Self-assembly techniques produce the perfect dermal scaffold by synthesizing ECM and facilitating assembly without the need for foreign proteins or nanomaterials. The procedure is usually followed by self-assembly methods: Fibroblasts are first cultivated to create cell sheets. The dermal layer is then formed by stacking these cell sheets. Ultimately, the epidermal layer and stratum corneum are formed by the seeding of epidermal cells onto the dermal layer. In vitro skin substitutes must closely resemble the structure of human skin in order to be functional, and the number of exogenous materials used must be minimized. Using the self-assembly technique, skin substitutes that lack artificial materials and exogenous ECM proteins can be created. These skin substitutes also exhibit many of the histological, physicochemical, and mechanical traits of healthy human skin in vivo [73, 74].

3.2.5 Examples/products

3.2.5.1 Integra

Integra, developed by Integra LifeSciences, is a groundbreaking synthetic acellular dermal regeneration template originally designed for covering severe burn wounds. This innovative product plays a crucial role in promoting revascularization and the formation of new dermal tissue, effectively addressing some of the most challenging aspects of burn treatment. The creation of Integra represents a major step forward in wound care and tissue engineering, offering a solution that facilitates dermal regeneration in cases where natural skin has been significantly damaged or lost.

Integra is constructed as a bilaminate sheet, consisting of two complementary layers that work in tandem to support the healing process. The inner layer is composed of highly purified bovine tendon collagen, which acts as a structural scaffold, similar to the ECM of human skin. This collagen layer is interwoven with glycosaminoglycans derived from shark, specifically chondroitin-6-sulfate, which are essential for maintaining the structural integrity and hydration of the ECM. Together, these components create an environment conducive to cellular infiltration and tissue regeneration, allowing the patient's own cells to migrate into the scaffold, promote revascularization, and begin the process of forming a new dermal layer.

The outer layer of Integra is a silicone sheet that serves as a temporary epidermal substitute, protecting the wound from infection and dehydration, while maintaining the moist environment necessary for optimal healing. This silicone barrier remains in place during the early stages of wound healing, providing essential support as the underlying dermal template integrates with the host tissue. As the new dermal layer forms and matures over time, the silicone sheet can be removed, allowing for the application of a thin skin graft or enabling the natural epithelialization process to complete wound closure.

Integra has revolutionized the treatment of extensive burns and other complex wounds, particularly in situations where traditional skin grafting methods are inadequate or not feasible. Its ability to support the regeneration of functional dermal tissue has led to better outcomes for patients, reducing the occurrence of scar contracture and improving the overall quality of the healed skin. Additionally, the biocompatibility of the materials used in Integra minimizes the risk of rejection and other complications, making it a highly effective option in managing severe skin injuries.

3.2.6 Applications and future perspectives

Polymer-based artificial skin systems have several uses in biomedical engineering [75–77]. In Yang et al.'s publication, the phrase "artificial skin" was first used to refer to wrinkle treatments [78]. Numerous varieties of bioengineered skin have been created for use as medicine delivery systems, wound dressings, and cosmetic purposes. Replicating the characteristics and functions of natural skin is the primary objective of manufactured human skin. It is mostly used as an in vitro model for disorders related to skin abnormalities and research. Different engineering concepts or approaches concentrate on different features of artificial skin, depending on the intended function. In vitro models employ artificial skin, such as patient skin-derived seed cells, to mimic particular components of the disease process. When applied medically, artificial skin seeks to reduce immunogenicity and enhance repair benefits, for example by including growth factors and other drugs. The sections that follow will give an overview of the specific artificial skins that are made according to their intended use [79, 80].

3.2.6.1 Skin cancer

In the context of melanoma in particular, skin reconstructions have shown to be extremely useful instruments for furthering our comprehension of the intricate interactions that take place within the skin's microenvironment. With the use of these advanced models, scientists can replicate the development and spread of melanoma cells in a 3D microenvironment that closely mimics the skin's natural environment. This feature is essential because it makes it possible to investigate the dynamic interactions between dermal fibroblasts, epidermal keratinocytes, and melanoma cells – all important participants in skin homeostasis and pathology – more precisely.

Observing the interactions between the initial melanoma cells and the surrounding fibroblasts within the dermis has been made possible by the use of manufactured artificial skin in melanoma research. Due to the fact that fibroblasts actively participate in the tumor microenvironment by reacting to signals from the melanoma cells, this contact is very important. These fibroblasts are drawn to the melanoma, where they secrete growth factors and change the deposition of ECM, which are vital survival signals [81, 82]. This procedure demonstrates how melanoma cells and the fibroblasts that surround them have a reciprocal connection. The melanoma cells receive structural support from the modified ECM, which also helps to create an environment that is more favorable for tumor growth and invasion [83, 84].

Furthermore, the formation of matrix metalloproteinases (MMPs) inside this milieu has been found to be upregulated in these skin reconstructions. Enzymes known as MMPs are critical for the ECM's breakdown, which is a necessary step in tissue remodeling and healing. But in the case of cancer, this rise in MMP activity makes it easier for tumor cells to invade by dissolving the surrounding ECM. As a result, the melanoma cells are able to enter the dermal layers more deeply and may even migrate to other areas of the body [85]. Examining these systems in a regulated 3D setting offers important insights into the progression and metastasis of melanoma, and provides useful data that may guide the creation of novel.

3.2.6.2 Reconstructive applications

Complex tissue abnormalities and congenital anomalies have also demonstrated significant promise for artificial skin repair. In situations where significant portions of the skin are absent or injured, a personalized artificial skin can be added or used in place of the missing skin tissue. These prosthetic skins offer mechanical stability, facilitate tissue integration, support the regrowth of functional skin, and can be tailored to the particular requirements of the defect. Scientists are working to create structures that closely resemble the structure, functionality, and aesthetic appearance of natural skin by combining various biomaterials, cells, and growth factors. For example, to repair a significant skin defect, researchers used the allogenic acellular dermal matrix

(ADM) [86]. The ADM functioned as a perfect artificial skin and successfully enhanced the scar quality in situations of abdominal wound, deep traumatic defect in the prepatellar region, and cranial defect. Here, early revascularization and decreased bacterial infection were also noted.

3.2.6.3 E-skin

In research and development, engineered artificial skin can be a useful tool for investigating wound healing mechanisms, assessing innovative therapeutic approaches, and evaluating potential drugs. They offer a consistent and controlled environment for examining tissue regeneration mechanisms, cellular behavior, and in understanding the effectiveness of different treatments. Through these uses, artificial skins help expand our knowledge of skin biology and wound healing, which will eventually lead to the creation of more potent therapeutic techniques.

Electronic skin (or "e-skin") technologies have also benefited from the use of artificial skin composites. By imitating the sensory properties of human skin, these "e-skins" have made it possible to create sophisticated robotic systems, prosthetic limbs, and wearable electronics that can sense touch, temperature, and other stimuli [87].The device's robustness and mechanics would be improved by an extremely thin e-skin that could be applied directly to the skin. Biocompatibility, biodegradable properties, self-repairing, temperature sensitivity, and self-powering are highly desirable qualities of e-skin [88].

3.2.6.4 Wound healing

Customized artificial skin has become a viable approach to enhance the results of wound healing, providing creative solutions that surpass conventional wound dressings. These painstakingly crafted artificial skins are intended to replace or augment absent or injured skin tissue, offering a nurturing environment that promotes the body's inherent healing ability. These state-of-the-art skin substitutes have shown great promise in wound healing, opening up a range of approaches to suit various wound types and the needs of patients Figure 3.2.3 shows the stages of the artificial skin's operation during wound healing.

Artificial skin, for example, is critical in the recovery process of burns because it efficiently covers the wound and promotes healing. Temporary artificial skin can be used to protect partial thickness burns, aiding natural healing and reducing the risk of complications. Permanent artificial skin grafts offer a long-term cure in more severe cases, such as full-thickness burns, by promoting tissue regeneration and restoring the skin's protective properties. This aids in both the physical and cosmetic rehabilitation of the wound by reducing scarring and improving the skin's overall appearance. It also be utilized to support split-thickness skin grafts, and is critical for scar correction, allowing patients to obtain even better functional and aesthetic outcomes.

Keratinocytes, melanocytes, and dermal fibroblasts are the three main cell types that are utilized to manufacture conventional skin constructions. Dermal fibroblasts are found in the dermis, while keratinocytes and melanocytes play significant roles in the epidermis [89]. Promising outcomes have been shown when cellular scaffolds, including fibroblasts, keratinocytes, or reprogrammed cells, are used to speed up in vivo wound healing and lessen the production of scars [90]. For rather thick burns, it can be employed as a temporary covering for protecting the area and speed up the healing process. In order to improve aesthetic results, it also makes split-thickness skin transplants easier and helps with scar remodeling [91].

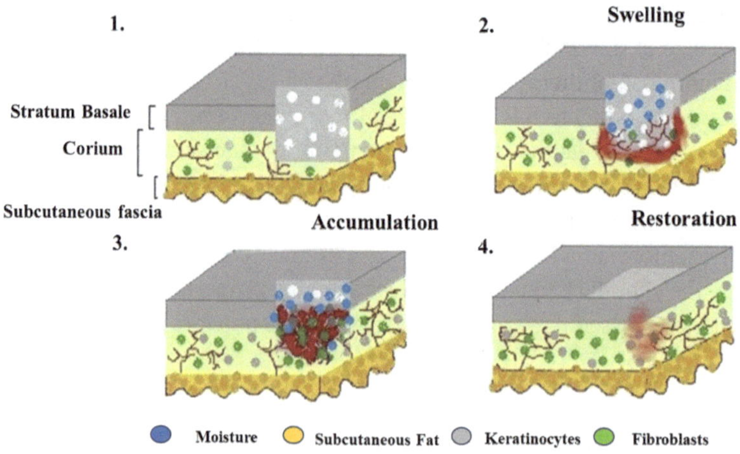

Figure 3.2.3: An overview of how artificial skin functions during wound healing [12].

3.2.7 Limitations and challenges

Many obstacles stand in the way of researchers' attempts to create successful skin substitutes. In addition to following stringent legal and ethical requirements, the production of these replacements necessitates tackling the difficulties involved in accurately recreating human skin. Commercially accessible artificial skin still does not offer patients in need of grafting, a complete and workable alternative, despite tremendous developments. The most formidable obstacle is imitating the complex architecture and physiology of the skin, especially with relation to ECM and vascularity. These elements are essential to ensure that the skin substitute can properly meld with the patient's body, offering the support and nutrition required for recovery and regeneration.

Enhancing these components is essential to creating skin substitutes that satisfy patients' various needs, particularly those with complicated or extensive injuries. It is necessary to carefully construct the ECM to closely resemble the natural ECM seen in human skin [92]. The ECM gives the surrounding cells structural and biochemical support. Furthermore, vascularity – the formation of blood vessels inside the skin substitute – is necessary to guarantee the transport of nutrients and oxygen to the tissue, thus fostering long-term viability and function.

Researchers are investigating novel strategies, such as automated tissue combination printing techniques and physiologically functional tissue printing, to solve these issues. These cutting-edge methods enable exact control over the content and structure of the tissue, which promises to improve the number and quality of skin substitutes. These techniques combine several cell types, growth hormones, and biomaterials to create skin substitutes that function and look quite similar to actual skin. They, therefore, have the potential to greatly enhance the results for those in need of skin grafts, providing a more dependable and efficient means of wound healing and tissue regeneration.

3.2.8 Future perspectives

Biomaterials with certain qualities have made it possible to create scaffolds that support tissue mechanically to promote tissue regeneration, by promoting cell adhesion and proliferation. Scaffold design is made more versatile through the application of both man-made and natural polymers, and the healing ability of artificial skins is improved with the inclusion of bioactive substances and growth factors. Complex and patient-specific structures may now be made, thanks to advanced manufacturing processes like 3D bioprinting and electrospinning [94, 95], which offer precise control over the growth factor supply, cellular dispersion, and scaffold architecture [93].

There is much potential for advancement in the field and to imagine synthetic skin in the future that is engineered. The goal of ongoing research is to improve artificial skin's long-term durability and functionality, while also strengthening its incorporation into the surrounding tissue. In order to optimize the flow of nutrients and oxygen within the constructions and facilitate quicker and more effective tissue regeneration, strategies to increase vascularization within them are being investigated. The integration of designed artificial skin into clinical practice will be facilitated by the creation of increasingly complex and in vitro models with physiological accuracy for preclinical testing [96].

For wound dressings and prosthetic skin, chitosan has long been the preferred polymer due to its unique hydrophilicity, solubility, and mechanical characteristics. These changes are typically made to derivatives with chemical alterations, blends, and composites with other appropriate polymers. Furthermore, chitosan polymer modifications by chemical means as well as biomanufacturing techniques involving the incorporation of medications, growth hormones. Chitosan is now a feasible bioresorbable skin substitute for tissue regeneration thanks to natural cells or DNA. Because of its special qualities, chitosan is extensively employed in biomedical applications, particularly in the field of tissue engineering. For example, the interaction between the negatively charged bacterial cell membrane and the positively charged chitosan structure inhibits the growth of bacteria. Because it creates a natural environment for cell growth and is essential for the movement of cells like keratinocytes, nutrients, and growth factors, chitosan's porosity and swelling behavior have also been noted and reported to have a significant impact on skin tissue regeneration [97]. Additionally, chitosan is said to possess sticky qualities that promote cell adhesion and facilitate wound closure, thus reducing the amount of blood lost from severe cuts or third-degree burns. In order to mimic the natural skin extracellular milieu, any of the inherent features of chitosan that have been previously discussed can also be customized by creating composites or blends with other appropriate polymers [98]. These can take the shape of films, membranes, sponges, micro/nanofibrous mats, matrix, or sponges.

Personalized medicine and tissue engineering are increasingly being used to develop artificial skin that perfectly matches each patient's unique characteristics. This involves using the patient's own cells and considering their specific genetic makeup to create skin grafts that can effectively heal and regenerate. Breakthroughs in stem cell and regenerative medicine have the potential to significantly improve the effectiveness and adaptability of artificial skin.

3.2.9 Summary

While tremendous progress has been achieved, an artificial skin substitute is still a long way from entirely replicating the anatomical hierarchy and functional qualities of genuine skin. An outline of the advancements in artificial skin is shown in Figure 3.2.4. Current research in biomaterials, stem cell biology, and advanced manufacturing methods shows promise for the future production of more functional and biomimetic artificial skin substitutes. However, several major obstacles remain to be addressed, including the creation of vascular networks to enable nutrient and oxygen transport, the integration of sensory and immunological systems, and the scalability of production methods to meet the growing demand for skin substitutes. Furthermore, the long-term durability and performance of these artificial skin structures in clinical settings necessitate additional research and modification. Despite these challenges, the potential of artificial skin to change a variety of sectors, including wound healing and prosthetics, continues to motivate researchers and doctors to push the boundaries of this rapidly growing field. Continued research into new biomaterial

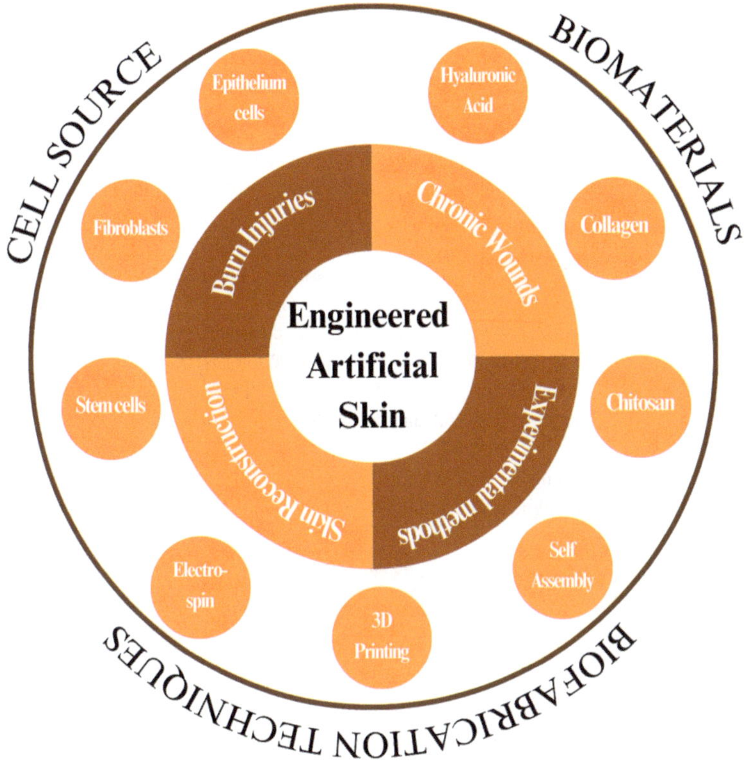

Figure 3.2.4: An outline of the advancements in artificial skin [15, 83].

composites and sophisticated manufacturing techniques will surely result in the creation of increasingly more effective and adaptable artificial skin substitutes capable of addressing the complex difficulties of skin regeneration and replacement.

References

[1] Low ZWK, *et al.* Using artificial skin devices as skin replacements: Insights into superficial treatment. 2019 Mar. 01; *Wiley-VCH Verlag*. doi: 10.1002/smll.201805453.

[2] Nguyen AV, Soulika AM. The dynamics of the skin's immune system. 2019 Apr. 02; *MDPI AG*. doi: 10.3390/ijms20081811.

[3] Sierra-Sánchez Á, Kim KH, Blasco-Morente G, Arias-Santiago S. Cellular human tissue-engineered skin substitutes investigated for deep and difficult to heal injuries. 2021 Dec. 01; *Nature Research*. doi: 10.1038/s41536-021-00144-0.

[4] Gao X, Su J-F, Wang S, Yang P. Smart self-nourishing and self-healing artificial skin composite using bionic microvascular containing liquid agent. 2022. doi: 10.3390/polym.

[5] Sheikholeslam M, Wright MEE, Jeschke MG, Amini-Nik S. Biomaterials for skin substitutes. 2018 Mar. 07; *Wiley-VCH Verlag*. doi: 10.1002/adhm.201700897.

[6] Lei Z, Zhu W, Zhang X, Wang X, Wu P. Bio-inspired ionic skin for theranostics. Advanced Functional Materials. 2021 Feb; 31(8). doi: 10.1002/adfm.202008020.
[7] Arjun Uppuluri VNV, Sathanantham ST, Bhimavarapu SK, Elumalai L. Polymeric hydrogel scaffolds: Skin tissue engineering and regeneration. 2022; *Tabriz University of Medical Sciences*. doi: 10.34172/apb.2022.069.
[8] Wang S, Li Q, Feng S, Lv Y, Zhang T. A water-retaining, self-healing hydrogel as ionic skin with a highly pressure sensitive properties. Journal of the Taiwan Institute of Chemical Engineers. 2019 Nov; 104: 318–329. doi: 10.1016/j.jtice.2019.09.005.
[9] Snyder DE, Sapper ED. See me, feel me, touch me, heal me: A contextual overview of conductive polymer composites as synthetic human skin. Journal of Composites Science. 2022 May; 6(5). doi: 10.3390/jcs6050141.
[10] Yu Y, *et al.* Biofuel-powered soft electronic skin with multiplexed and wireless sensing for human-machine interfaces. Science Robotics. 2020 Apr; 5(41). doi: 10.1126/SCIROBOTICS.AAZ7946.
[11] Park S, Shin BG, Jang S, Chung K. Three-dimensional self-healable touch sensing artificial skin device. ACS Applied Materials and Interfaces. 2020 Jan; 12(3): 3953–3960. doi: 10.1021/acsami.9b19272.
[12] Sun Q, Gao X, Wang S, Shao RY, Wang XY, Su JF. Microstructure and self-healing capability of artificial skin composites using biomimetic fibers containing a healing agent. Polymers (Basel). 2023; 15(1). doi: 10.3390/polym15010190.
[13] Sharma S, Batra S. Recent Advances of Chitosan Composites in Artificial Skin: The Next Era for Potential Biomedical Application. In: Eds: Holban A-M, Grumezescu AM, Materials for Biomedical Engineering: Nanobiomaterials in Tissue Engineering, Elsevier; 2019, pp. 97–119. doi: 10.1016/B978-0-12-816909-4.00005-1.
[14] MacNeil S. Biomaterials for tissue engineering of skin. Materials Today, 2008; 11(5):26–35.
[15] Dash BC, Hsia HC. Composite Scaffolds for Skin Repair and Regeneration. In: Ed: García-Gareta E, Biomaterials for Skin Repair and Regeneration, Northwood, UK: Woodhead Publishing, Elsevier; 2019, pp. 193–223. doi: 10.1016/B978-0-08-102546-8.00007-8.
[16] Xu Y, Wu X, Zhang Y, Yu Y, Gan J, Tan Q. Engineered artificial skins: Current construction strategies and applications. 2023 Dec. 01; *KeAi Communications Co*. doi: 10.1016/j.engreg.2023.09.001.
[17] Sivanesan I, *et al.* Exploring the impact of chitosan composites as artificial organs. Polymers. 2022; 14(8):1587, doi: 10.3390/polym14081587.
[18] Joseph J, Deshmukh K, Tung T, Chidambaram K, Khadheer Pasha SK. 3D Printing Technology of Polymer Composites and Hydrogels for Artificial Skin Tissue Implementations. In: Sadasivuni K, Ponnamma D, Rajan M, Ahmed B, Al-Maadeed M, editors. Polymer Nanocomposites in Biomedical Engineering. Lecture Notes in Bioengineering. Cham, Switzerland: Springer; 2019, 205–233. doi: 10.1007/978-3-030-04741-2_7.
[19] Chen H, *et al.* Covalently antibacterial alginate-chitosan hydrogel dressing integrated gelatin microspheres containing tetracycline hydrochloride for wound healing. Materials Science and Engineering: C. 2017; 70(Part 2): 287–295. doi: 10.1016/j.msec.2016.08.086.
[20] Koide M, *et al.* A new type of biomaterial for artificial skin: Dehydrothermally cross-linked composites of fibrillar and denatured collagens. Journal of Biomedical Materials Research, 1993; 27(1): 79–87.
[21] Simionescu BC, Ivanov D. Natural and Synthetic Polymers for Designing Composite Materials. In: Ed: Antoniac I, Handbook of Bioceramics and Biocomposites, Cham, Switzerland: Springer; 2015, pp. 233–286. doi: 10.1007/978-3-319-09230-0_11-1.
[22] Sionkowska A, Skrzyński S, Śmiechowski K, Kołodziejczak A. The review of versatile application of collagen. 2017 Jan. 01; *John Wiley and Sons Ltd*. doi: 10.1002/pat.3842.

[23] Bae SO, Shoda M. Production of bacterial cellulose by Acetobacter xylinum BPR2001 using molasses medium in a jar fermentor. Applied Microbiology and Biotechnology. 2005 Apr; 67(1): 45–51. doi: 10.1007/s00253-004-1723-2.

[24] Kucińska-Lipka J, Gubanska I, Janik H. Bacterial cellulose in the field of wound healing and regenerative medicine of skin: Recent trends and future prospectives. 2015 Sep. 19; *Springer Verlag*. doi: 10.1007/s00289-015-1407-3.

[25] Miao J, *et al.* Lysostaphin-functionalized cellulose fibers with antistaphylococcal activity for wound healing applications. Biomaterials. 2011; 32(36): 9557–9567. doi: 10.1016/j.biomaterials.2011.08.080.

[26] Abazari MF, *et al.* Recent advances in cellulose-based structures as the wound-healing biomaterials: A clinically oriented review. 2021 Sep. 01; *MDPI AG*. doi: 10.3390/app11177769.

[27] Venkatesan J, Nithya R, Sudha PN, Kim SK. Role of Alginate in Bone Tissue Engineering. Advances in Food and Nutrition Research, 2014; 73: 45–57. doi: 10.1016/B978-0-12-800268-1.00004-4.

[28] Lee KY, Mooney DJ. Alginate: Properties and biomedical applications. 2012; *Elsevier Ltd*. doi: 10.1016/j.progpolymsci.2011.06.003.

[29] Venkatesan J, Bhatnagar I, Manivasagan P, Kang KH, Kim SK. Alginate composites for bone tissue engineering: A review. 2015 Jan. 01; *Elsevier B.V.* doi: 10.1016/j.ijbiomac.2014.07.008.

[30] Zdiri K, Cayla A, Elamri A, Erard A, Salaun F. Alginate-based bio-composites and their potential applications. 2022 Sep. 01; *MDPI*. doi: 10.3390/jfb13030117.

[31] Dovedytis M, Liu ZJ, Bartlett S. Hyaluronic acid and its biomedical applications: A review. 2020 Jan. 01; *KeAi Communications Co*. doi: 10.1016/j.engreg.2020.10.001.

[32] Mineo A, Suzuki R, Kuroyanagi Y. Development of an artificial dermis composed of hyaluronic acid and collagen. Journal of Biomaterials Science, Polymer Edition. 2013 Apr; 24(6): 726–740. doi: 10.1080/09205063.2012.708190.

[33] Sionkowska A, Gadomska M, Musiał K, Piatek J. Hyaluronic acid as a component of natural polymer blends for biomedical applications: A review. 2020 Sep. 01; *MDPI AG*. doi: 10.3390/molecules25184035.

[34] Yasin A, Ren Y, Li J, Sheng Y, Cao C, Zhang K. Advances in hyaluronic acid for biomedical applications. 2022 Jul. 04; *Frontiers Media S.A*. doi: 10.3389/fbioe.2022.910290.

[35] Della Sala F, Longobardo G, Fabozzi A, Di Gennaro M, Borzacchiello A. Hyaluronic acid-based wound dressing with antimicrobial properties for wound healing application. 2022 Mar. 01; *MDPI*. doi: 10.3390/app12063091.

[36] Romanò CL, De Vecchi E, Bortolin M, Morelli I, Drago L. Hyaluronic acid and its composites as a local antimicrobial/antiadhesive barrier. 2017 Jan. 01; *Copernicus GmbH*. doi: 10.7150/jbji.17705.

[37] Sundaramurthi D, Krishnan UM, Sethuraman S. Electrospun nanofibers as scaffolds for skin tissue engineering. 2014 Apr. 03; *Taylor and Francis Inc*. doi: 10.1080/15583724.2014.881374.

[38] Kohli N, Sharma V, Brown SJ, García-Gareta E. Synthetic Polymers for Skin Biomaterials. In: Biomaterials for Skin Repair and Regeneration, Ed: García-Gareta E, Woodhead Publishing, Elsevier, Northwood, UK; 2019, pp. 125–149. doi: 10.1016/B978-0-08-102546-8.00005-4.

[39] Xu R, *et al.* Recent advances in biodegradable and biocompatible synthetic polymers used in skin wound healing. 2023 Aug. 01; *Multidisciplinary Digital Publishing Institute (MDPI)*. doi: 10.3390/ma16155459.

[40] Hilmy N. Poly(N-vinylpyrrolidone) hydrogels 2. Hydrogel composites as wound dressing for tropical environment. Radiation Physics and Chemistry. 1993; 42(4-6): 911–914.

[41] De Lima GG, *et al.* Synthesis and in vivo behavior of PVP/CMC/agar hydrogel membranes impregnated with silver nanoparticles for wound healing applications. ACS Applied Bio Materials. 2018; 1(6): 1842–1852. doi: 10.1021/acsabm.8b00369.

[42] Avinc O, Khoddami A. Overview of Poly(lactic acid) (PLA) fibre Part I: Production, properties, performance, environmental impact, and end-use applications of poly(lactic acid) fibres. Fibre Chemistry. 2009; 41(6): 391–401.

[43] Ulery BD, Nair LS, Laurencin CT. Biomedical applications of biodegradable polymers. 2011 Jun. 15. doi: 10.1002/polb.22259.

[44] Sin LT, Rahmat AR, Rahman WAWA. Applications of Poly(lactic Acid). In: Ed: Sina Ebnes, Willium Andrew, Elsevier, Handbook of Biopolymers and Biodegradable Plastics: Properties, Processing and Applications, MA, USA; 2013, 55–69. doi: 10.1016/B978-1-4557-2834-3.00003-3.

[45] Liu S, Qin S, He M, Zhou D, Qin Q, Wang H. Current applications of poly(lactic acid) composites in tissue engineering and drug delivery. 2020 Oct; 15: *Elsevier Ltd*. doi: 10.1016/j.compositesb.2020.108238.

[46] Sheikholeslam M, Wright MEE, Jeschke MG, Amini-Nik S. Biomaterials for skin substitutes. 2018 Mar. 07; *Wiley-VCH Verlag*. doi: 10.1002/adhm.201700897.

[47] Xie J, MacEwan MR, Ray WZ, Liu W, Siewe DY, Xia Y. Radially aligned, electrospun nanofibers as dural substitutes for wound closure and tissue regeneration applications. ACS Nano. 2010 Sep; 4(9): 5027–5036. doi: 10.1021/nn101554u.

[48] Woodruff MA, Hutmacher DW. The return of a forgotten polymer – Polycaprolactone in the 21st century. 2010; *Elsevier Ltd*. doi: 10.1016/j.progpolymsci.2010.04.002.

[49] Prete S, Dattilo M, Patitucci F, Pezzi G, Parisi OI, Puoci F. Natural and synthetic polymeric biomaterials for application in wound management. 2023 Sep. 01; *Multidisciplinary Digital Publishing Institute (MDPI)*. doi: 10.3390/jfb14090455.

[50] Savoji H, Godau B, Hassani MS, Akbari M. Skin tissue substitutes and biomaterial risk assessment and testing. 2018 Jul. 26. doi: 10.3389/fbioe.2018.00086.

[51] Fujie T, Furutate S, Niwa D, Takeoka S. A nano-fibrous assembly of collagen-hyaluronic acid for controlling cell-adhesive properties. Soft Matter. 2010 Oct; 6(19): 4672–4676. doi: 10.1039/c0sm00527d.

[52] Park S-N, Jung Lee H, Hoon Lee K, Suh H. Biological characterization of EDC-crosslinked collagen-hyaluronic acid matrix in dermal tissue restoration. Biomaterials. 2003; 24(9): 1631–1641.

[53] Chung EJ, Jakus AE, Shah RN. In situ forming collagen-hyaluronic acid membrane structures: Mechanism of self-Assembly and applications in regenerative medicine. Acta Biomaterial. 2013; 9(2): 5153–5161. doi: 10.1016/j.actbio.2012.09.021.

[54] Sionkowska A, Gadomska M, Musiał K, Piatek J. Hyaluronic acid as a component of natural polymer blends for biomedical applications: A review. 2020 Sep. 01; *MDPI AG*. doi: 10.3390/molecules25184035.

[55] Ma L, *et al.* Collagen/chitosan porous scaffolds with improved biostability for skin tissue engineering. Biomaterials. 2003; 24(26): 4833–4841. doi: 10.1016/S0142-9612(03)00374-0.

[56] Romanova OA, *et al.* Chitosan as a modifying component of artificial scaffold for human skin tissue engineering. Bulletin of Experimental Biology and Medicine. 2015; 159(4): 557–566. doi: 10.1007/s10517-015-3014-6.

[57] Bonvallet PP, Culpepper BK, Bain JL, Schultz MJ, Thomas SJ, Bellis SL. Microporous dermal-like electrospun scaffolds promote accelerated skin regeneration. Tissue Engin (Part A). 2014 Sep; 20 (17–18): 2434–2444. doi: 10.1089/ten.tea.2013.0645.

[58] Huang C, Fu X, Liu J, Qi Y, Li S, Wang H. The involvement of integrin β1 signaling in the migration and myofibroblastic differentiation of skin fibroblasts on anisotropic collagen-containing nanofibers. Biomaterials. 2012 Feb; 33(6): 1791–1800. doi: 10.1016/j.biomaterials.2011.11.025.

[59] Yang Y, Zhu X, Cui W, Li X, Jin Y. Electrospun composite mats of poly[(D, Llactide)-co-glycolide] and collagen with high porosity as potential scaffolds for skin tissue engineering. Macromolecular Materials and Engineering. 2009 Sep; 294(9): 611–619. doi: 10.1002/mame.200900052.

[60] Huang S, *et al.* Functional bilayered skin substitute constructed by tissue-engineered extracellular matrix and microsphere-incorporated gelatin hydrogel for wound repair. Tissue Engineering Part A. 2009; 15(9): 2617–2624.

[61] Ziai Y, *et al.* Chameleon-inspired multifunctional plasmonic nanoplatforms for biosensing applications. NPG Asia Mater. 2022 Dec; 14(1). doi: 10.1038/s41427-022-00365-9.
[62] Si Y, Yu J, Tang X, Ge J, Ding B. Ultralight nanofibre-assembled cellular aerogels with superelasticity and multifunctionality. Nature Communications. 2014; 5. doi: 10.1038/ncomms6802.
[63] Yang M, *et al.* High-performance flexible pressure sensor with a self-healing function for tactile feedback. Advanced Science. 2022 Jul; 9(20). doi: 10.1002/advs.202200507.
[64] Sun B, *et al.* Advances in three-dimensional nanofibrous macrostructures via electrospinning. 2014; *Elsevier Ltd.* doi: 10.1016/j.progpolymsci.2013.06.002.
[65] Liu X, Xu H, Zhang M, Yu DG. Electrospun medicated nanofibers for wound healing: Review. 2021 Oct. 01; *MDPI.* doi: 10.3390/membranes11100770.
[66] Liu M, Duan XP, Li YM, Yang DP, Long YZ. Electrospun nanofibers for wound healing. 2017 Jul. 01; *Elsevier Ltd.* doi: 10.1016/j.msec.2017.03.034.
[67] Bishop ES, *et al.* 3-D bioprinting technologies in tissue engineering and regenerative medicine: Current and future trends. 2017 Dec. 01; *Chongqing University.* doi: 10.1016/j.gendis.2017.10.002.
[68] Wang C, *et al.* 3D printing of bone tissue engineering scaffolds. 2020 Mar. 01; *KeAi Communications Co.* doi: 10.1016/j.bioactmat.2020.01.004.
[69] Kamolz LP, Kotzbeck P, Schintler M, Spendel S. Skin regeneration, repair, and reconstruction: Present and future. European Surgery – Acta Chirurgica Austriaca. 2022 Jun; 54(3): 163–169. doi: 10.1007/s10353-022-00757-9.
[70] Choi WS, *et al.* Development of a multi-layer skin substitute using human hair keratinic extract-based hybrid 3d printing. Polymers (Basel). 2021 Aug; 13(16). doi: 10.3390/polym13162584.
[71] Wistner SC, Rashad L, Slaughter G. Advances in tissue engineering and biofabrication for in vitro skin modeling. 2023 Nov. 01; *Elsevier B.V.* doi: 10.1016/j.bprint.2023.e00306.
[72] Gao C, *et al.* 3D bioprinting for fabricating artificial skin tissue. 2021 Dec. 01; *Elsevier B.V.* doi: 10.1016/j.colsurfb.2021.112041.
[73] Jean J, Lapointe M, Soucy J, Pouliot R. Development of an in vitro psoriatic skin model by tissue engineering. Journal of Dermatological Science. 2009 Jan; 53(1): 19–25. doi: 10.1016/j.jdermsci.2008.07.009.
[74] Jean J. Bioengineered skin: The self-assembly approach. Journal of Tissue Science & Engineering. 2013; 03(05). doi: 10.4172/2157-7552.s5-001.
[75] Nicoletti G, *et al.* Long-term in vivo assessment of bioengineered skin substitutes: A clinical study. Journal of Tissue Engineering and Regenerative Medicine. 2015; 9(4): 460–468. doi: 10.1002/term.1939.
[76] Kenry JCY, Lim CT. Emerging flexible and wearable physical sensing platforms for healthcare and biomedical applications. 2016; *Nature Publishing Group.* doi: 10.1038/micronano.2016.43.
[77] Han Y, Hu J, Jiang L. Collagen skin, a water-sensitive shape memory material. Journal of Materials Chemical B. 2018; 6(31): 5144–5152. doi: 10.1039/c8tb01346b.
[78] Brohem CA, Da Silva Cardeal LB, Tiago M, Soengas MS, De Moraes Barros SB, Maria-Engler SS. Artificial skin in perspective: Concepts and applications. 2011 Feb. doi: 10.1111/j.1755-148X.2010.00786.x.
[79] Bhowmick S, Thanusha AV, Kumar A, Scharnweber D, Rother S, Koul V. Nanofibrous artificial skin substitute composed of mPEG-PCL grafted gelatin/hyaluronan/chondroitin sulfate/sericin for 2nd degree burn care: In vitro and in vivo study. RSC Advances. 2018; 8(30): 16420–16432. doi: 10.1039/c8ra01489b.
[80] Ma Z, *et al.* Advanced electronic skin devices for healthcare applications. Royal Society of Chemistry. 2019. doi: 10.1039/c8tb02862a.
[81] Bell E, Ivarsson B, Merrill C. Production of a tissue-like structure by contraction of collagen lattices by human fibroblasts of different proliferative potential in vitro (tissue structure/hydrated collagen lattice/fibroblast function/cell aging/contractility). 1979; 76(3): 1274–1278.

[82] Haass NK, Smalley KSM, Li L, Herlyn M. Adhesion, migration and communication in melanocytes and melanoma. 2005; 18(3): 150–159. doi: 10.1111/j.1600-0749.2005.00235.x.

[83] Smalley KSM, Lioni M. Life ins't flat: Taking cancer biology to the next dimension. In Vitro Cellular & Developmental Biology-Animal. 2006; 42:242–247.

[84] Li L, Fukunaga-Kalabis M, Herlyn M. The three-dimensional human skin reconstruct model: A tool to study normal skin and melanoma progression. Journal of Visualized Experiments. 2011 Aug: 54:2937. doi: 10.3791/2937.

[85] Sarkozyova N, *et al.* Preparation and processing of human allogenic dermal matrix for utilization in reconstructive surgical procedures. Bratislava Medical Journal. 2020; 121(6): 386–394. doi: 10.4149/BLL_2020_063.

[86] Dolbashid AS, Mokhtar MS, Muhamad F, Ibrahim F. Potential applications of human artificial skin and electronic skin (e-skin): A review. 2018 Jun. 26; *ICE Publishing*. doi: 10.1680/jbibn.17.00002.

[87] Hammock ML, Chortos A, Tee BCK, Tok JBH, Bao Z. 25th anniversary article: The evolution of electronic skin (E-Skin): A brief history, design considerations, and recent progress. Advances in Materials. 2013 Nov; 25(42): 5997–6038. doi: 10.1002/adma.201302240.

[88] Cai R, Gimenez-Camino N, Xiao M, Bi S, Divito KA. Technological advances in three-dimensional skin tissue engineering. 2023 Jan. 01; *Walter de Gruyter GmbH*. doi: 10.1515/rams-2022-0289.

[89] Kim HS, Sun X, Lee JH, Kim HW, Fu X, Leong KW. Advanced drug delivery systems and artificial skin grafts for skin wound healing. 2019 Jun. 01; *Elsevier B.V.* doi: 10.1016/j.addr.2018.12.014.

[90] Albanna M, *et al.* In situ bioprinting of autologous skin cells accelerates wound healing of extensive excisional full-thickness wounds. 2019 *Scientific Reports*; 9(1): Dec. doi: 10.1038/s41598-018-38366-w.

[91] Böttcher-Haberzeth S, Biedermann T, Reichmann E. Tissue engineering of skin. 2010 Jun. doi: 10.1016/j.burns.2009.08.016.

[92] Zeng Q, *et al.* 6.20 Skin Tissue Engineering. In: Comprehensive Biomaterials II, Ed: Ducheyne P; Elsevier; 2017, pp. 334–382. 10.1016/B978-0-12-803581-8.10157-2.

[93] Xu Y, Wu X, Zhang Y, Yu Y, Gan J, Tan Q. Engineered artificial skins: Current construction strategies and applications. 2023 Dec. 01; *KeAi Communications Co.* doi: 10.1016/j.engreg.2023.09.001.

[94] Hosseinpour M, Khalili tanha G, Forouzanfar F, Moghbeli M, Enderami SE, Saburi E. Application of natural polymers in skin tissue engineering using 3D scaffold. International Journal of Polymeric Materials and Polymeric Biomaterials. 2025; 74(4):285–296.

[95] Gupta V, Singla R, Singh G, Chanda A. Development of soft composite based anisotropic synthetic skin for biomechanical testing. Fibers. 2023 Jun; 11(6). doi: 10.3390/fib11060055.

[96] Parvez S, *et al.* Preparation and characterization of artificial skin using chitosan and gelatin composites for potential biomedical application. Polymer Bulletin. 2012; 69(6): 715–731. doi: 10.1007/s00289-012-0761-7.

[97] Wang J, Duan X, Zhong D, Zhang M, Li J, Hu Z, Han F. Pharmaceutical applications of chitosan in skin regeneration: A review. International Journal of Biological Macromolecules. 2024 1; 261:129064.

[98] Zhang M, Xing J, Zhong Y, Zhang T, Liu X, Xing D. Advanced function, design and application of skin substitutes for skin regeneration. 2024 Feb. 01; *Elsevier B.V.* doi: 10.1016/j.mtbio.2023.100918.

[99] Joseph J, Deshmukh K, Tung T, Chidambaram K, Khadheer Pasha SK. 3D Printing Technology of Polymer Composites and Hydrogels for Artificial Skin Tissue Implementations. in Lecture Notes in Bioengineering, Springer, 2019, pp. 205–233. doi: 10.1007/978-3-030-04741-2_7.

[100] Yue K, Trujillo-de Santiago G, Alvarez MM, Tamayol A, Annabi N, Khademhosseini A. Synthesis, properties, and biomedical applications of gelatin methacryloyl (GelMA) hydrogels. Elsevier Ltd. Dec. 01, 2015. doi: 10.1016/j.biomaterials.2015.08.045.3

Hossein Sahbafar*, Fatemeh Ahmadvand, and Narmin Zarrinabadi

Chapter 3.3 Development of nerve tissue replacement using composites

Abstract: Since currently available treatments are not very effective in restoring nerve function, nerve tissue engineering and replacement are considered important approaches for addressing nerve-related conditions. Composites are among the material types extensively employed in tissue engineering applications. These materials are known for their tailorable properties, which are crucial for tissue regeneration. Examples of composite structures for nerve tissue replacement include conduits, scaffolds, hydrogels, films, fibers, and membranes, which are used to mimic the target nerve tissue. This chapter discusses the most significant literature that have employed composite structures and materials for the purpose of nerve repair and replacement. Through the creation of bioengineered scaffolds and grafts, scientists hope to promote cell survival and neuron regeneration. Scientific advancements, such as stem cell therapy and biomaterial fabrication, offer new ways to enhance cognitive and physical performance. These findings suggest the possibility of successfully simulating autologous grafts using customized scaffolds and conduits, providing viable substitutes for nerve tissue engineering and recovery in various animal models, and opening the door for potential therapeutic applications.

Keywords: Nerve tissue replacement, nerve tissue engineering, composites, conduit, scaffold, hydrogel

3.3.1 Introduction

Nerve tissue engineering and replacement is important because nerve injuries and neurodegenerative conditions may cause irreversible problems in patients' lifestyles and nerve–related activities [1–3]. Since current therapies are thought to be inadequate in restoring nerve function, bioengineered neural scaffolds and grafts must be developed [4–7]. Due to the fact that current therapies are thought to be inadequate

***Corresponding author: Hossein Sahbafar**, School of Mechanical Engineering, College of Engineering, University of Tehran, Tehran, Iran, e-mail: hosseinsahba@gmail.com, hsahbafar@ut.ac.ir
Fatemeh Ahmadvand, Department of Experimental Medicine, University of Genova, Genova, Italy, e-mail: Fateme.ahmadvand74@gmail.com, 6017408@studenti.unige.it
Narmin Zarrinabadi, Mechanical Engineering Department, Tarbiat Modares University, Tehran, Iran, e-mail: Narminzarrinabadi@gmail.com, Narmin.zarinabadi@modares.ac.ir

https://doi.org/10.1515/9783111386027-015

in restoring nerve function, bioengineered neural scaffolds and grafts need to be developed [8–11]. Researchers can provide solutions via techniques like stem cell therapy and the development of appropriate biomaterials and bio-structures. These solutions contribute to the improvement of sensory and cognitive capacities, in addition to restoring physical function [12, 13]. Developments in nerve tissue engineering have the potential to transform therapies and significantly enhance patient outcomes as the world's population ages and the prevalence of nerve-related diseases rises.

Composites are created from two or more unique components to improve their overall characteristics, such as strength, flexibility, and biocompatibility [14–16]. Composites play a major role in the engineering and replacement of nerve tissue by fostering the cellular connections and structural support necessary for nerve repair [17–19]. For instance, combining synthetic materials with natural polymers like collagen may result in the development of scaffolds that imitate the extracellular matrix's function, which is crucial for promoting cell attachment and proliferation in the target region [20–22]. Additionally, these composites may be designed to carry bioactive chemicals that promote nerve regeneration and aid in the development of new neurons [23, 24]. Because of their adaptable qualities, materials may be designed to match the mechanical and biological features of the intended neural tissue [25–27]. Composite materials

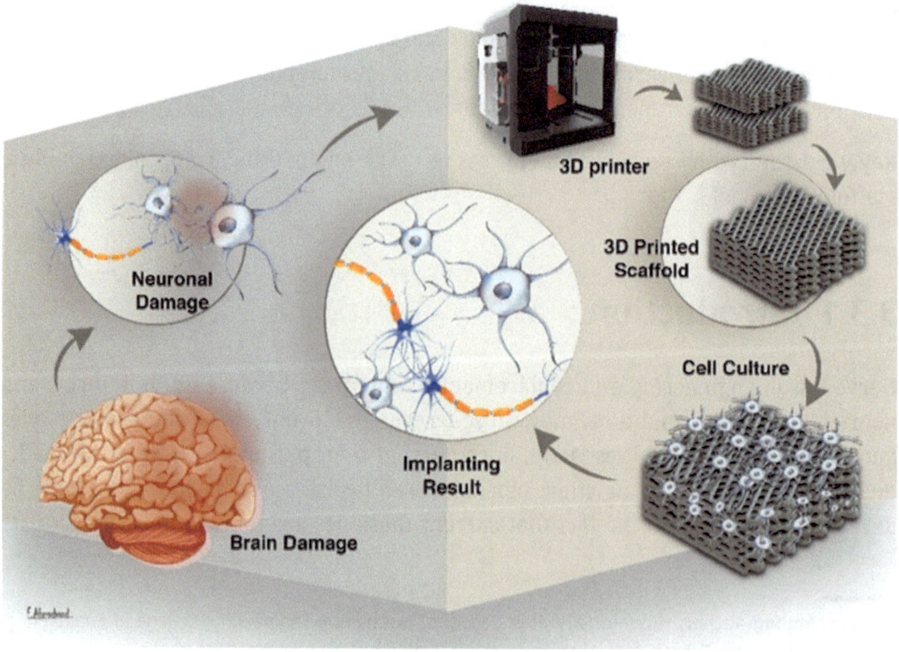

Scheme 3.3.1: Overall illustration of the development of composite structures for nerve tissue replacement and engineering [some items were reused freely from Freepik. In addition, the image related to the 3D printer device was adapted from the website of TechnoZoo].

considerably improve the environment for nerve cells, which enhances the efficacy of regeneration techniques and boosts recovery outcomes for patients with nerve-related injuries. Scheme 3.3.1 is the general illustration of the development of composite structures for nerve tissue replacement.

In this chapter, the most significant types of composite materials, including conduits, scaffolds, hydrogels, films, fibers, membranes, etc., for nerve tissue engineering and replacements are addressed in detail.

3.3.2 Development of nerve tissue replacement using composites

3.3.2.1 Composite conduits

Composite conduits are tubular constructions that facilitate nerve regeneration by fusing several materials together. They provide channels for the development of axons and may include bioactive substances to promote the regeneration of nerve tissue [28–30]. Sun et al. used graphene with chitosan and dental pulp stem cells (DPSCs) in an attempt to create a new nerve conduit for facial nerve restoration. They developed synthetic graphene/chitosan tubes (G/CSTs) and assessed the biological, chemical, and physical characteristics of the materials. Later, using a rabbit model, they investigated how G/CST and DPSCs affected the regeneration of the facial nerve. In comparison to CSTs alone, the findings demonstrated that G/CST displayed advantageous qualities and improved nerve regeneration. Nevertheless, the absence of long-term follow-up and the small sample size hampered the research. Subsequent investigations need to concentrate on refining the composition of G/CST and investigating its possibilities for therapeutic uses [31]. In another study, Zhang et al. used a chitosan nerve conduit, loaded with ZIF-8-functionalized guide microfibers, in an attempt to enhance nerve regeneration. The conduit was developed, and its capacity to stimulate Schwann cell growth in vitro, as well as its biocompatibility, were evaluated. Rats with sciatic nerve abnormalities using the conduit with ZIF-8-modified fibers demonstrated better axonal elongation and functional recovery in vivo compared to control groups. The healing results resembled those of autologous nerve transplants. The absence of long-term follow-up and the study's concentration on a specific kind of nerve damage were the main limitations. Subsequent studies might examine the conduit's impact on various nerve injuries and assess ZIF-8's potential for therapeutic agent delivery [32]. Moreover, Tang et al. investigated the efficacy of a composite conduit for treating long-gap peripheral nerve damage in rats injected with exosomes derived from human umbilical cord mesenchymal stem cells. The conduit, both with and without exosomes, was compared with an autograft control group. To evaluate nerve regeneration and functional recovery, the research included muscle mass measurement, histological staining, electrical recordings, and the sciatic functional index. Comparing the conduit filled with exosomes to the conduit alone, the findings demonstrated a considerable improvement in nerve regeneration, motor function, and muscle mass, which was almost as effective as autograft. Future studies should concentrate on examining the treatment's long-term effects, identifying the precise molecular mechanisms underlying exosome-mediated regeneration, and refining the exosome loading technique for use in clinical settings [33]. The potential of multilayered conductive PVP/Col/CS composite, doped with nanoparticles as nerve guide conduits, was studied by Sierakowska-Byczek et al. (Figure 3.3.1). These composites were produced by combining electrospraying and electrospinning (ELS) methods, and conductivity tests, FT-IR and SEM/EDS, were used to characterize the materials. The biomaterials underwent biodegradation and cytotoxicity testing. The composites showed high biocompatibility and biodegradability, and the addition of nanoparticles improved the composites' conductivity, according to the scientists. To confirm the effectiveness of these materials as nerve guide conduits, further in vivo research is required since the current work was restricted to in vitro testing. Subsequent investigations need to concentrate on enhancing the composition and structure of the composites to elevate their functionality and investigate their potential for therapeutic applications [34].

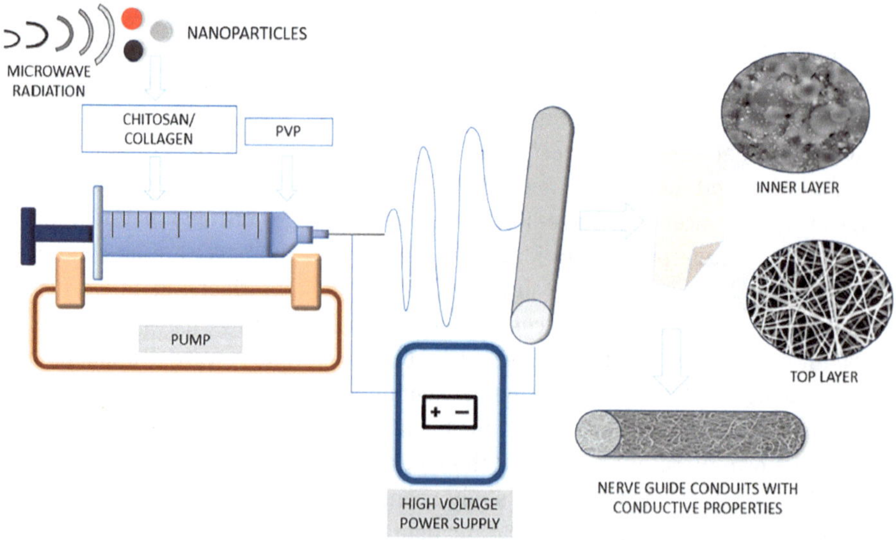

Figure 3.3.1: Schematic of the composite chitosan/collagen/PVP conduit developed in [34]. Reused with permission under CC BY 4.0.

Through the ELS of poly(3-hydroxybutyrate) (PHB) with 8 wt% magnetite nanoparticles, modified by citric acid (Fe_3O_4-CA), Shlapakova et al. developed magnetoactive nerve guidance conduits (NGCs). Compared to pure PHB NGCs, the composite NGCs exhibited improved ductility and surface-free energy. The conduits promoted the growth of SH-SY5Y cells and rat mesenchymal stem cells in vitro when exposed to a low-frequency magnetic field. The conduits were inserted into injured sciatic nerves in rats in vivo, demonstrating no signs of acute inflammation, and encouraging nerve regeneration. The lack of an in vivo examination was a limitation of the study. Future research should examine the potential of these magnetoactive NGCs for treating different peripheral nerve lesions, with an emphasis on long-term in vivo assessments [35]. In addition, for peripheral nerve regeneration, Nabipour et al. developed a bilayer nerve conduit using multiwall carbon nanotubes (MWCNTs), polylactic acid (PLA), and polyurethane (PU). The nanofibrous scaffold was produced using ELS, and the authors went on to describe its morphology, hydrophilicity, mechanical strength, and rate of degradation. The results of the study demonstrated that the conduit facilitated cell adhesion and proliferation in vitro; however, further in vivo research is required to determine how well it promotes neuron regeneration. The authors recommend that in order to improve nerve regeneration, future studies concentrate on improving scaffold design and adding growth factors [36]. In a rat model of peripheral nerve damage, Li et al. examined the effectiveness of a BDNF-loaded chitosan-based mimetic mussel polymer conduit (MPC@BDNF) for healing. The conduit was developed by the researchers, who also assessed its adhesive and mechanical properties as well as its biocompatibility and antibacterial activity in vitro. Comparing MPC@BDNF to a commercial control group, in vivo experiments employing a 15 mm sciatic nerve damage model demonstrated substantial improvements in nerve regeneration, myelination, and functional recovery. Nevertheless, a small sample size and a brief follow-up period hampered the research. Future investigations should concentrate on longer-term trials with larger sample numbers, and investigate MPC@BDNF's potential for treating other nerve injuries [37]. In addition, Johnson et al. attempted to enhance peripheral nerve regeneration through the development of porous NGCs by emulsion-templating methods. The authors studied the mechanical properties, porosity, and biocompatibility of the composite structures they developed by adding poly(glycerol sebacate)-methacrylate (PGS-M) to a high internal phase emulsion (HIPE). The improved mechanical characteristics and increased porosity of the composite NGCs allowed for better nutrition transfer and inhibited fibroblast infiltration. Studies conducted in vitro and in vivo demonstrated that the conduit surface is neuronally growing and cytocompatible. In vivo investigations to evaluate the long-term effectiveness of these porous NGCs in stimulating nerve regeneration should be the main focus of future research [38]. To encourage peripheral nerve regeneration, Huang et al. developed a dual-layer nerve conduit with an exterior layer made of PLGA and an inner layer made of GelMA that encapsulated Schwann cells treated with VEGF-A. In a rat model of nerve damage, they assessed the conduit's effectiveness, biocompatibility, and mechanical properties. The outcomes demonstrated that the conduit induced nerve regeneration and angiogenesis, producing results similar to those of autograft. To evaluate the conduit's long-term effectiveness and safety in humans, further research is necessary, since this study was limited to a single-animal model. Subsequent investigations have to concentrate on refining the conduit's design, investigating distinct cell types and growth factors, and carrying out clinical studies to verify its feasibility for the restoration of peripheral nerves [39].

In a rat sciatic nerve damage model, Hoveizi et al. examined the possibility of promoting nerve regeneration using a 3D-printed polyacrylonitrile (PAN) conduit, filled with fibrin hydrogel and graphene quantum dots (GQDs). Mesenchymal stem cells from Wharton's jelly (WJMSCs) were developed into Schwann cells and then added to the conduit. Rats treated with the composite hydrogel-filled PAN conduit with GQDs had considerably greater sensory recovery, larger axon numbers, and remyelination, compared to other groups, indicating that the conduit/Schwann method greatly enhanced nerve regeneration. However, the short trial period and small sample size were the limitations of the research. Future studies should examine the potential of this strategy for therapeutic applications and concentrate on longer-term studies with larger sample numbers [40]. In another study, Dai et al. used electrospun PLLA-PPSB nanofibers to develop a biocompatible and bioactive nerve conduit for the repair of peripheral nerve defects. The researchers developed and examined PLLA-PPSB nanofibers with different mass ratios, tested their biocompatibility with Schwann cells in vitro, and evaluated their potential for in vivo repairing in a model of a sciatic nerve lesion in rats. The findings demonstrated that PLLA-PPSB nanofibers improved myelin formation, axonal regeneration, angiogenesis, and functional recovery in vivo, and stimulated SC proliferation, glial expression, and adhesion in vitro. The use of a single-animal model and the absence of long-term in vivo investigations were limitations of the study. Subsequent studies should concentrate on refining the nanofibers' shape and content, examining their long-term biocompatibility and effectiveness in larger animal models, and figuring out the fundamental principles of SC growth [41].

The objective of Xu et al. was to develop a biocompatible nerve conduit that could be used to fix peripheral nerve abnormalities with lengthy segments. They created GelMA-CNTF/IGF-1, a composite hydrogel, and placed it into PCL electrospun nerve conduits. Research conducted in vitro demonstrated that the GelMA-CNTF/IGF-1 hydrogel maintained the release of CNTF and IGF-1, stimulated the development of Schwann cells, and improved neuronal axonal growth. Research conducted on rats in vivo demonstrated that PCL-GelMA-CNTF/IGF-1 efficiently sped up neuron regeneration, encouraged the creation of myelin sheaths, and enhanced the recovery of motor function. Nevertheless, the lack of long-term follow-up and the small sample size were the limitations of the research. Future research will need to concentrate on refining the hydrogel's composition and examining its long-term effectiveness and safety in larger animal models [42]. Moreover, Tagandurdyyeva et al. used chitosan fibers and poly(L-lactide) nanofibers to develop resorbable conduits for peripheral nerve regeneration (Figure 3.3.2). While in vivo investigations revealed enhanced sciatic nerve regeneration with the inclusion of chitosan fibers, in vitro studies exhibited

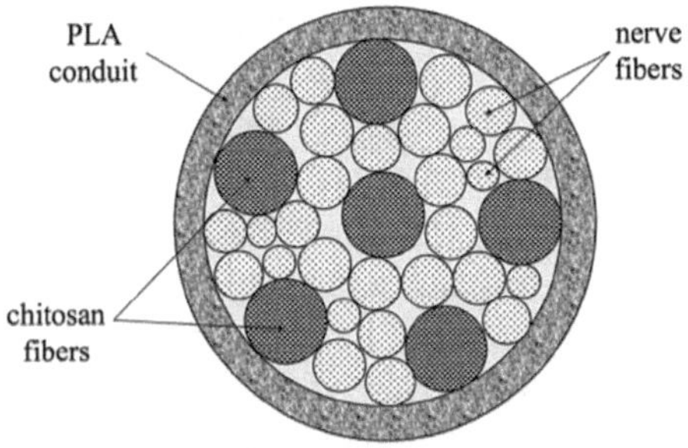

Figure 3.3.2: Schematic illustration of the conduit developed in [43]. Reused with permission under CC BY 4.0.

excellent cell adherence and differentiation on the conduits. The conduits guided nerve development and facilitated electrical transmission; however, the study's small sample size meant that more investigation was required to assess the filler composition's long-term effectiveness [43]. The efficacy of a Gel-Ppy-modified nerve conduit in repairing sciatic nerve abnormalities in rats was studied by Jiang et al. Gelatin (Gel) for biocompatibility and polypyrrole (Ppy) for conductivity were used in the conduit's construction. The Gel-Ppy conduit exhibited increased Schwann cell proliferation in in vitro tests as compared to the control, Gel, and Ppy groups. The Gel-Ppy group exhibited enhanced nerve regeneration, axon diameter, myelin sheath thickness, muscle atrophy, nerve conduction velocity, and motor function recovery in rat sciatic nerve damage model in vivo experiments. The scientists hypothesized that by activating the Rap1 pathway, the Gel-Ppy conduit promoted neuron regeneration. A small sample size and a lack of long-term follow-up were among the limitations. Further studies need to concentrate on examining the extended effectiveness and reliability of Gel-Ppy conduits in larger animal models and clinical trials [44].

In order to repair 10 mm peripheral nerve lesions in rats, Jiang et al. investigated the effectiveness of a multichannel nerve conduit, made of an internally oriented N-succinyl-chitosan (NS-chitosan) fiber scaffold and warp-knitted chitosan scaffold. They proved that the conduit was biocompatible and had sufficient mechanical strength. It was discovered that the NS-chitosan fibers enhanced the migration and survival of Schwann cells, while blocking apoptosis via the Bcl-2/Bax/caspase-3 pathway. The conduit demonstrated nerve regeneration effectiveness that was on par with autografts. The study, however, was restricted to a particular model of nerve damage, and further investigation is required to examine the long-term impacts and applicability in other forms of nerve injury. Further studies may focus on refining the conduit's structure and adding more biomolecules to improve nerve regeneration and functional recovery [45]. The development of conductive, biodegradable, 3D-printed PGSA composites for nerve tissue regeneration was studied by Huang et al. In a PGSA matrix, they incorporated PVP, AgNPs, and graphene. They next evaluated the composites' electrical conductivity, biodegradability, 3D printability, and ability to stimulate cell proliferation. The PGSA composites demonstrated biodegradability by enzymatic breakdown and hydrolysis, were highly conductive in a swelled state, and facilitated cell proliferation and guidance through electrical stimulation and 3D-printed microstructures, according to the research. By combining electrical stimulation with Schwann cell culture, the scientists were able to demonstrate directed cell development and enhanced proliferation. Only in vitro trials were conducted for this study; more in vivo research is required to validate the results. Future investigations should concentrate on refining the composite structure and examining the long-term biocompatibility and effectiveness of the 3D-printed nerve conduits in animal models [46]. Dong et al. examined the efficacy of a unique composite nerve conduit for bridging long-gap peripheral nerve lesions in canine models. The conduit was made to resemble the natural brain microarchitecture, with an anisotropic microfiber grid core and a nanofiber sheath. The composite conduit was compared with Neurolac® conduits that can be purchased commercially and conduits that merely had a sheath. At 12 months after implantation, the composite conduit exhibited greater nerve regeneration, as proven by enhanced histology recovery, improved gait recovery, increased electrophysiological performance, and decreased muscle atrophy. The limited sample size and the need for more research on long-term effectiveness were among the limitations [47]. In addition, Dong et al. used platelet-rich plasma (PRP) in a composite nerve conduit, in an effort to enhance peripheral nerve regeneration. In vitro and in vivo tests were performed on GelMA/SA-3/PRP-20 NGCs to assess their mechanical properties, rate of degradation, and biological efficacy. In a rat sciatic nerve lesion model, the GelMA/SA-3/PRP-20 NGCs increased nerve regeneration and functional recovery, sustained growth factor release, Schwann cell migration, and angiogenesis. Nevertheless, the study only investigated the impact of PRP at a single concentration; additional investigation is required to find the ideal PRP concentration and analyze the GelMA/SA-3/PRP-20 NGCs' long-term effects [48].

Table 3.3.1 is a summary of the investigations that developed composite conduit structures for nerve tissue repair and engineering.

3.3.2.2 Composite scaffolds

Composite scaffolds are multimaterial and three-dimensional constructs that facilitate the development of new tissues and the proliferation of cells. They enhance axonal connection in the replacement of nerve tissue and provide an environment that is favorable for nerve regeneration [57–59]. Wang et al. examined the possibility of stimulating sciatic nerve regeneration in rats using a 3D-printed neural scaffold that included reduced graphene oxide (rGO) and pulsed electromagnetic field (PEMF) stimulation. They assessed the scaffold's biocompatibility in vitro as well as its material characteristics. Studies conducted in vivo revealed that compared to a control group, the combination method dramatically enhanced neuron regeneration and functional recovery. The structure of the scaffold should be optimized, the fundamental processes of PEMF-induced nerve regeneration should be investigated, and larger-

Table 3.3.1: Summary of literature on the development of composite conduits for nerve tissue replacement.

Objectives	Composite type or structure	Fabrication method	Composition and materials	Results	Limitations	Future recommendations	Reference
Repair facial nerve injury with biocompatible materials	Graphene/chitosan composite tubes	ELS and chemical modification	Graphene, chitosan, and DPSCs	G/CST enhances nerve regeneration and functional recovery	Microstructure of the conduit, the loading mode of cells, and the mechanism of the composite conduit	Explore the effect of the electrical conductivity of graphene	[31]
Enhance nerve regeneration and sensory function recovery	Chitosan nerve conduit with ZIF-8-functionalized guide microfibers	Mold casting and in situ ZIF-8 loading	Chitosan, ZIF-8 nanoparticles	Improved nerve regeneration, sensory function, and muscle recovery, compared to control groups	Limited to in vitro and in vivo studies on rats	Clinical trials are exploring other MOFs for nerve repair	[32]
Repair long-gap peripheral nerve injury in rats	3D composite conduit with Col/HA inner sponge and PLGA outer shell	ELS for PLGA shell, collagen/hyaluronic acid sponge filling	Collagen, hyaluronic acid, poly(lactic-co-glycolic acid) (PLGA)	Exosome-loaded conduit promotes nerve regeneration and functional recovery in rats	Frequent measures to closely study the correlation between the degree of behavioral improvement and axon growth	Investigate the long-term efficacy and safety of the exosome-loaded conduit in humans. Anti-inflammatory or anti-pyroptotic effects must be investigated. The number of released exosoms after implantation must be studied	[33]

(continued)

Table 3.3.1 (continued)

Objectives	Composite type or structure	Fabrication method	Composition and materials	Results	Limitations	Future recommendations	Reference
Develop conductive, biocompatible nerve guide conduits	Multilayered, gradient, and conductive nerve guide conduits	ELS and electrospraying	PVP, collagen, chitosan, and metallic nanoparticles (Ag, Au, and Pt)	Biocompatible, conductive, and potentially effective for nerve regeneration	Further in vivo studies needed to confirm effectiveness	Further research on long-term biodegradation and in vivo efficacy	[34]
Develop magnetoactive NGCs for peripheral nerve repair	Biodegradable, electrospun, magnetic nerve guidance conduit	ELS	PHB and magnetite nanoparticles	Composite conduits exhibit enhanced mechanical properties, biocompatibility, and promote nerve regeneration in vitro and in vivo	Further in vivo studies with longer time points are needed	Explore the potential of magnetic stimulation for enhanced nerve regeneration	[35]
Develop bilayer nerve conduit for peripheral nerve regeneration	Bilayer PLA/PU/ MWCNT nerve conduit	ELS and hand-rolling	PLA, PU, MWCNT, and human endometrial stem cells	Conduit supports cell attachment and proliferation of hEnSCs	In vitro study only, no in vivo testing	In vivo testing and optimization of conduit design	[36]
Repair peripheral nerve injury with a novel conduit	MPC@BDNF	Chitosan-based mimetic mussel polymer conduit fabrication	Chitosan, gelatin, polydopamine, and BDNF	Improved nerve regenerat on and functional recovery in rats	Further clinical trials needed to confirm efficacy	Explore long-term effects and optimize BDNF delivery	[37]

Develop porous NGCs for peripheral nerve regeneration	Polymer-polyHIPE composite NGCs	Emulsion templating with photocross-linking	Poly(glycerol sebacate)-methacrylate (PGS-M) and Hypermer	Porous NGCs with tunable mechanical properties; favorable for nerve regeneration	In vivo testing not yet performed, limited sample size	Further in vivo studies and clinical trials	[38]
Enhance peripheral nerve regeneration via angiogenesis	Dual-layer nerve conduit	3D printing and injection molding	PLGA, GelMA, VEGF-A transfected Schwann cells	Composite conduit promotes angiogenesis and nerve regeneration, comparable to autograft	Limited to in vitro and animal studies, further human trials needed	Further clinical trials are needed to evaluate long-term efficacy and safety, replacing the insertion material to achieve optimal bioactivity and utilizing more growth factors	[39]
Enhance sciatic nerve regeneration using 3D-printed conduit	3D-printed PAN conduit with fibrin hydrogel and GQDs	3D printing and solution printing method	PAN, fibrin hydrogel, GQDs, WJMSCs, RA, bFGF, PDGF, GGF-2	Conduit/Schwann cell approach improves nerve regeneration	Study conducted on rats, limited sample size	Further clinical trials with larger sample size	[40]
Repair peripheral nerve defects with composite nanofibers	Electrospun PLLA-PPSB nanofiber nerve conduits	ELS and curing	PLLA, PPSB (poly (PCL triol-co-sebacic acid-co-BES sodium salt))	PLLA-PPSB nanofibers enhance SC proliferation, myelin formation, axonal regeneration, angiogenesis, and functional recovery in a rat sciatic nerve defect model	Limited in vivo study duration, further investigation needed on long-term effects and clinical translation	Explore clinical trials and investigate the long-term effects of PLLA-PPSB nanofibers on nerve regeneration. More diverse experimental methods such as electrophysiologic assessments	[41]

(continued)

Table 3.3.1 (continued)

Objectives	Composite type or structure	Fabrication method	Composition and materials	Results	Limitations	Future recommendations	Reference
Repair long-segment peripheral nerve defects	PCL electrospun nerve conduit filled with GelMA-CNTF/IGF-1 hydrogel	ELS and photopolymerization	PCL, GelMA, CNTF, and IGF-1	Improved nerve regeneration, myelin formation, and motor function recovery in rats	Limited to sciatic nerve injury, further in vivo studies needed	Investigate other nerve injuries, clinical trials for human applications	[42]
Develop a resorbable conduit for peripheral nerve regeneration	Tubular conduit with nanofiber and fiber fillers	ELS for nanofibers and wet-spinning for fibers	PLA nanofibers, chitosan fibers, and chitin nanofibrils	Improved nerve regeneration, increased SFI values, and better M-response	Limited to defects larger than 3 cm, and further research needed	Explore long-term effects, clinical trials, and optimization	[43]
Enhance nerve regeneration after injury	Gel-Ppy modified nerve conduit	ELS	PLLA, Gelatin, and Ppy	Improved Schwann cell proliferation, axon diameter, and myelin sheath thickness	Study only on rats, limited time follow-up	Clinical trials, long-term studies	[44]
Repair peripheral nerve defects using a novel chitosan-based conduit	Multichannel nerve conduit with warp-knitted chitosan scaffold and NS-chitosan fibers	Warp knitting for the chitosan scaffold, internal NS-chitosan fibers	NS-chitosan	Conduit promotes Schwann cell survival, axonal regeneration, and functional recovery in rats	Study limited to 10 mm nerve defects in rats, and further research needed for clinical applications	Investigate long-term efficacy, explore clinical trials, and optimize conduit design for various nerve defect sizes	[14]

Develop conductive, biodegradable nerve conduits for regeneration	3D-printed, microstructured PGSA composites	Digital light processing additive manufacturing (DLP-AM)	PGSA, PVP, AgNPs, and graphene	Improved conductivity, biodegradability, and cell proliferation with electrical stimulation	Limited in vivo testing, and further optimization needed	In vivo studies, clinical trials, and further material optimization	[46]
Promote long-gap peripheral nerve regeneration	Microfiber grid cores with nanofiber sheaths	Melt electroprinting and ELS	PDS for core and PLCL for sheath	Composite conduits outperform controls in nerve regeneration	Limited to canine models, and long-term effects unknown	Clinical trials and larger animal studies	[47]
Enhance peripheral nerve regeneration	GelMA/SA/PRP nerve conduit	Photo-cross-linking and ionic crosslinking	GelMA, sodium alginate, and PRP	Sustained growth factor release, improved cell migration and angiogenesis, and enhanced nerve regeneration in vivo	Limited to in vitro and in vivo studies, and further clinical trials needed	Clinical trials, and explore other growth factors and delivery methods	[48]
Develop a bioactive nerve conduit for peripheral nerve repair	Warp-knitted chitosan tube with radial C-CM -CTS sponge	Warp knitting, freeze drying, and cross-linking	Medical-grade chitosan fiber and carboxymethyl chitosan	Promotes axonal regeneration and improves functional recovery in rats	Further long-term studies needed and clinical trials required	Explore different cross-linking methods and optimize material properties	[49]
Peripheral nerve repair using a double-layer composite hydrogel conduit	Double-layer composite hydrogel nerve conduit	Electrodeposition and Schiff base reaction	Chitosan, polyaniline, Pluronic F-127, and 7,8-dihydroxyflavone	Improved nerve regeneration compared to traditional conduits, comparable to autologous transplantation	Further in vivo studies needed for long-term efficacy	Clinical trials to assess effectiveness and safety in humans	[50]

(continued)

Table 3.3.1 (continued)

Objectives	Composite type or structure	Fabrication method	Composition and materials	Results	Limitations	Future recommendations	Reference
Promote peripheral nerve regeneration	Conductive nerve guide conduit	ELS and soft-templating	PLA, PPy, VEGF, PF, and CS	Improved conductivity, mechanical properties, biocompatibility, and drug release	In vitro studies only, and no in vivo testing	Further optimization of drug loading and release. More in-depth studies, such as in vivo experiments, are needed to better demonstrate the advantages of the NGC	[51]
Evaluate PDLLA/β-TCP/HA/CHS/NGF conduits for nerve regeneration	Multilayered, sustained-release nerve conduit	Layer-by-layer assembly and cross-linking	PDLLA, β-TCP, HA, CHS, and NGF	Conduits promote nerve regeneration and similar to autograft	Limited sample size and short-term study duration	Long-term studies, clinical trials, and optimize NGF release	[52]
Develop stable HA nanocomposite conduits for neural regeneration	Hyaluronan-based nanocomposite conduits	Cross-linked HA with MWCNTs or mesoporous silica	Hyaluronic acid, MWCNTs and mesoporous silica	Improved stability, mechanical properties, and cell adhesion	Limited in vivo testing	In vivo studies, optimization of nanomaterial dosage. Further use of MWCNTs in prosthetic neurotechnological devices for electric stimulation	[53]

Repair large gap peripheral nerve injuries	Nerve regeneration conduit	3D printing and photo-cross-linking	GelMA, bFGF, DPSCs, and CSM	Functional nerve regeneration in rat model	Limited to rat model, and further clinical trials needed	Clinical trials and long-term efficacy evaluation	[54]
Repair sciatic nerve defects in rats	Tissue-engineered nerve graft	Chitin biological conduit with SANs and PRP	Chitin, SANs, and PRP	Improved nerve regeneration and functional recovery	Small sample size and limited long-term follow-up	Further clinical trials and investigation of long-term effects	[55]
Promote peripheral nerve regeneration	Multilayered nerve guidance conduit	ELS	PCL, MLT, and Fe_3O_4-MNPs	Improved nerve regeneration compared to controls	Long-term effects of nanoparticles not fully investigated	Further study of long-term nanoparticle effects	[56]

scale clinical trials should be carried out to evaluate the effectiveness and safety of this intriguing strategy [60]. The objective of Soltani Gerdefaramarzi et al. was to develop and assess PLA/PCL/G nanocomposite scaffolds for the purpose of tissue engineering peripheral nerves. They produced four distinct scaffold compositions using FDM 3D printing. The scaffolds' mechanical properties, electrical conductivity, and biodegradability were assessed as well as their characterization using SEM, XRD, and FTIR. With its regular shape, appropriate mechanical properties, and degradation rate that closely matched the rates of regeneration of peripheral nerve tissue, the PLA-PCL-G scaffold demonstrated the most promising findings. The study's shortcomings were emphasized, namely the absence of in vivo testing and the need for more graphene dispersion optimization. In vivo investigations to evaluate the scaffolds' biocompatibility and ability to support nerve regeneration should be the main focus of future research [61]. Moreover, Aseer et al. used electrospun collagen/PLA (Coll/PLA) nanofibers, together with retinoic acid (RA), to develop a biocompatible and bioactive nerve conduit. In vitro biocompatibility was studied using PC12 cells, and the mechanical properties, hydrophilicity, and degradation rate of the Coll/PLA and Coll/PLA/RA nanofibers were analyzed by the authors. The potential of Coll/PLA/RA nanofibers for nerve repair was shown by the results, which indicated that they enhanced PC12 cell adhesion and proliferation. The use of a single cell line and the absence of in vivo testing presented limitations to the research. Future studies should concentrate on examining these scaffolds' long-term impact on nerve regeneration in animal models as well as their potential for use in clinical settings [62]. By adding VEGF and gastrodin liposomes (GAS-LP) to electrospun silk fibroin/graphene oxide (SF/GO) nanofibrous and SF sponge, respectively, Liu et al. developed a new composite scaffold. This scaffold allowed for the combination of electrical stimulation and intraneural vascularization medications, which improved RSC96 cell proliferation and provided protection from an inflammatory milieu. The research proved that electrical stimulation and sequential medication release work synergistically to promote peripheral nerve regeneration. Nevertheless, because the research was carried out in vitro, further in vivo investigations are required to verify the safety and effectiveness of this scaffold for the regeneration of peripheral nerves. Subsequent studies could concentrate on refining the scaffold's design and examining the scaffold's long-term impact on nerve regeneration [63].

In order to encourage peripheral nerve regeneration, Zhang et al. developed a piezoelectric SF/poly(vinylidene fluoride-co-hexafluoropropylene)/Ti_3C_2Tx (SF/PVDF-HFP/MXene) composite scaffold by ELS. The physical and biological characteristics of the scaffold were described. Research conducted in vitro revealed that Schwann cell growth and proliferation were enhanced by piezoelectric stimulation. Axonal myelination and functional recovery were facilitated by the SF/PVDF-HFP/MXene nerve conduit, according to in vivo investigations conducted in a rat sciatic nerve injury model. The study was restricted to only one animal model; thus, further research using several animal models as well as clinical trials is required. Future studies may concentrate on improving the scaffold's design and investigating piezoelectric stimulation's long-term impact on nerve regeneration [64]. Moreover, Yang et al. aimed to create an electroconductive scaffold that could operate wirelessly to encourage the regeneration of peripheral nerves. The shell of the composite scaffold that the scientists developed was made of aligned electrospun PLLA nanofibers, while the core was made of CNTs@GelMA hydrogel. According to in vitro research, the scaffold increased the number of neurites in dorsal root ganglions, axonal outgrowth, and Schwann cell adhesion and elongation. According to in vivo research, the scaffold improved motor functional recovery by myelination and axon outgrowth of peripheral nerve regeneration in rats with a 10 mm sciatic nerve lesion [65]. To repair spinal cord injury (SCI), Song et al. used 3D bioprinting to fabricate a conductive composite hydrogel (CCH) scaffold filled with neural stem cells (NSCs). NSC astrocytic differentiation is inhibited, and neuronal development is promoted in vivo by the CCH scaffold design. The CCH scaffold was 3D bioprinted by the researchers using a linear filament-based layered-stacking geometry, and they were able to demonstrate that it was beneficial in fostering the growth of nerve tissue and the restoration of locomotor function in SCI rats. Nevertheless, the lack of long-term follow-up and the small sample size were the limitations of the research. Future studies should concentrate on enhancing the printing settings and bioink composition, as well as in examining the CCH scaffold's long-term effects on nerve regeneration and functional recovery [66].

In another investigation, the development of multifunctional nanofibrous scaffolds for nerve regeneration was the objective of Al-Hadeethi et al. They developed electrospun nanofibrous scaffolds of PCL and PCL-MWCNT with varying morphologies (random and aligned) and investigated their potential for nerve regeneration in Schwann cells. In order to encourage cell proliferation, the scientists further covered these scaffolds with an acellular matrix released by Schwann cells. According to the research, as compared to pure PCL, the mechanical strength of the nanocomposite scaffolds was decreased by adding MWCNT. However, they did not find any conclusive evidence linking the surface characteristics or scaffold shape to Schwann cells' capacity for nerve regeneration. The authors recommended that further research be conducted to determine the scaffolds' long-term impact on nerve regeneration [67]. In addition, the objective of Guan et al. was to develop a multilevel, multiscale anisotropic SF/metformin scaffold for the regeneration of peripheral nerves. The scaffold was created by micromolding ELS, and its shape, wettability, mechanical properties, degradability, and drug release were evaluated. Dorsal root ganglia axon extension and Schwann cell orientation were both enhanced by the scaffold, according to in vitro research. Genes linked to adhesion, migration, myelination, and Schwann cell proliferation have been shown to be upregulated, according to gene expression analysis. The absence of in vivo testing placed limitations on the research. Further studies might focus on in vivo trials to assess the scaffold's effectiveness in fostering peripheral nerve regeneration in animal models [68]. Table 3.3.2 is a summary of the investigations that developed composite scaffolds for nerve tissue repair and engineering.

Table 3.3.2: Summary of literature on the development of composite scaffolds for nerve tissue replacement.

Objectives	Composite type or structure	Fabrication method	Composition and materials	Results	Limitations	Future recommendations	Reference
Promote sciatic nerve regeneration in rats	Reduced GO composite 3D-printed nerve scaffold	3D bioprinting	Gelatin, sodium alginate, SF, and rGO	Enhanced cell adhesion, increased nerve regeneration-related mRNA expression, and functional recovery of sciatic nerves in rats	Limited in vivo study duration; further investigation of long-term effects needed	Further explore the potential of the combined application of rGO and PEMF for other types of nerve injuries	[60]
Develop 3D-printed scaffolds for peripheral nerve regeneration	PLA/PCL/graphene nanocomposite scaffolds	Fused deposition modeling (FDM) 3D printing	PLA, PCL, and graphene nanoparticles	PLA-PCL-G scaffold showed promising properties for nerve tissue engineering	Limited in vivo testing, potential for graphene particle release	Further in vivo studies, and optimize graphene dispersion	[61]
Develop biocompatible scaffold for nerve repair	Electrospun collagen/PLA nanofibers	ELS and cross-linking	Collagen, PLA, and RA	Scaffold supports PC12 cell adhesion, proliferation, and potential for nerve repair	In vitro study only, further in vivo testing needed	Investigate in vivo efficacy and clinical translation	[62]

Accelerated peripheral nerve regeneration	Double-layer conductive drug carrier	ELS and freeze-drying	SF, GO, VEGF, and GAS-LP	Synergistic drug release, enhanced cell growth, anti-inflammatory effects.	In vitro studies only, not tested in vivo	In vivo studies, clinical trials	[63]
Promote peripheral nerve regeneration using piezoelectric scaffolds	Electrospun SF/PVDF-HFP/MXene composite scaffold	ELS technology	SF, PVDF-HFP, and MXene	Piezoelectric scaffold enhances nerve regeneration and functional recovery in rats	In vivo study conducted only on rats	Further research on clinical applications and long-term effects	[64]
Promote peripheral nerve regeneration	Wireless self-powered scaffold	ELS and UV cross-linking	CNTs@GelMA/PLLA	Improved nerve function and regeneration	In vivo study only 12 weeks post-implantation	Long-term in vivo studies andclinical trials	[65]
Repair SCI with NSCs	CCH scaffold	3D bioprinting	Gelatin, polyethylene glycol, PEDOT:CSMA, and TA	Improved neuronal differentiation, reduced astrocytic differentiation, and enhanced locomotor function recovery	In vivo performance of transplanted stem cells is still a major concern	Investigate application in other nervous system diseases	[66]

(continued)

Table 3.3.2 (continued)

Objectives	Composite type or structure	Fabrication method	Composition and materials	Results	Limitations	Future recommendations	Reference
Develop multifunctional scaffolds for nerve regeneration	PCL-MWCNT nanofibrous scaffolds	ELS	Polycaprolactone (PCL), multiwalled carbon nanotubes (MWCNTs), and Schwann cell acellular matrix	Improved PMP22 expression with Schwann cell matrix coating, but MWCNT reduced mechanical strength	MWCNT reduced mechanical strength, and unclear influence of morphology on nerve regeneration	Further investigation of morphology's impact, in vivo studies to validate findings	[67]
Enhance nerve tissue regeneration with PCL/PGS/MWCNT scaffold	Aligned electrospun PCL/PGS/MWCNT fibers	ELS with varying MWCNT concentrations	PCL, polyglycerol sebacate (PGS), and MWCNTs	Improved mechanical properties, biocompatibility, and cell adhesion with increasing MWCNTs.	Limited in vivo testing, further optimization of MWCNT loading needed	In vivo studies, explore long-term effects, investigate different MWCNT types	[69]
Develop biocompatible scaffolds for neural tissue engineering	Chitosan/polyethylene glycol/MWCNT	Freeze-drying	Chitosan, polyethylene glycol, MWCNTs	MWCNTs enhance scaffold properties, and support PC12 cell growth and differentiation	Biocompatibility of CNTs still needs further investigation	Further research on in vivo applications and long-term biocompatibility	[70]

Repair long-segment sciatic nerve defects in rats	Biomimetic chitosan scaffold	ELS and freeze-drying	Chitosan, nerve growth factor (NGF), and collagen	Improved motor and sensory function, nerve regeneration, and functional circuit reconstruction	Study only on rats, limited sample size, short-term follow-up	Clinical trials, long-term follow-up, investigate other nerve injuries	[71]
Develop antibacterial nerve scaffold for regeneration	Porous NDM-CS scaffold	Freeze-drying technique	Nerve decellularized matrix and chitosan	High biocompatibility, antibacterial activity, promotes Schwann cell growth.	In vitro study, and further in vivo studies needed	Clinical trials for nerve regeneration	[72]
Develop nerve regeneration scaffold	PLA/HA nanofibrous scaffold	ELS	PLA, HA, and methylprednisolone	Improved mechanical and biological properties and enhanced cell viability	In vitro study, limited to Schwann cells	In vivo studies, explore other cell types	[73]
Develop biocompatible chitosan-based composite scaffolds for neural tissue engineering	3D porous scaffolds with interconnected network	Freeze-drying with cross-linking	Chitosan, hyaluronic acid, gelatin, collagen, and polyethylene glycol	Composite scaffolds exhibit good biocompatibility and promote cell adhesion, proliferation, and differentiation of PC12 cells	Electrophysiological behavior of chitosan-based 3D neuron cultures is unknown	Further investigate the electrophysiological behavior of the scaffolds and explore their application in in vivo models	[74]

(continued)

Table 3.3.2 (continued)

Objectives	Composite type or structure	Fabrication method	Composition and materials	Results	Limitations	Future recommendations	Reference
Enhance nerve tissue regeneration using melatonin-functionalized scaffolds	Electrospun fibrous scaffolds	ELS	PCL, gelatin, and melatonin	Melatonin promotes PC12 cell proliferation and differentiation	In vitro study, and limited in vivo data	Further in vivo studies, and optimize melatonin concentration	[75]
Nerve tissue regeneration using biocompatible scaffolds	Polyvinyl alcohol (PVA)-graphene nanosheet nanocomposite	ELS technique	PVA, graphene nanosheets (Gr-NS) and omega-3 fatty acids (O3FA)	Scaffolds with 0.5% Gr-NS show promising properties for nerve tissue engineering	Further in vivo studies are needed to confirm the effectiveness of the scaffolds	Investigate the long-term biocompatibility and efficacy of the scaffolds in vivo.	[76]
Develop 3D-printed scaffold for nerve regeneration	Gellan gum/starch composite scaffold	3D printing technology	Gellan gum, starch, and YIGSR polypeptide	Scaffold exhibits porous structure, good biocompatibility, and promotes Schwann cell growth	Limited in vivo testing, and further optimization needed	In vivo studies, clinical trials, and explore other biomaterials	[77]

Evaluate collagen/heparin scaffold with VEGF for TBI repair	Collagen/heparin scaffold loaded with VEGF	Chemical cross-linking of collagen and heparin	Collagen, heparin, and vascular endothelial growth factor	Improved neurological function, angiogenesis, and tissue regeneration in TBI rats	Further studies needed to confirm long-term effects and clinical translation	Investigate clinical application and long-term efficacy of the scaffold. Develop more scaffolds in combination with cell therapy, gene therapy, and exosome therapy	[78]
Influence of rGO on stem cells for nerve regeneration	rGO-based scaffolds	Chemical reduction of GO	rGO, biopolymers, and stem cells	rGO promotes stem cell differentiation and nerve regeneration	Toxicity of rGO, biodegradability concerns, unclear mechanisms	Further research on rGO biocompatibility and mechanisms	[79]
Investigate GO and reduced GO in nerve tissue engineering	Micro/nanofibrous scaffolds	ELS with in situ postreduction	SF, GO, and rGO	Enhanced cell viability, metabolic activity, and neurite outgrowth on conductive scaffolds	Limited comparative studies on graphene family materials in nerve regeneration	Further explore the molecular mechanisms of graphene's effects on nerve cells	[80]
Peripheral nerve regeneration using a novel scaffold	Graphene mesh-supported double network hydrogel	Chemical vapor deposition, ion exchange, UV polymerization	Graphene, alginate, GelMA, and netrin-1	Improved nerve regeneration and muscle restoration compared to autologous grafts.	Further investigation needed on long-term effects and specific mechanisms	Clinical trials and exploration of other growth factors. Investigation to confirm its therapeutic effects	[81]

(continued)

Table 3.3.2 (continued)

Objectives	Composite type or structure	Fabrication method	Composition and materials	Results	Limitations	Future recommendations	Reference
Promote peripheral nerve regeneration using graphene-based materials	Graphene-based hybrid materials as scaffold	Chemical and physical methods for graphene and its derivatives	Graphene, GO, rGO, and polymers	Promising in vitro and in vivo results for nerve regeneration	Toxicity concerns, need for further optimization of material properties	Continued research on biocompatibility, clinical trials, and exploring applications for specific nerve injuries	[82]
Enhance nerve regeneration via electrical stimulation	Conductive PPy/SF composite scaffold	3D bioprinting and ELS	Ppy and SF	Enhanced SC viability, proliferation, migration, and axonal regeneration in vivo.	Further investigation needed on long-term effects and clinical trials	Optimize stimulation parameters and explore clinical applications	[83]
Enhance neuronal cell affinity of composite scaffolds	Bioactive glass/ polyhydroxyalkanoate blend composites	Solvent casting method	Bioactive glass 45S5, bioactive glass 1393, and poly(3-hydroxyoctanoate/ poly3-hydroxybutyrate) blend	Improved neuronal growth and differentiation on composites	Limited to in vitro studies, and further in vivo testing required	Investigate long-term biocompatibility and in vivo efficacy. The measurement of calcium release from PHA blend/ BG composites and investigation of calcium homeostasis	[84]

Enhance peripheral nerve regeneration using a biocompatible scaffold	Electrospun nanofiber scaffold with RGO-enfolded AuNPs	ELS of PCL solution with incorporated RGO-AuNPs	PCL, rGO, and gold nanoparticles (AuNPs)	Improved cell viability, neurite growth, and differentiation of PC12 and S42 cells	Limited in vivo testing and long-term degradation studies	Further in vivo studies and optimization for clinical applications	[85]
Peripheral nerve regeneration using electrospun scaffold	PLA/GO/NGF conductive scaffold	Improved emulsion ELS	PLA, GO, NGF, and BSA	Scaffold promotes cell proliferation and differentiation	In vitro study only, and long-term in vivo effects are unknown	Further in vivo testing and clinical trials	[86]
Develop a 3D scaffold for peripheral nerve regeneration	Core-shell structure with collagen-hyaluronic acid sponge core and electrospun PCL nanofiber shell	ELS and freeze-drying	Collagen, hyaluronic acid, and PCL	Scaffold supports Schwann cell proliferation and axon growth	Lower Schwann cell proliferation with higher HA content due to increased degradation rate and pore size	Further in vivo studies to evaluate the scaffold's efficacy in nerve regeneration	[87]
Enhance peripheral nerve regeneration using piezoelectric scaffolds	3D porous PVDF/PCL composite nerve scaffold	Cast/annealing-solvent displacement method	Polyvinylidene fluoride (PVDF) and PCL	Improved biocompatibility, mechanical properties, and nerve regeneration in a rat model	Short-term in vivo study (4 months) and limited sample size	Long-term in vivo studies, clinical trials, and explore applications in other tissues. Investigation of piezoelectric tissue engineering scaffolds	[88]

(continued)

Table 3.3.2 (continued)

Objectives	Composite type or structure	Fabrication method	Composition and materials	Results	Limitations	Future recommendations	Reference
Enhance 3D neurite outgrowth and elongation	Multiscale composite scaffold	3D bioprinting with modular bioinks	GelMA/chitosan microspheres, and hydrogel	Enhanced neurite growth and Schwann cell proliferation	Limited to in vitro studies, and further in vivo studies needed	In vivo studies, clinical applications, and investigation of the underlying mechanisms with optimization	[89]
Develop conductive nerve tissue engineering materials	Zero-valent iron nanoparticle-containing PCL nanofiber scaffold	ELS	PCL and zero-valent iron nanoparticles	Increased conductivity and mechanical properties, and improved cell adhesion and neurite outgrowth	Cytotoxicity at higher Fe nanoparticle concentrations, and limited in vivo studies	Optimize nanoparticle ratio, in vivo testing, and clinical applications	[90]

3.3.2.3 Composite hydrogels/films/fibers/membranes

Composite hydrogels, films, fibers, and membranes are biomaterials that support cell growth in nerve tissue replacement. They enhance biocompatibility, provide structural support, and can release growth factors to promote regeneration [91–95]. For peripheral nerve interface, Trueman et al. aimed to develop an aligned bioelectronic architecture. The electrical and mechanical properties of the aligned collagen hydrogels with embedded Ppy nanoparticles were studied. Comparing the PPy-loaded structures with the collagen control constructs, cultured neurons exhibited increased neurite length supported by the PPy-loaded constructs; moreover, electrical stimulation promoted neurite development. The research, however, did not investigate the long-term effects of PPy on nerve regeneration, since it was restricted to in vitro tests. To evaluate the biocompatibility and effectiveness of these constructions in bridging nerve gaps and boosting functional recovery, future research should concentrate on in vivo experiments [96]. Moreover, Javanmardi et al. used gelatin-based hydrogel and hyaluronic acid (HA) microparticles to develop a sustained release method for dexamethasone (Dex) to treat sciatic nerve damage. Using a microfluidic device, they produced Dex-loaded HA microparticles by cross-linking horseradish peroxidase (HRP), which they subsequently added to a proanthocyanidin (PA)-cross-linked gelatin hydrogel. Sustained Dex release and enhanced mechanical properties were shown by the composite system. The encapsulated cells exhibited excellent growth and viability in vitro. In a rat sciatic nerve lesion paradigm, the composite system greatly enhanced nerve regeneration in vivo. Nevertheless, just one animal model was used in the investigation. Future studies should investigate the composite system's long-term impacts and assess its effectiveness in other animal models before moving on to clinical trials [97]. The potential of iron oxide magnetic nanoparticles (IONPs)-loaded fibrin nanofiber hydrogel to stimulate peripheral nerve regeneration in rats was examined by Hong et al. They synthesized fibrin/IONP hydrogels at different concentrations and assessed the breakdown and biocompatibility of the hydrogels both in vitro and in vivo. In a rat sciatic nerve defect model, they discovered that 2% IONPs/fibrin hydrogel demonstrated ideal biocompatibility and encouraged nerve regeneration, resulting in better nerve fiber form and motor function recovery. Nevertheless, the study's sample size was constrained, and the underlying processes were not thoroughly examined. Further research will need focus on examining the enduring impacts of IONPs/fibrin hydrogel, refining the formulation and mode of administration, and delving into the particular pathways of nerve regeneration augmentation [98]. Using a microfluidic system and laccase-mediated cross-linking, Zolfagharzadeh et al. produced insulin-loaded biomimetic microspheres (Figure 3.3.3). They examined the microspheres' mechanical resistance, insulin release rate, and swelling rate, among other physical characteristics. The microspheres' capacity to stimulate sciatic nerve regeneration in rats was examined by the researchers, along with their biocompatibility. The outcomes demonstrated that the insulin-containing microspheres markedly

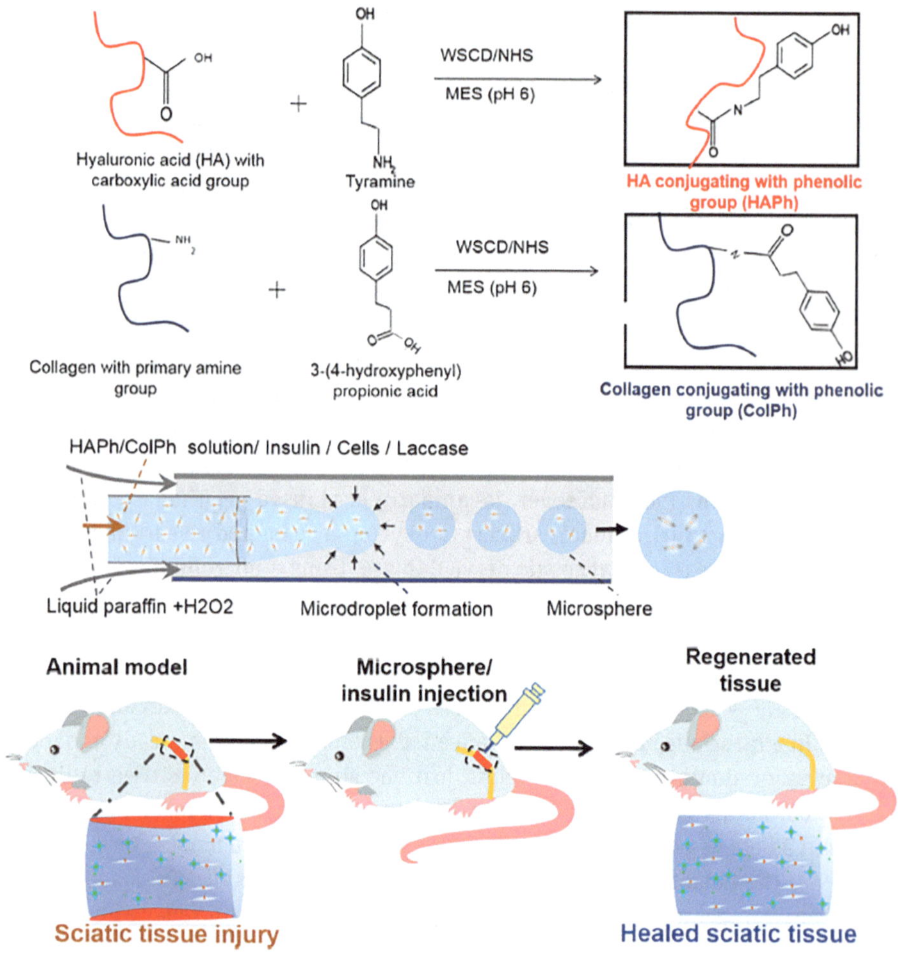

Figure 3.3.3: Schematic illustration of the study conducted by Zolfagharzadeh et al. [99]. Reprinted with permission from [99]. Copyright (2023) Elsevier.

enhanced the sciatic nerve's axonal regeneration, remyelination, and functional recovery. Nevertheless, the investigation was limited to a rat model, and more study is required to extrapolate these results to humans. Further studies need to concentrate on refining the microsphere composition and investigating the possibilities of these microspheres for additional nerve damage and tissue regeneration uses [99].

In another investigation, Yuan et al. investigated the possibility of treating peripheral nerve damage (PNI) in rats using collagen/chitosan composite films, injected with PRP gel. They developed films with different ratios of collagen to chitosan and assessed their effectiveness, biocompatibility, and physicochemical characteristics using a rat sciatic nerve defect model. The findings demonstrated the excellent swelling, mechanical strength, degradability, and hydrophilicity of the composite films. Studies conducted in vitro demonstrated that PRP gel loading increased biocompatibility and stimulated the growth of Schwann cells. PRP-loaded films considerably sped up functional recovery and nerve regeneration in vivo compared to control groups. Further research will need to focus on refining the composition and structure of the composite films, delving into the fundamental principles of nerve regeneration, and carrying out clinical studies to evaluate the safety and effectiveness of this methodology in human subjects [100]. The objective of Tang et al. was to develop a biomimetic scaffold for the repair of SCI. Using a hydrogel coating and ELS technique, they produced a composite fiber scaffold that allowed the release of BDNF and SDF-1α. In a rat model of SCI, this scaffold attracted NSCs, stimulated macrophage polarization toward the anti-inflammatory M2 subtype, and improved nerve regeneration. The investigation proved this composite fiber's promise as a therapy for SCI. To improve the scaffold's design and assess its long-term effectiveness and safety in larger animal models, further study is necessary [101]. In order to promote corneal nerve regeneration, Liu et al. set out to develop a new substance based on collagen. They created a methacryloyl-collagen composite film (ColMA) that was photo-cross-linked; then, they grafted taurine molecules onto the surface of the film (ColMA-Tr). The evaluation of the films' physicochemical characteristics and biocompatibility revealed that, in comparison to ColMA, ColMA-Tr considerably enhanced the proliferation of nerve cells and ocular epithelial cells. The in vitro testing limited the study, and further in vivo research is needed. Further studies need to focus on refining the production procedure and investigating the enduring impacts of ColMA-Tr in fostering corneal nerve regeneration [102].

Developing a unique composite membrane to encourage peripheral nerve regeneration was the objective of Liu et al. Using ELS, they developed a multilayer nanofiber composite membrane made of PCL and amniotic membrane (AM). According to in vitro research, SH-SY5Y cell axon development and differentiation were stimulated by the PCL/AM composite. The PCL/AM combination enhanced macrophage polarization toward the M2 phenotype, increasing nerve regeneration and lowering inflammation, according to in vivo tests conducted in rats. The use of a single animal model was the limitation of the study, and further research is required to determine the PCL/AM composite's long-term impact on nerve regeneration. Further studies may focus on refining the composite membrane's composition and structure and investigating its possible uses in healthcare settings [103]. In addition, the potential of electrically conductive, biocompatible, and bioresorbable fibers for peripheral nerve regeneration was examined by Dresvyanina et al. They synthesized chitosan and single-walled carbon nanotubes (SWCNTs) to create composite fibers, and then assessed the mechanical, electrical, biological, and structural characteristics of the materials. According to the research, adding 0.5 wt% of SWCNTs increased the fiber's flexibility and strength. Additionally, the fibers were biocompatible and did not significantly increase inflammation in the in vivo experiments. To assess the fibers' long-term impact on nerve regeneration, further research is necessary since the study's sample size was limited. In order to improve the fibers' biocompatibility and promote nerve regeneration, future studies might focus on improving the structure and material [104]. In addition, Chen et al. combined decellularized amniotic membrane (dAM) with gelatin (Gel) nanofiber membrane in an attempt to develop a composite membrane for peripheral nerve injury (PNI) repair. Through interfacial bonding, they developed the Gel-dAM composite membrane and assessed its mechanical properties and rate of degradation. The biocompatibility of the Gel-dAM was then evaluated in vitro with macrophages and Schwann cells (SCs). Finally, using a rat sciatic nerve transection model, they assessed the pro-regenerative repair mediated by the Gel-dAM. While maintaining the bioactivity and biocompatibility of dAM, the Gel-dAM demonstrated improved mechanical properties and a decreased rate of degradation. In PNI repair, it enhanced functional recovery, decreased fibrotic response, and encouraged axon regrowth. The lack of long-term follow-up and the small sample size were the limitations of the research. Further studies should focus on refining the Gel-dAM's composition and manufacturing procedure and assessing its effectiveness in other PNI models [105].

Table 3.3.3 is a summary of the investigations that developed composite hydrogels, films, fibers, and membranes for nerve tissue repair and engineering.

3.3.2.4 Other types of composite structures

There are also some other types of composite structures, such as tubes, patches, electrodes, etc., that have been developed and utilized for nerve tissue engineering and replacement. The objective of Zhang et al. was to develop an electrically conductive, biocompatible composite for spinal cord repair. They developed a composite of collagen, Ppy, and quercetin (Figure 3.3.4) and evaluated its mechanical strength, electrical conductivity, shape, and chemical functioning. The composite demonstrated appropriate mechanical strength, electrical conductivity, and acceptable biocompatibility. Additionally, it demonstrated promise in stimulating the development of neurons in human astrocyte cells. To confirm the composite's effectiveness in spinal cord regeneration, further in vivo research is required since the current study was limited to in vitro tests. Future studies might focus on refining the characteristics of the composite and assessing its enduring impacts in animal models [113].

Table 3.3.4 is a summary of the investigations that developed other composite types for nerve tissue repair and engineering.

Table 3.3.3: Summary of literature on the development of composite hydrogels, films, fibers, and membranes for nerve tissue replacement.

Objectives	Composite type or structure	Fabrication method	Composition and materials	Results	Limitations	Future recommendations	Reference
Develop aligned bioelectronic constructs for nerve regeneration	Aligned Ppy/ collagen constructs	Gel-aspiration-ejection (GAE) technique	Ppy nanoparticles, and collagen hydrogel	Enhanced neurite length with PPy and electrical stimulation	Conductivity lower than cell culture medium	Further investigate the mechanism of increased conductivity	[96]
Sustained Dex delivery for sciatic nerve regeneration	Dex-loaded HA microparticles in gelatin hydrogel	Microfluidic device and enzymatic cross-linking	Hyaluronic acid, gelatin, proanthocyanidin, and dexamethasone	Improved hydrogel properties, sustained Dex release, and enhanced nerve regeneration	In vivo study only in rats, and further clinical trials needed	Explore clinical applications, and optimize drug loading and release	[97]
Promote peripheral nerve regeneration using IONPs/ fibrin hydrogel	IONP-loaded fibrin nanofiber hydrogel	Direct mixing of fibrinogen, thrombin, and IONPs	Fibrinogen, thrombin, iron oxide magnetic nanoparticles (IONPs), and chitosan	2% IONPs/fibrin hydrogel showed promising results in promoting nerve regeneration in rats, comparable to autografts	Further in-depth studies needed to assess long-term effects and clinical translation	Investigate the potential of IONPs/ fibrin hydrogel for various nerve injuries and explore clinical applications	[98]
Develop insulin-loaded biomimetic microspheres for sciatic nerve regeneration	Biomimetic hydrogel microspheres	Laccase-mediated cross-linking using a microfluidic device	Phenol-substituted hyaluronic acid (HAPh), collagen (ColPh), insulin, and laccase	Improved nerve regeneration in a rat model through sustained insulin release	Study conducted only in a rat model, and further clinical trials are needed	Investigate the long-term efficacy and safety of the microspheres in humans	[99]

(continued)

Table 3.3.3 (continued)

Objectives	Composite type or structure	Fabrication method	Composition and materials	Results	Limitations	Future recommendations	Reference
Improve sciatic nerve injury repair	Collagen/chitosan composite film	Physical doping	Collagen, chitosan, and PRP	PRP-loaded COL/CS film promotes nerve regeneration and functional recovery	Further in vivo studies needed to confirm long-term efficacy	Investigate other growth factors and delivery methods	[100]
Promote nerve regeneration after SCI	Hydrophilic GelMA-coated electrospun fibers	Microsol ELS with GelMA coating	PLLA, HA, GelMA, SDF-1α, and BDNF	Improved immune microenvironment, enhanced NSC recruitment and differentiation, and improved nerve function recovery	In vivo study only in rats, and further human trials needed	Explore clinical applications and optimize the composite for long-term efficacy	[101]
Promote corneal nerve regeneration	Photo-cross-linked collagen composite film	Surface grafting of taurine molecules	Collagen, methacryloyl groups and taurine	Improved mechanical properties and enhanced nerve cell growth	In vitro study, and further in vivo testing needed	In vivo studies and clinical trials	[102]
Promote peripheral nerve regeneration	Multilayer nanofiber composite membrane	ELS	PCL and amniotic membrane (AM)	PCL/AM composite promotes nerve regeneration by regulating macrophage polarization and inflammatory microenvironment	In vivo study sample size is limited	Further in vivo studies with larger sample sizes and clinical trials are needed	[103]

Develop conductive fibers for nerve regeneration	Chitosan/SWCNT composite fibers	Wet spinning	Chitosan, single-walled carbon nanotubes	Improved strength, elasticity, biocompatibility, and electrical conductivity	Limited in vivo data, and small sample size	Further in vivo studies and optimize fiber properties	[104]
Promote axon regeneration and functional recovery in PNI repair	Gelatin nanofiber-reinforced decellularized amniotic membrane	Interfacial bonding of gelatin nanofiber membrane and dAM	Gelatin nanofiber membrane and decellularized amniotic membrane	Gel-dAM enhances mechanical properties and stability, promotes SC adhesion and macrophage polarization, and improves axon regeneration and functional recovery in PNI repair	Further in vivo studies are required to confirm the long-term effects of Gel-dAM.	Clinical trials to evaluate the safety and efficacy of Gel-dAM for PNI repair	[105]
Investigate degradation of PHB/CS nanofibers in different pH conditions	Electrospun PHB/chitosan composite fibrous membrane	ELS	Polyhydroxybutyrate (PHB) and chitosan (CS)	Increased CS content enhances hydrophilicity, reduces degradation temperature and crystallinity, and impacts fiber morphology	Limited to in vitro degradation study, and no in vivo testing	Further in vivo studies and exploration of potential applications in peripheral nerve regeneration	[106]
Support neuron differentiation and alignment	Directional submicrofiber hydrogel composite	ELS and enzymatic cross-linking	Gelatin hydrogel and PCL fibers	Composite supports cell attachment, differentiation, and neurite alignment	No chemical bonding between fibers and hydrogel	Investigate in vivo applications and therapeutic efficacy	[107]
Accelerate peripheral nerve regeneration using piezoelectric stimulation	Electrospun composite nanofibers	ELS	PCL and zinc oxide (ZnO)	PZNF demonstrated faster and superior sciatic nerve repair in vivo than PNF and nerve bridging in situ	ZnO toxicity needs further investigation	Further explore the mechanism of piezoelectric stimulation in nerve regeneration	[108]

(continued)

Table 3.3.3 (continued)

Objectives	Composite type or structure	Fabrication method	Composition and materials	Results	Limitations	Future recommendations	Reference
Develop conductive collagen/MWNT films for nerve tissue engineering	Conductive collagen/ multiwalled carbon nanotube composite films	Solution-evaporation method	Collagen, multiwalled carbon nanotubes and hexadecyl trimethyl ammonium bromide	Improved mechanical and electrical properties, and good biocompatibility	High MWNT concentrations may lead to cytotoxicity	Further investigate long-term biocompatibility and in vivo studies	[109]
Peripheral nerve regeneration using chitin hydrogel	Conductive chitin hydrogel film	Electrostatic interaction and partial deacetylation	Chitin, PEDOT nanoparticles, and CRGD peptide	Improved nerve regeneration and enhanced cell adhesion	Limited in vivo study duration and potential for inflammation	Long-term in vivo studies and clinical trials	[110]
Develop composite nanofibers for neuroprotection after peripheral nerve injury	Electrospun composite nanofibers with embedded nanoparticles	ELS and nanopray drying	PLGA, chitosan, alpha lipoic acid, and atorvastatin	Composite nanofibers improved motor and sensory function recovery in a rat sciatic nerve crush injury model	Study focused on a single nerve injury model; further research is needed for broader applications	Investigate the long-term effects and biocompatibility of the composite nanofibers	[111]
Enhance peripheral nerve regeneration	rGO-GelMA-PCL hybrid nanofibers	ELS	rGO, GelMA, and PCL	Improved electrical conductivity, biocompatibility, and cell proliferation; promoted nerve regeneration and functional recovery in rats	Cytotoxicity of rGO at higher concentrations	Further research on electrical stimulation for nerve regeneration	[112]

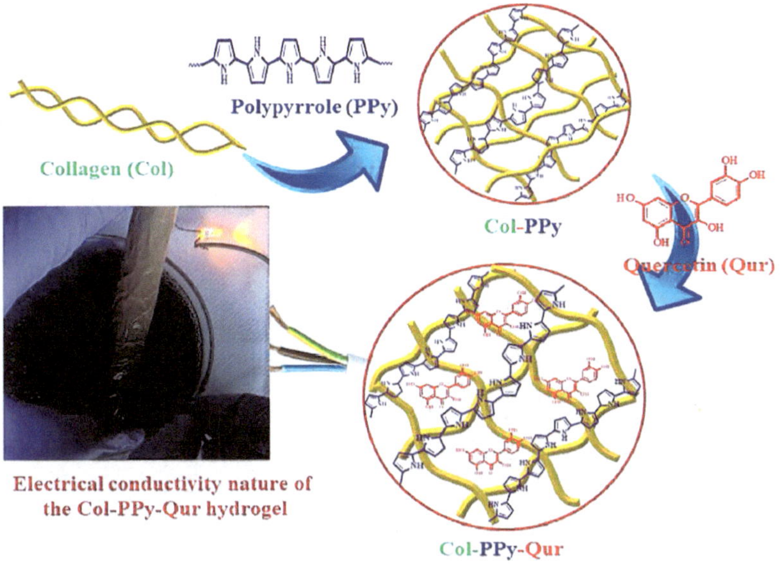

Figure 3.3.4: Schematic illustration of Col-PPy-Qur composite developed in [113]. Reprinted with permission from [113]. Copyright (2023) Elsevier.

3.3.3 Discussion and future research directions

To overcome the limitations of existing therapies for nerve injuries and neurodegenerative conditions, nerve tissue engineering and replacement are crucial. Through the development of bioengineered scaffolds and grafts, scientists hope to promote cell survival and neuron regeneration. Scientific advancements like stem cell treatment and biomaterial fabrication provide ways to improve cognitive and physical performance. Composite materials, which combine synthetic and natural components, provide the necessary cellular connections and structural support for nerve repair. By delivering bioactive molecules and imitating the extracellular matrix, these materials improve patient outcomes and greatly increase the efficacy of regeneration techniques.

The findings demonstrate that a variety of cutting-edge substances and methods considerably improve nerve regeneration and functional recovery in experimental animals. Piezoelectric materials, composite scaffolds, and exosome-loaded conduits all exhibited enhanced mechanical properties and biocompatibility, which encouraged cell adhesion, proliferation, and nerve repair. Notably, Schwann cell proliferation, myelin production, and axonal regeneration were increased by PLLA-PPSB nanofibers and MWCNT-enriched scaffolds. Methods such as electrical stimulation and prolonged growth factor release improved the results of nerve regeneration even further. Fur-

Table 3.3.4: Summary of literature on the development of other types of composite structures for nerve tissue replacement.

Objectives	Composite type or structure	Fabrication method	Composition and materials	Results	Limitations	Future recommendations	Reference
SCI regeneration using a biocompatible conductive composite	Collagen- Ppy-quercetin composite	Lyophilization of a stirred mixture	Collagen, Ppy, and quercetin	Composite exhibits electrical conductivity, mechanical strength, and promotes neuronal differentiation	In vivo studies not yet performed	Further in vivo studies to assess efficacy in SCI models	[113]
Efficient corneal nerve regeneration via electrical stimulation	Transparent graphene-based corneal stimulation electrode	Electroactive nitrogen-containing conducting polymer-functionalized graphene	Polyaniline-functionalized graphene (PAG) on ITO membrane	Promoted axon regeneration of trigeminal ganglion (TG) neurons in vitro and in vivo	In vivo study only on rabbit model, further human trials needed	Explore the long-term effects of electrical stimulation on corneal nerve regeneration	[114]
Promote optic nerve regeneration in traumatic optic neuropathy	Implantable patches with gelatin/SF microspheres	W/O emulsion technique with genipin cross-linking	Gelatin, SF, genipin, mesenchymal stem cells	MSC@MS patches promote RGC survival and axon regeneration compared to bolus injection	Small sample size in some experiments, further in vivo studies needed	Explore delivery of therapeutic cargoes within microspheres	[115]
Promote neurite outgrowth for nerve regeneration	Ppy-assisted LIG flexible electrodes	Electrochemical deposition of Ppy on LIG	Ppy, laser-induced graphene, and polyimide	Electrical stimulation enhances PC-12 cell proliferation and neurite outgrowth	Limited by in vitro testing, and further in vivo studies needed	Develop practical conductive platforms for clinical use	[116]

Controlled NGF release for nerve repair	TPP ionically cross-linked chitosan/PLGA microspheres	Re-emulsification-ionic cross-linking	Chitosan, PLGA, TPP, and NGF	Controlled NGF release and enhanced nerve regeneration in rats	Limited in vivo study duration, and potential for burst release	Further in vivo studies, and optimize release kinetics	[117]
Promote early osseointegration and nerve regeneration	NGF-CS/HA-coating composite implant	Modified biomimetic method	Titanium, hydroxyapatite, chondroitin sulfate, and NGF	NGF-CS/HA composite implants promote early osseointegration and nerve regeneration	Study conducted on Beagle dogs, and limited sample size	Further clinical trials with larger sample size, and explore long-term effects	[118]

thermore, PRP-loaded films and PCL/AM composites enhanced neuronal differentiation and the immunological microenvironment. When everything is considered, these results imply that customized conduits and scaffolds may accurately replicate autologous grafts, providing viable substitutes for nerve tissue development and recovery in diverse scenarios.

There were several drawbacks, such as the heavy dependence on in vitro and in vivo research, which centered primarily on rat models and required further clinical trials to verify effectiveness in humans. A number of studies had small sample numbers and little follow-up periods, which raised questions about the materials' long-term safety and efficacy. Further research is necessary to address potential problems, including cytotoxicity, especially at greater concentrations of certain nanoparticles, and the unexplained processes behind neuron regeneration. Furthermore, the generalizability of the findings was limited since, while several studies revealed encouraging outcomes, they often concentrated on particular damage models or defect sizes. The evaluation of long-term results is further complicated by the variation in research settings and length. To ensure the safety and effectiveness of new nerve regeneration techniques, extensive long-term research and wider applications are still necessary to convert these discoveries into clinical practice.

Further research should focus on carrying out comprehensive clinical studies in order to evaluate the enduring effectiveness and safety of diverse nerve regeneration approaches, specifically those pertaining to exosome-loaded conduits and PLLA-PPSB nanofibers. To improve the results of nerve repair, researchers should investigate the potential of various growth factors, medication delivery strategies, and stimulation approaches, such as magnetic and electrical stimulation. In vivo investigations are essential to optimize conduit designs and assess the biocompatibility of materials such as rGO and MWCNTs across a range of nerve lesions. For clinical translation, it is crucial to investigate these materials' long-term impacts, including their toxicity and biodegradation. Important insights will also be obtained by investigating the processes causing enhanced conductivity and the effects of various scaffolds on nerve cell activity. It will be possible to confirm research results and promote wider applications by expanding the study to encompass more animal models and various damage types. Nerve regeneration treatments must be advanced via a multimodal strategy that optimizes material qualities, delivery methods, and stimulation settings.

3.3.4 Conclusion

To sum up, nerve tissue engineering and replacement are significant breakthroughs that will help overcome the drawbacks of the existing therapies for neurodegenerative illnesses and nerve injuries. Together with cutting-edge techniques like stem cell treatment and composite materials, the development of bioengineered scaffolds and grafts offers promises for promoting nerve regeneration and improved patient outcomes. The findings show that in experimental settings, a variety of new materials, such as conduits filled with exosomes and PLLA-PPSB nanofibers, dramatically enhance nerve repair and functional recovery. To establish effectiveness in humans, more clinical trials are necessary, as confirmed by the dependence on in vitro and animal research. Limitations like limited sample numbers and brief follow-up periods highlight how crucial long-term research is for determining safety and efficacy. Further studies should give precedence to comprehensive clinical assessments, delve into a variety of growth components and supply strategies, as well as optimize material attributes and stimulation methods. Transforming these encouraging discoveries into clinical practice will need a thorough and multidisciplinary approach, which will eventually result in better therapy alternatives for patients with nerve-related disorders.

3.3.5 Declarations

3.3.5.1 Conflict of interests

The authors declare that there is no conflict of interest.

3.3.5.2 Data availability

No data was used in this study.

3.3.5.3 Funding

No funding source is available for this study.

3.3.5.4 Declaration of generative AI and AI-assisted technologies in the writing process

During the preparation of this work, the authors used ChatGPT 3.5, Assistant, QuillBot, and Grammarly to rephrase and edit the text grammatically and in terms of typos. After using these tools/services, the authors reviewed and edited the content as needed and take full responsibility for the content of the publication.

References

[1] Zhang P-X, et al. Tissue engineering for the repair of peripheral nerve injury. Neural Regeneration Research. 2019; 14(1): 51–58.

[2] Maresova P, et al. Activities of daily living and associated costs in the most widespread neurodegenerative diseases: A systematic review. Clinical Interventions in Aging. 2020; 15(null): 1841–1862.

[3] Galts CPC, et al. Depression in neurodegenerative diseases: Common mechanisms and current treatment options. Neuroscience and Biobehavioral Reviews. 2019; 102: 56–84.

[4] Di Paolo M, et al. Natural products in neurodegenerative diseases: A great promise but an ethical challenge. International Journal of Molecular Sciences. 2019; 20(20): 5170.

[5] Lamptey RNL, et al. A review of the common neurodegenerative disorders: Current therapeutic approaches and the potential role of nanotherapeutics. International Journal of Molecular Sciences. 2022; 23(3): 1851.

[6] Akhtar A, et al. Neurodegenerative diseases and effective drug delivery: A review of challenges and novel therapeutics. Journal of Controlled Release. 2021; 330: 1152–1167.

[7] De Gioia R, et al. Neural stem cell transplantation for neurodegenerative diseases. International Journal of Molecular Sciences. 2020; 21(9): 3103.

[8] Li R, et al. Growth factors-based therapeutic strategies and their underlying signaling mechanisms for peripheral nerve regeneration. Acta Pharmacologica Sinica. 2020; 41(10): 1289–1300.

[9] Sarker MD, et al. Regeneration of peripheral nerves by nerve guidance conduits: Influence of design, biopolymers, cells, growth factors, and physical stimuli. Progress in Neurobiololy. 2018; 171: 125–150.

[10] Cen L-P, et al. Human periodontal ligament-derived stem cells promote retinal ganglion cell survival and axon regeneration after optic nerve injury. Stem Cells. 2018; 36(6): 844–855.

[11] Jessen KR, Mirsky R. The success and failure of the Schwann cell response to nerve injury. Frontiers in Cellular Neuroscience. 2019; 13: 33.

[12] Jiao Y, et al. Neuroregeneration and functional recovery after stroke: Advancing neural stem cell therapy toward clinical application. Neural Regeneration Research. 2021; 16(1): 80–92.

[13] Zakerinia M, et al. Intrathecal autologous bone marrow-derived hematopoietic stem cell therapy in neurological diseases. International Journal of Organ Transplantation Medicine. 2018; 9(4): 157–167.

[14] Egbo MK. A fundamental review on composite materials and some of their applications in biomedical engineering. Journal of King Saud University – Engineering Sciences. 2021; 33(8): 557–568.

[15] Zagho MM, Hussein EA, Elzatahry AA. Recent overviews in functional polymer composites for biomedical applications. Polymers. 2018; 10(7): 739.

[16] Ramakrishna S, et al. Biomedical applications of polymer-composite materials: A review. Composites Science and Technology. 2001; 61(9): 1189–1224.

[17] Niemczyk-Soczynska B, et al. Hydrogel, electrospun and composite materials for bone/cartilage and neural tissue engineering. Materials. 2021; 14(22): 6899.

[18] Cristina Chircov, Daniela-Alina Ioniță, Andra-Maria Sîrmon, Ionela Andreea Neacșu, Anton Ficai, Chapter 3 – Natural, synthetic, and hybrid and composite biomaterials for neural tissue engineering, Editor(s): Oguzhan Gunduz, Cem Bulent Ustundag, Mustafa Sengor, In Woodhead Publishing Series in Biomaterials, Biomaterials for Neural Tissue Engineering, Woodhead Publishing, 2023, Pages 21–58, ISBN 9780323905541, https://doi.org/10.1016/B978-0-323-90554-1.00008-2.

[19] McCullen SD, et al. Nanofibrous composites for tissue engineering applications. WIREs Nanomedicine and Nanobiotechnology. 2009; 1(4): 369–390.

[20] Kijeńska E, et al. Electrospun bio-composite P(LLA-CL)/collagen I/collagen III scaffolds for nerve tissue engineering. Journal of Biomedical Materials Research Part B: Applied Biomaterials. 2012; 100B(4): 1093–1102.
[21] Si J, et al. Controlled degradable chitosan/collagen composite scaffolds for application in nerve tissue regeneration. Polymer Degradation and Stability. 2019; 166: 73–85.
[22] Radhakrishnan, S., Nagarajan, S., Bechelany, M., Kalkura, S.N. (2020). Collagen Based Biomaterials for Tissue Engineering Applications: A Review. In: Frank-Kamenetskaya, O., Vlasov, D., Panova, E., Lessovaia, S. (eds) Processes and Phenomena on the Boundary Between Biogenic and Abiogenic Nature. Lecture Notes in Earth System Sciences. Springer, Cham. https://doi.org/10.1007/978-3-030-21614-6_1
[23] Grimaudo MA, et al. Bioactive injectable hydrogels for on demand molecule/cell delivery and for tissue regeneration in the central nervous system. Acta Biomaterialia. 2022; 140: 88–101.
[24] Elliott Donaghue I, et al. Cell and biomolecule delivery for tissue repair and regeneration in the central nervous system. Journal of Controlled Release. 2014; 190: 219–227.
[25] Subramanian A, Krishnan UM, Sethuraman S. Development of biomaterial scaffold for nerve tissue engineering: Biomaterial mediated neural regeneration. Journal of Biomedical Science. 2009; 16(1): 108.
[26] Sahbafar H. Mechanical and Biological Properties of Polylactic Acid (PLA)-hardystonite–fe3o4 Scaffolds Made by 3D Printing for Bone Tissue Engineering, Iran: University of Tehran; 2023.
[27] Noroozi R, et al. 3D and 4D bioprinting technologies: A game changer for the biomedical sector?. Annals of Biomedical Engineering. 2023; 51(8): 1683–1712.
[28] Kang NU, Lee SJ, Gwak SJ. Fabrication techniques of nerve guidance conduits for nerve regeneration. Yonsei Medical Journal. 2022; 63(2): 114–123.
[29] Regas I, et al. Functionalized nerve conduits for peripheral nerve regeneration: A literature review. Hand Surgery and Rehabilitation. 2020; 39(5): 343–351.
[30] Zhang X, et al. Functional polymer-based nerve guide conduits to promote peripheral nerve regeneration. Advanced Materials Interfaces. 2020; 7(14): 2000225.
[31] Sun J, et al. Graphene/chitosan tubes inoculated with dental pulp stem cells promotes repair of facial nerve injury. Frontiers in Chemistry. 2024; 12: 1417763 https://doi.org/10.3389/fchem.2024.1417763.
[32] Zhang X, et al. Chitosan nerve conduit filled with ZIF-8-functionalized guide microfibres enhances nerve regeneration and sensory function recovery in sciatic nerve defects. Chemical Engineering Journal. 2024; 480: 147933.
[33] Tang H, et al. Human umbilical cord mesenchymal stem cell-derived exosomes loaded into a composite conduit promote functional recovery after peripheral nerve injury in rats. Neural Regeneration Research. 2024; 19(4): 900–907. DOI: 10.4103/1673-5374.380911
[34] Sierakowska-Byczek A, et al. Environment-friendly preparation and characterization of multilayered conductive PVP/Col/CS composite doped with nanoparticles as potential nerve guide conduits. Polymers. 2024; 16: 10.3390/polym16070875.
[35] Shlapakova LE, et al. Magnetoactive composite conduits based on poly(3–hydroxybutyrate) and magnetite nanoparticles for repair of peripheral nerve injury. ACS Applied Bio Materials. 2024; 7(2): 1095–1114.
[36] Nabipour M, et al. Preparation of bilayer tissue-engineered polyurethane/poly-L-lactic acid nerve conduits and their in vitro characterization for use in peripheral nerve regeneration. Journal of Biological Engineering. 2024; 18(1): 16.
[37] Li L, et al. BDNF-loaded chitosan-based mimetic mussel polymer conduits for repair of peripheral nerve injury. Frontiers in Cell and Developmental Biology. 2024; 12. https://doi.org/10.3389/fcell.2024.1431558.

[38] D. V. Johnson L, et al. Emulsion templated composites: Porous nerve guidance conduits for peripheral nerve regeneration. Materials and Design. 2024; 239: 112779.

[39] Huang Y, et al. Dual-layer conduit containing VEGF-A – transfected Schwann cells promotes peripheral nerve regeneration via angiogenesis. Acta Biomaterialia. 2024; 180: 323–336.

[40] Hoveizi E. Enhancement of nerve regeneration through Schwann cell-mediated healing in a 3D printed polyacrylonitrile conduit incorporating hydrogel and graphene quantum dots: A study on rat sciatic nerve injury model. Biomedical Materials. 2024; 19(1): 015012.

[41] Dai Y, et al. Electrospun composite PLLA-PPSB nanofiber nerve conduits for peripheral nerve defects repair and regeneration. Advanced Healthcare Materials. 2024; 13(10): 2303539.

[42] Xu H, et al. Electrospun PCL nerve conduit filled with GelMA gel for CNTF and IGF-1 delivery in promoting sciatic nerve regeneration in rat. ACS Biomaterials Science and Engineering. 2023; 9(11): 6309–6321.

[43] Tagandurdyyeva NA, et al. Properties of resorbable conduits based on poly(L-Lactide) nanofibers and chitosan fibers for peripheral nerve regeneration. Polymers. 2023; 15: 10.3390/polym15153323.

[44] Jiang L, et al. The role of Gel-Ppy-modified nerve conduit on the repair of sciatic nerve defect in rat model. The FASEB Journal. 2023; 37(5): e22921.

[45] Jiang Z, et al. Multichannel nerve conduit based on chitosan derivates for peripheral nerve regeneration and Schwann cell survival. Carbohydrate Polymers. 2023; 301: 120327.

[46] Huang W-J, Wang J. Development of 3D-printed, biodegradable, conductive PGSA composites for nerve tissue regeneration. Macromolecular Bioscience. 2023; 23(3): 2200470.

[47] Dong X, et al. Micro-nanofiber composite biomimetic conduits promote long-gap peripheral nerve regeneration in canine models. Bioactive Materials. 2023; 30: 98–115.

[48] Dong Q, et al. Composite hydrogel conduit incorporated with platelet-rich plasma improved the regenerative microenvironment for peripheral nerve repair. ACS Applied Materials and Interfaces. 2023; 15(20): 24120–24133.

[49] Zhang Y, et al. Nerve regeneration effect of a composite bioactive carboxymethyl chitosan-based nerve conduit with a radial texture. Molecules. 2022; 27: 10.3390/molecules27249039.

[50] Deng P, et al. Multifunctional double-layer composite hydrogel conduit based on chitosan for peripheral nerve repairing. Advanced Healthcare Materials. 2022; 11(13): 2200115.

[51] Yan X, et al. Preparation and characterization of conductive nerve guide conduit filled with dual drug-loaded nanofibers. Journal of Bioactive & Compatible Polymers. 2021; 36(6): 531–547.

[52] Yan X, et al. PDLLA/β-TCP/HA/CHS/NGF sustained-release conduits for peripheral nerve regeneration. Journal of Wuhan University of Technology-Mater Sci Ed. 2021; 36(4): 600–606.

[53] Ruiz IM, et al. Development and evaluation of hyaluronan nanocomposite conduits for neural tissue regeneration. Journal of Biomaterial Science, Polymer Edition. 2021; 32(17): 2227–2245.

[54] Luo L, et al. Application of bioactive hydrogels combined with dental pulp stem cells for the repair of large gap peripheral nerve injuries. Bioactive Materials. 2021; 6(3): 638–654.

[55] Lu C-F, et al. Combining chitin biological conduits with small autogenous nerves and platelet-rich plasma for the repair of sciatic nerve defects in rats. CNS Neuroscience and Therapeutics. 2021; 27(7): 805–819.

[56] Chen X, et al. Electrospinning multilayered scaffolds loaded with melatonin and Fe3O4 magnetic nanoparticles for peripheral nerve regeneration. Advanced Functional Materials. 2020; 30(38): 2004537.

[57] Zha F, et al. Electrospun natural polymer and its composite nanofibrous scaffolds for nerve tissue engineering. Journal of Biomaterials Science, Polymer Edition. 2020; 31(4): 519–548.

[58] Yi S, Xu L, Gu X. Scaffolds for peripheral nerve repair and reconstruction. Experimental Neurology. 2019; 319: 112761.

[59] Wang ZZ, Sakiyama-Elbert SE. Matrices, scaffolds & carriers for cell delivery in nerve regeneration. Experimental Neurology. 2019; 319: 112837.

[60] Wang Z, et al. Pulsed electromagnetic field-assisted reduced graphene oxide composite 3D printed nerve scaffold promotes sciatic nerve regeneration in rats. Biofabrication. 2024; 16(3): 035013.

[61] Soltani Gerdefaramarzi R, Ebrahimian-Hosseinabadi M, Khodaei M. 3D printed poly(lactic acid)/poly (ε-caprolactone)/graphene nanocomposite scaffolds for peripheral nerve tissue engineering. Arabian Journal of Chemistry. 2024; 17(9): 105927.

[62] Aseer M, et al. Potential of collagen/PLA-based nanofibrous scaffold to support PC12 cells and neural repair. Journal of Contemporary Medical Sciences. 2024; 10(1): 13–18

[63] Liu J, et al. Electrospun SF/GO/VEGF/GAS–LP scaffolds with synergistic sequential drug release and electrical stimulation for accelerated peripheral nerve regeneration. International Journal of Polymeric Materials and Polymeric Biomaterials. 2024; 73(13): 1136–1148.

[64] Zhang H, et al. Electrospun piezoelectric scaffold with external mechanical stimulation for promoting regeneration of peripheral nerve injury. Biomacromolecules. 2023; 24(7): 3268–3282.

[65] Yang Y, et al. Engineering a wirelessly self-powered and electroconductive scaffold to promote peripheral nerve regeneration. Nano Energy. 2023; 107: 108145.

[66] Song S, et al. Inhibited astrocytic differentiation in neural stem cell-laden 3D bioprinted conductive composite hydrogel scaffolds for repair of spinal cord injury. Biomaterials Advances. 2023; 148: 213385.

[67] Al-Hadeethi Y, et al. Schwann cell-matrix coated PCL–MWCNT multifunctional nanofibrous scaffolds for neural regeneration. RSC Advances. 2023; 13(2): 1392–1401.

[68] Guan W, et al. Multi-scale, multi-level anisotropic silk fibroin/metformin scaffolds for repair of peripheral nerve injury. International Journal of Biological Macromolecules. 2023; 246: 125518.

[69] Saudi A, et al. A study on the role of multi-walled carbon nanotubes on the properties of electrospun poly(Caprolactone)/poly(Glycerol sebacate) scaffold for nerve tissue applications. Materials Chemistry and Physics. 2022; 282: 125868.

[70] Sang S, et al. Biocompatible chitosan/polyethylene glycol/multi-walled carbon nanotube composite scaffolds for neural tissue engineering. Journal of Zhejiang University–SCIENCE B. 2022; 23(1): 58–73.

[71] Liu F-D, Duan HM, Hao F, et al. Biomimetic chitosan scaffolds with long-term controlled release of nerve growth factor repairs 20-mm-long sciatic nerve defects in rats. Neural Regeneration Research. 2022; 17(5): 1146–1155. doi:10.4103/1673-5374.324860

[72] Kong Y, et al. Nerve decellularized matrix composite scaffold with high antibacterial activity for nerve regeneration. Frontiers in Bioengineering and Biotechnology. 2022; 9: 840421. https://doi.org/10.3389/fbioe.2021.840421.

[73] Keikha M, et al. Fabrication and characterization of methylprednisolone-loaded polylactic acid/hyaluronic acid nanofibrous scaffold for soft tissue engineering. Journal of Industrial Textiles. 2022; 52: 15280837221116593.

[74] Cheng R, et al. Fabrication and characterization of chitosan–based composite scaffolds for neural tissue engineering. International Journal of Polymeric Materials and Polymeric Biomaterials. 2022; 71(11): 831–841.

[75] Chen T, et al. Proliferation and differentiation study of melatonin functionalized polycaprolactone/gelatin electrospun fibrous scaffolds for nerve tissue engineering. International Journal of Biological Macromolecules. 2022; 197: 103–110.

[76] Bitaraf P, Asefnejad A, Hassanzadeh Nemati N. Fabrication of polyvinyl alcohol-graphene nanosheet nanocomposite loading of omega–3 fatty acids for ceramic engineering. Iranian Journal of Basic Medical Sciences. 2022; 25(3): 330–340.

[77] Zhang L, et al. Fabrication and characterization of 3D–printed gellan gum/starch composite scaffold for Schwann cells growth. Nanotechnology Reviews. 2021; 10(1): 50–61.

[78] Zhang J, et al. Collagen/heparin scaffold combined with vascular endothelial growth factor promotes the repair of neurological function in rats with traumatic brain injury. Biomaterials Science. 2021; 9(3): 745–764.

[79] Yao X, et al. The influence of reduced graphene oxide on stem cells: A perspective in peripheral nerve regeneration. Regenerative Biomaterials. 2021; 8(4): rbab032.

[80] Magaz A, et al. Graphene oxide and electroactive reduced graphene oxide-based composite fibrous scaffolds for engineering excitable nerve tissue. Materials Science and Engineering: C. 2021; 119: 111632.

[81] Huang Q, et al. Aligned graphene mesh-supported double network natural hydrogel conduit loaded with netrin-1 for peripheral nerve regeneration. ACS Applied Materials & Interfaces. 2021; 13(1): 112–122.

[82] Grijalvo S, Díaz DD. Graphene-based hybrid materials as promising scaffolds for peripheral nerve regeneration. Neurochemistry International. 2021; 147: 105005.

[83] Zhao Y, et al. Application of conductive PPy/SF composite scaffold and electrical stimulation for neural tissue engineering. Biomaterials. 2020; 255: 120164.

[84] Lizarraga-Valderrama LR, et al. Modulation of neuronal cell affinity of composite scaffolds based on polyhydroxyalkanoates and bioactive glasses. Biomedical Materials. 2020; 15(4): 045024.

[85] Jaswal R, et al. Nanographene enfolded AuNPs sophisticatedly synchronized polycaprolactone based electrospun nanofibre scaffold for peripheral nerve regeneration. Materials Science and Engineering: C. 2020; 116: 111213.

[86] Xu H, et al. Preparation and characterization of poly lactic acid/graphene oxide/nerve growth factor scaffold with electrical stimulation for peripheral nerve regeneration in vitro. Journal of Wuhan University of Technology-Mater. Sci. Ed. 2020; 35(6): 1149–1161.

[87] Entekhabi E, et al. Fabrication and in vitro evaluation of 3D composite scaffold based on collagen/ hyaluronic acid sponge and electrospun polycaprolactone nanofibers for peripheral nerve regeneration. Journal of Biomedical Materials Research, Part A. 2021; 109(3): 300–312.

[88] Cheng Y, et al. 3D structured self-powered PVDF/PCL scaffolds for peripheral nerve regeneration. Nano Energy. 2020; 69: 104411.

[89] Chen J, et al. 3D bioprinted multiscale composite scaffolds based on gelatin methacryloyl (GelMA)/ chitosan microspheres as a modular bioink for enhancing 3D neurite outgrowth and elongation. Journal of Colloid and Interface Science. 2020; 574: 162–173.

[90] Aydemir Sezer U, et al. Zero-valent iron nanoparticles containing nanofiber scaffolds for nerve tissue engineering. Journal of Tissue Engineering and Regenerative Medicine. 2020; 14(12): 1815–1826.

[91] Sheffield C, et al. Application of composite hydrogels to control physical properties in tissue engineering and regenerative medicine. Gels. 2018; 4(2): 51.

[92] Moradi S, et al. Chitosan/graphene oxide composite films and their biomedical and drug delivery applications: A review. Applied Sciences. 2021; 11(17): 7776.

[93] Wang Y, et al. The progress of biomaterials in peripheral nerve repair and regeneration. Journal of Neurorestoratology. 2020; 8(4): 252–269.

[94] Zhang J, et al. Conductive composite fiber with optimized alignment guides neural regeneration under electrical stimulation. Advanced Healthcare Materials. 2021; 10(3): 2000604.

[95] Cheng M, et al. Study on physical properties and nerve cell affinity of composite films from chitosan and gelatin solutions. Biomaterials. 2003; 24(17): 2871–2880.

[96] Trueman RP, et al. Aligned bioelectronic polypyrrole/collagen constructs for peripheral nerve interfacing. Advanced Engineering Materials. 2024; 26(6): 2301488.

[97] Javanmardi K, et al. Dexamethasone release from hyaluronic acid microparticle and proanthocyanidin-gelatin hydrogel in sciatic tissue regeneration. Journal of Materials Science Materials in Medicine. 2024; 35(1): 5.

[98] Hong J, et al. Magnetic fibrin nanofiber hydrogel delivering iron oxide magnetic nanoparticles promotes peripheral nerve regeneration. Regenerative Biomaterials. 2024; 11: rbae075.
[99] Zolfagharzadeh V, et al. Sustain release of loaded insulin within biomimetic hydrogel microsphere for sciatic tissue engineering in vivo. International Journal of Biological Macromolecules. 2023; 225: 687–700.
[100] Yuan B, et al. Platelet-rich plasma gel–loaded collagen/chitosan composite film accelerated rat sciatic nerve injury repair. ACS Omega. 2023; 8(3): 2931–2941.
[101] Tang Y, et al. Architecture-engineered electrospinning cascade regulates spinal microenvironment to promote nerve regeneration. Advanced Healthcare Materials. 2023; 12(12): 2202658.
[102] Liu Y, et al. Preparation and characterization of a photo-crosslinked methacryloyl-collagen composite film to promote corneal nerve regeneration via surface grafting of taurine molecules. International Journal of Molecular Sciences. 2023; 24: 10.3390/ijms241411248.
[103] Liu C, et al. Nanofibrous polycaprolactone/amniotic membrane facilitates peripheral nerve regeneration by promoting macrophage polarization and regulating inflammatory microenvironment. International Immunopharmacology. 2023; 121: 110507.
[104] Dresvyanina EN, et al. Structure and properties of composite fibers based on chitosan and single-walled carbon nanotubes for peripheral nerve regeneration. Polymers. 2023; 15: 10.3390/polym15132860.
[105] Chen L, et al. Gelatin nanofiber-reinforced decellularized amniotic membrane promotes axon regeneration and functional recovery in the surgical treatment of peripheral nerve injury. Biomaterials. 2023; 300: 122207.
[106] Zhou Y, et al. Electrospun PHB/chitosan composite fibrous membrane and its degradation behaviours in different pH conditions. Journal of Functional Biomaterials. 2022; 13: 10.3390/jfb13020058.
[107] Mungenast L, et al. Directional submicrofiber hydrogel composite scaffolds supporting neuron differentiation and enabling neurite alignment. International Journal of Molecular Sciences. 2022; 23: 10.3390/ijms231911525.
[108] Mao R, et al. Piezoelectric stimulation from electrospun composite nanofibers for rapid peripheral nerve regeneration. Nano Energy. 2022; 98: 107322.
[109] Li Q, et al. Construction and characterization of conductive collagen/multiwalled carbon nanotube composite films for nerve tissue engineering. AIP Advances. 2022; 12(5): 055124.
[110] Huang L, et al. Biocompatible chitin hydrogel incorporated with PEDOT nanoparticles for peripheral nerve repair. ACS Applied Materials & Interfaces. 2021; 13(14): 16106–16117.
[111] Haidar MK, et al. Composite nanofibers incorporating alpha lipoic acid and atorvastatin provide neuroprotection after peripheral nerve injury in rats. European Journal of Pharmaceutics and Biopharmaceutics: Official Journal of Arbeitsgemeinschaft Fur Pharmazeutische Verfahrenstechnik eV. 2020; 153: 1–13.
[112] Fang X, et al. Reduced graphene oxide–GelMA–PCL hybrid nanofibers for peripheral nerve regeneration. Journal of Materials Chemical B. 2020; 8(46): 10593–10601.
[113] Zhang S, Li Q, Zhang S. Neural regeneration ability of Polypyrrole-Collagen–Quercetin composite in the spinal cord injury. Regenerative Therapy. 2023; 24: 85–93.
[114] Zheng Z, et al. The efficient regeneration of corneal nerves via tunable transmembrane signaling channels using a transparent graphene-based corneal stimulation electrode. Advanced Healthcare Materials. 2022; 11(10): 2101667.
[115] Wang J-J, et al. Implantable patches assembled with mesenchymal stem cells and gelatin/silk fibroin composite microspheres for the treatment of traumatic optic neuropathy. Applied Materials Today. 2022; 26: 101278.

[116] Liu R, et al. Mediate neurite outgrowth of PC-12 cells using polypyrrole-assisted laser-induced graphene flexible composite electrodes combined with electrical stimulation. European Polymer Journal. 2022; 181: 111634.
[117] Zeng W, et al. TPP ionically cross-linked chitosan/PLGA microspheres for the delivery of NGF for peripheral nerve system repair. Carbohydrate Polymers. 2021; 258: 117684.
[118] Ye J, Huang B, Gong P. Nerve growth factor-chondroitin sulfate/hydroxyapatite-coating composite implant induces early osseointegration and nerve regeneration of peri-implant tissues in Beagle dogs. Journal of Orthopaedic Surgery and Research. 2021; 16(1): 51.

R. C. Kishor and Sumit Pramanik*

Chapter 3.4
Manufacturing of advanced prosthetic limbs using composites

Abstract: The limbs of over 30 million people are affected for various reasons and they need a prosthetic limb. Prosthetic limbs of different materials are manufactured. There has been significant improvement in manufacturing of prosthetic limbs that have high strength, are lightweight, cost-effective and that score over traditional materials for sustainability [1]. The manufacturing process of prosthetic limbs through composites by hand layup, vacuum bagging, filament winding, compression molding and resin transfer molding (RTM) are explored in this study. Hand layup and vacuum bagging methods are highlighted due to their flexibility in creating custom-fit sockets. At the same time, RTM and compression molding are preferred for high-strength, complex components like knee joints and prosthetic feet. Filament winding is particularly effective for producing cylindrical components such as pylons, offering exceptional strength-to-weight ratios. Additive manufacturing (3D printing) developing prosthetic limb design, allowing for customization and enhanced comfort when combined with composite materials, produces light weight, strong prosthetics. To meet the exact requirements of prosthetic materials, selecting a suitable manufacturing process is important, which is discussed in this study. The prosthetic limbs are produced using composites that are more comfortable, strong, and personalized, based on the needs of an individual.

Keywords: Prosthetic limbs, composites, manufacturing, hand layup, RTM, vacuum molding, filament winding, compression molding

3.4.1 Introduction

In biomedical engineering, prosthetics is one of the oldest inventions. During the Second World War, a large number of soldiers were injured and disabled; prosthetics are advances in medicine and the assistive technology became a true engineering in itself

***Corresponding author: Sumit Pramanik,** Functional and Biomaterials Engineering Lab, Department of Mechanical Engineering, College of Engineering and Technology, SRM Institute of Science and Technology, Kattankulathur, Kancheepuram, Chennai 603203, Tamil Nadu, India, e-mail: sumitprs@srmist.edu.in or prsumit@gmail.com
R. C. Kishor, Functional and Biomaterials Engineering Lab, Department of Mechanical Engineering, College of Engineering and Technology, SRM Institute of Science and Technology, Kattankulathur, Kancheepuram, Chennai 603203, Tamil Nadu, India, email: kr9264@srmist.edu.in

https://doi.org/10.1515/9783111386027-016

in the period. Internal or external devices that replace the lost parts or functions of the neuroskeletomotor system are defined as prosthetics. A distinct power supply or myoelectricity controls external devices powered by the body.

The prosthetic limbs are classified into upper limbs and lower limbs. The upper prosthetic limb majorly consists of transradial and transhumeral regions. The amputation that occurs between the wrist and elbow is transradial and the amputation that occurs between the shoulder and elbow is transhumeral. The lower limb majorly consists of transtibial and transfemoral regions. The amputation that occurs between the knee and foot is transtibial and the amputation that occurs between the hip and knee is transfemoral [2]. The upper and lower limb parts classification is shown in Figure 3.4.1.

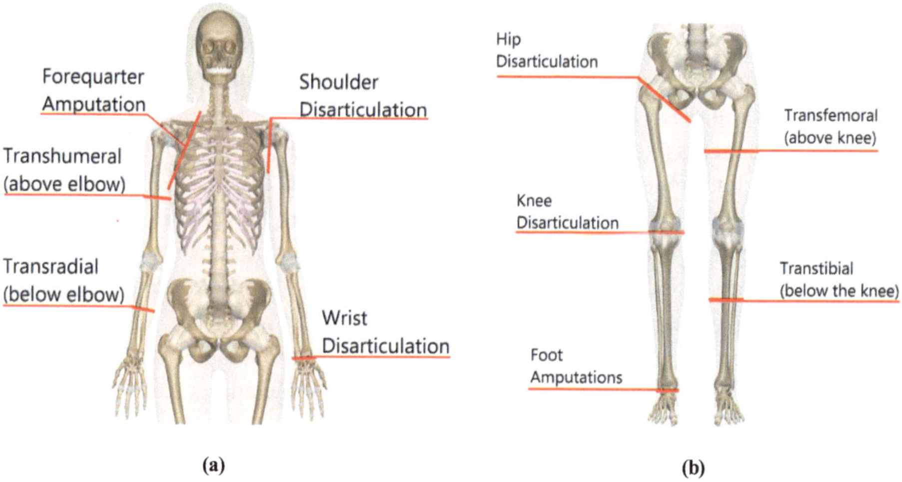

Figure 3.4.1: Classification of (a) upper limb parts and (b) lower limb parts (adapted from [2] – Open access under CC BY 4.0 license).

The prosthetic limbs majorly made from wood, metal, and leather, which was used earlier were to address the basic needs to complex functionality. The advanced methods of prosthetic limbs in material science and engineering have transformed the design and effectiveness of prosthetics. The combination of composite materials is the most important advancement in prosthetic technology [3]. Figure 3.4.2 shows the categorization of prosthetics and accessories.

Two or more materials are added to prepare components with better properties to create a product called composite material. For prosthetic limbs, composites contain a reinforced fiber like carbon fiber, glass fiber, Kevlar fiber, or natural fibers combined in a polymer matrix. Prosthetic limbs made of composite materials are very useful in everyday functions as they are lightweight, durable, strong, and are capable of withstanding stress. The advantages of using prosthetic limbs made from

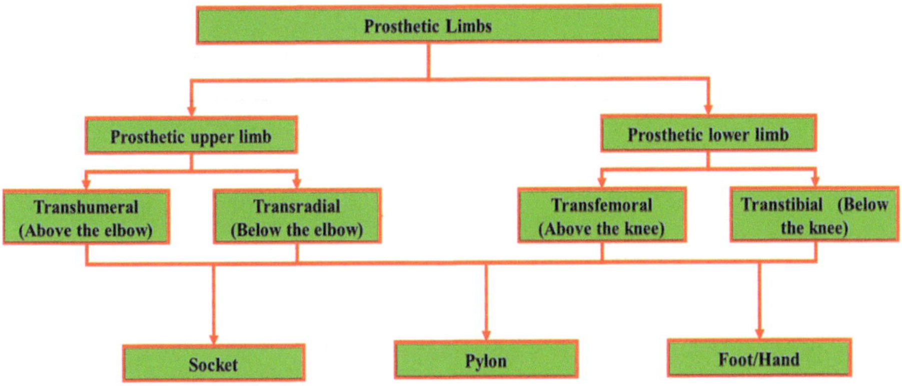

Figure 3.4.2: Categorization of prosthetics and accessories.

composite materials over traditional materials are improved durability, strength-to-weight ratio, improved comfort and fit, and biomechanical efficiency.

The advanced technology in prosthetic limbs is made from composite materials with the incorporation of nanomaterials like graphene, silicon dioxide, silicon carbide, and carbon nanotubes to improve the material strength. Another development in fabrication of prosthetics with composite materials is the 3D printing technology, which is a cost–effective and accurate production method based on the user measurement needs.

In medical prosthetics, composite materials offer significant improvement in the prosthetic limbs, contributing supreme benefits in terms of lightweight, durability, strength, and customization. They help improve the quality of life of persons who depend on prosthetic limbs, allowing them to lead more independent, active, and fulfilling lives through the improvement of advanced technology [4].

3.4.2 Classification of prosthetic limb materials

Common materials, which are used in prosthetic limbs with their properties and applications are listed in Table 3.4.1.

3.4.3 Advantages of using composites in prosthetic limbs

Biocompatibility is the need of the hour as it is not injurious to the active tissues. At the time of manufacturing, surgical tools or any other tools that interrelate with the human organ will be considered frequently. One more feature of biocompatibility is in

Table 3.4.1: Materials used in prosthetic limbs.

S. no.	Category	Materials	Properties	Applications	References
1	Composite	Carbon fiber	Stiffness and high strength-to-weight ratio	Prosthetic socket, pylon, and foot	[5–7]
2		Kevlar fiber	Impact resistant and high tensile strength	High wear areas and protective prosthetics	[8–10]
3		Glass fiber	Affordable, flexible, and impact resistant	Prosthetic socket and pylon	[11–13]
4	Plastic	Polyethylene	Shock absorbing and flexible	Liners and prosthetic sockets	[14]
5		Polypropylene	Flexible, durable, and lightweight	Limb covers and prosthetic sockets	[15]
6		Polyurethane	Elastic, cushioning, and soft	Liners and cushioning	[16, 17]
7	Metal	Stainless steel	Strong, corrosion-resistant, and durable	Structural components and weight-bearing joints	[18]
8		Aluminum	Corrosion- resistant and lightweight	Pylons, joints, and structural components	[19]
9		Titanium	Biocompatible and high strength-to-weight ratio	Prosthetic foot and load-bearing joints	[20, 21]

the way a material relates to the surface of the skin of external body. When prosthetics are involved in the outside limb and constant movement occurs, the patient may feel uncomfortable about the skin and there may be painful side effects [22]. Composite materials have transformed the design and functionality of prosthetic limbs by providing strength, flexibility, and durability and are lightweight. These materials are planned by combining two or more different materials to make a composite with superior properties. The properties of composites have high tensile strength and stiffness, low weight and thermal expansion, high chemical resistance, high specific strength, and high temperature tolerance and specific modulus [23]. High elastic modulus material is not a ductile material. Carbon fiber-reinforced composites have better modulus – three times compared to stainless steel, titanium and aluminum, and high tensile and compressive strength, in addition to high elastic deformation for CFRC [24].

Prosthetic limbs are manufactured by composite materials because human bones themselves are composites. Human bones are hard and soft materials. An innovative technique is a natural fiber-reinforced polymer within a polymer matrix incorporated as a reinforcing agent. Jute, flax, bamboo, hemp, and sisal are the natural fibers commonly used in these composites. Natural fibers are selected because of their unique properties like flexibility, high strength-to-weight ratio, and environmental sustainability [25]. Natural fiber-reinforced polymers (NFRP) have advantages in prosthetics like better comfort for users, eco-friendliness, and possible customization. The advancement in NFRP is concentrated on functional, satisfying, and sustainable solutions [26]. The incorporation of natural fibers with polymer matrix composites into prosthetics with eco-friendly healthcare, and personalized medical care benefits the environment and improves user quality of life.

3.4.4 Design considerations for prosthetic limbs using composites

Designing composite prosthetic limbs that aims to reproduce the functionality of natural limbs while ensuring durability, aesthetic appeal, and comfort involves a complex process. Composite materials, such as Kevlar fiber, carbon fiber, and glass fiber, provide a sole mixture of flexibility, lightweight properties, and strength making it appropriate for prosthetic applications. The design of composite prosthetic limbs requires particular consideration of various elements to ensure user satisfaction and maximum performance [27].

Important considerations in designing the material selection are the maintenance of structural integrity, weight optimization, and load distribution, range of motion and flexibility, fit and comfort, aesthetics, and cosmetics. Carbon fiber, glass fiber, and Kevlar fiber are the materials commonly used due to their flexibility, impact resistance, and strength-to-weight ratio. The design structure must confirm that the prosthetic limb withstands stresses and forces that come with regular activities such as walking, running, and carrying weights [28].

Strength and durability are the major performance factors for balancing the weight. To improve the distribution of composite materials, computational modeling can be used while designing the prosthetic material, ensuring that the limb is light without compromising strength and functionality. Flexibility is important for the natural movement of a limb and composite materials can be modeled to deliver specific stages of stiffness and flexibility, depending on the projected usage of the prosthetic.

Comfort and fit of composite prosthetic limbs are the most important characteristics that directly affect the standards of the user's life. To avoid damage and discomfort, the socket, which is directly connected with the residual limb, should be designed ergonomically. Sockets made of composite materials should be flexible and supportive to confirm enough support for the user's limb structure. Another important characteristic of composite prosthetic limbs is the ability to choose them from a variety of finishes, including glossy, patterned surfaces, and matte based on the users' personal style or preferences. Table 3.4.2 represents the design considerations for prosthetic limbs.

Table 3.4.2: Design considerations for prosthetic limbs.

S. no.	Design consideration	Materials/ techniques	Key factors	Applications	References
1	Material selection	Carbon fiber, glass fiber, Kevlar fiber	Strength, flexibility, and impact resistance	Socket, pylon, foot	[29, 30]
2	Structural integrity	Fiber orientation and layering	Load distribution and stress minimization	Load-bearing components	[31]
3	Weight optimization	Computational modeling and lightweight composites	Balance between strength and weight	Prosthetic limbs and foot	[32]
4	Flexibility and range of motion	Adjustable stiffness and tailored flexibility	Representing natural limb movement, and energy storage and return	Running blades and prosthetic foot	[33]
5	Comfort and fit	Flexible sockets and custom sockets	Pressure distribution and adaptation to anatomy	Prosthetic sockets	[34]
6	Aesthetic considerations	Personalization and custom finishes	User satisfaction, personal style, and confidence	External surfaces and cosmetic covers	[35, 36]

3.4.5 Manufacturing processes for prosthetic limbs using composites

Prosthetic limbs using composites are manufactured by advanced procedures that confirm the final product is strong, durable, and lightweight [37]. Composite materials are naturally made from fibers like carbon or glass implanted in a polymer matrix and offer greater mechanical properties compared to traditional materials. Hand layup, vacuum bagging, resin transfer molding (RTM), compression molding, and injection molding are some of the most important composite prosthetic production methods [38].

In hand layup technique, the fabric layers of reinforcing fibers are physically placed in a mold followed by the application of a resin matrix over the fibers/fabric layers one by one, is depicted in Figure 3.4.3. This method is widely used for prosthetic sockets that need specific shapes and sizes [39]. It is appropriate for low-volume manufacturing and has many benefits, including design and customization freedom, cheap cost, simplicity, and minimum equipment requirements. However, it may be difficult and time-consuming.

Vacuum bagging is an addition to the hand layup method that increases composite quality and standardization by adding pressure during curing (see Figure 3.4.4).

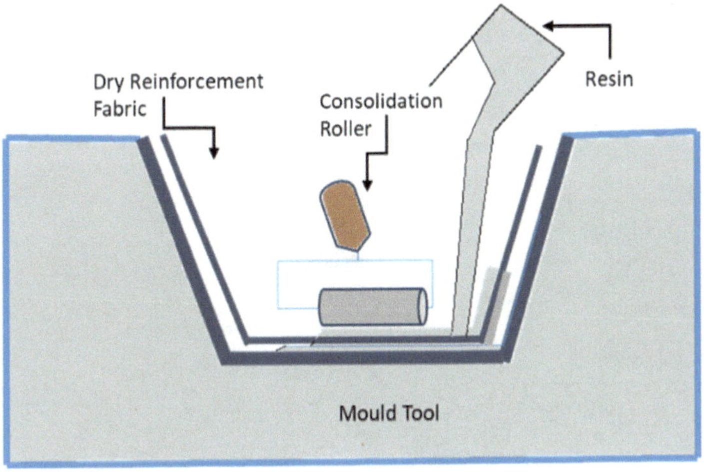

Figure 3.4.3: Hand layup process.

Following the fiber and resin layup, the complete assembly is sealed in a flexible film or bag and the air is evacuated using a vacuum pump to compress layers and ensure uniform resin distribution [40, 41]. The component is subsequently cured under pressure to remove voids and improve the fiber-to-resin ratio.

RTM is a more automated procedure that produces high quality composite prostheses with complicated geometries (see Figure 3.4.5). It requires inserting a dry fiber preform into a closed mold, injecting resin under pressure, saturating the fibers, and heating the mold to cure the resin, resulting in the finished composite component [40–42]. Compression molding is another procedure that uses a particular type of resin, such as polyurethane, to form a composite structure. However, it demands careful management of the resin flow and curing conditions.

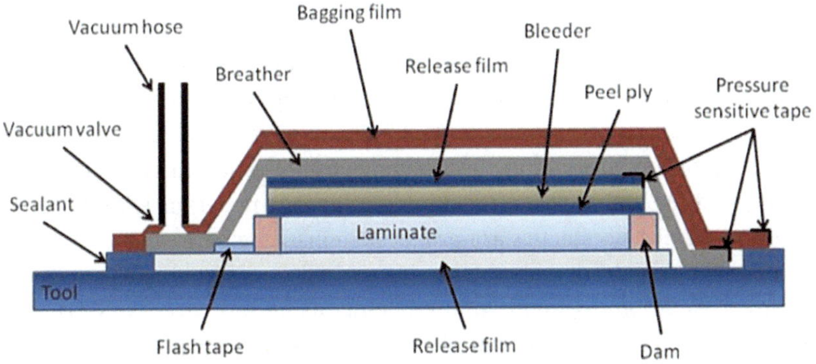

Figure 3.4.4: Vacuum bagging process (adapted from [41] – Open access under CC BY 4.0 license).

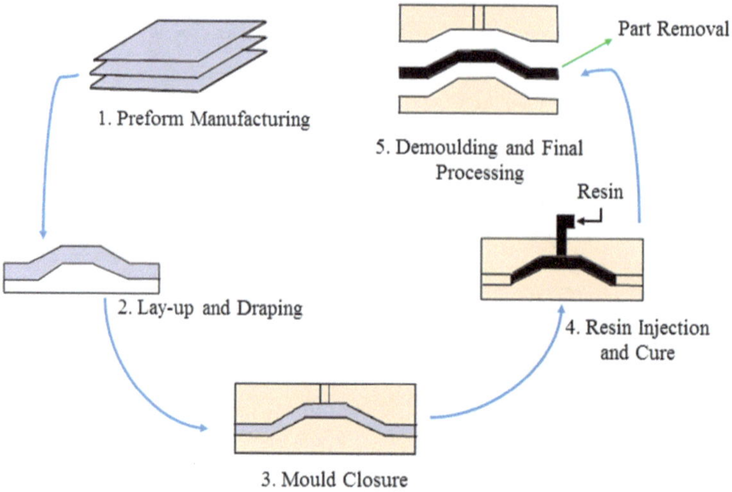

Figure 3.4.5: Resin transfer molding process (adapted from [41] – Open access under CC BY 4.0 license).

Compression molding is a heat- and pressure-based procedure for shaping composite materials into the required form (see Figure 3.4.6). It requires inserting a determined quantity of fiber and resin material into a head mold cavity, sealing the mold, and exerting pressure until the material confirms the mold shape [43]. The method is used to mass produce tiny, high-strength components such as prosthetic connections and joints. The benefits of compression molding are constant quality and dimensional precision, suitability for high volume manufacturing with shorter cycle times, greater surface polish, and little post processing. However, because of mold limits, it can only produce basic designs and has significant initial tooling costs [44, 45].

Filament winding is another method for producing cylindrical or tubular composite components by wrapping resin-impregnated fibers around a revolving mandrel (see Figure 3.4.7) [41]. Continuous fibers are dragged through a resin solution and coiled around a spinning mandrel in a predetermined pattern, with the number of layers and winding angle adjusted to provide the appropriate strength and stiffness. After winding, the assembly is cured to solidify the resin, resulting in a hollow composite structure [46, 47].

Additive manufacturing also known as 3D printing involves developing the design and manufacturing of prosthetic limbs by allowing customization to an individual's unique anatomy and functional needs. Combining composite materials with 3D printing, lightweight yet strong prosthetics are produced, enhancing performance and durability [48, 49] A typical additive manufacturing process is depicted in Figure 3.4.8. The process involves designing and digital modeling using software like CAD, and combining fibers with a matrix material like thermoplastic [50, 51]. Advantages include customization, complex geometries, rapid prototyping, cost efficiency, and lightweight properties. However, challenges include material limitations, surface finish, and mechanical properties [52, 53].

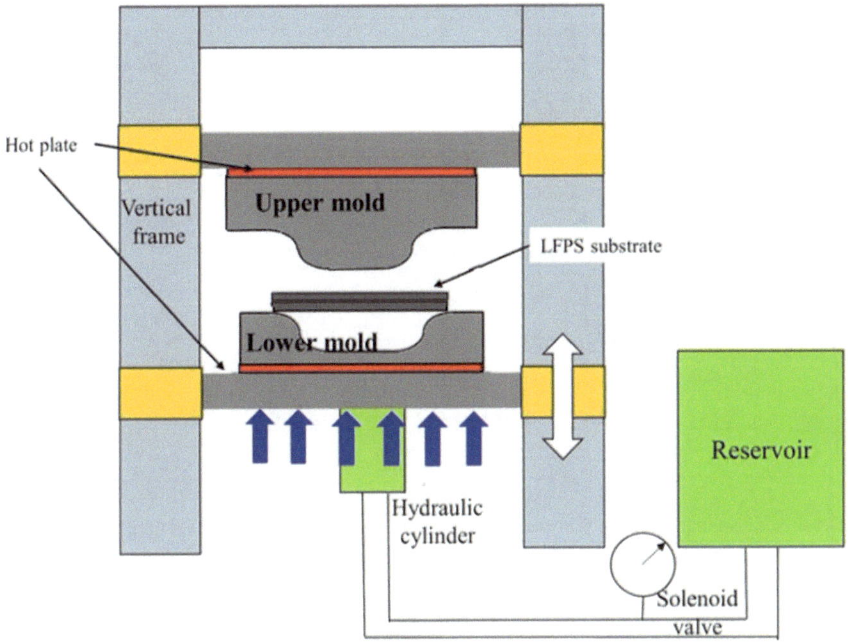

Figure 3.4.6: Compression molding process (adapted from [45] – Open access under CC BY 4.0 license).

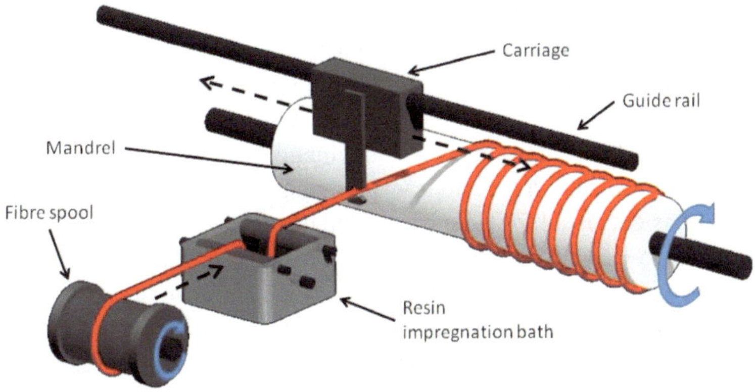

Figure 3.4.7: Filament winding process (adapted from [41] – Open access under CC BY 4.0 license).

Factors of composite prosthetics include hand layup, vacuum bagging, RTM, filament winding, and compression molding. Hand layup offers flexibility in design, low cost, simplicity, and custom prosthetic sockets; vacuum bagging improves resin distribution, reduces voids, and offers better mechanical properties, while vacuum bagging ensures a consistent resin-to-fiber ratio, resulting in lightweight, yet strong components [54].

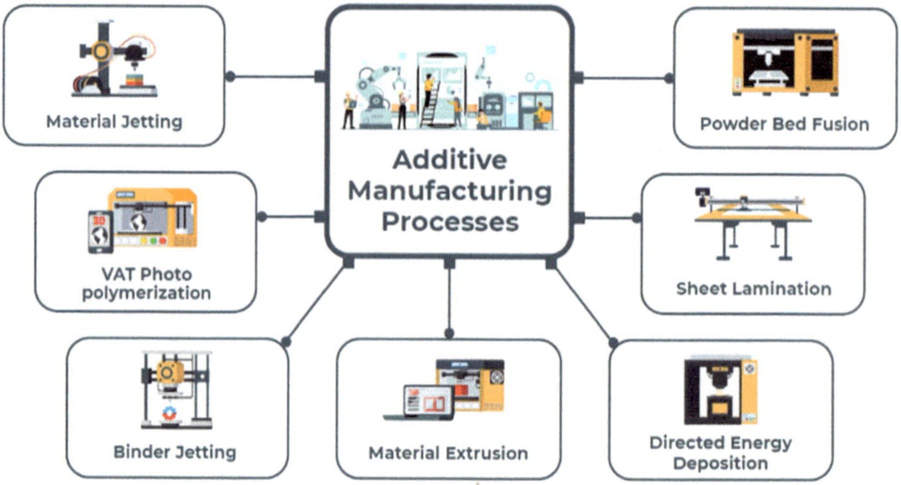

Figure 3.4.8: Additive manufacturing process (adapted from [49] – Open access under CC BY 4.0 license).

3.4.6 Conclusions

Manufacturing composite prosthetics is a multifaceted process that integrates advanced engineering techniques and materials science to produce extremities that are robust, lightweight, and customized [55]. The manufacturing process comprises hand layup, vacuum bagging, RTM, filament winding, and compression molding, each of which offers its benefits. Custom-fit sockets are manufactured using vacuum bagging and hand layup, while high strength intricate components such as knee joints and prosthetic foot are manufactured using resin transfer molding and compression molding. Filament winding is employed for cylindrical components such as pylons due to its exceptional strength-to-weight ratio. 3D printing with composite materials is transforming the field of prosthetics, allowing for the creation of lighter, stronger, and more personalized limbs. The progress of these manufacturing processes will result in more efficient production methods and higher performance prosthetics as composite technology continues to advance. The quality of life for individuals who require prosthetics will be improved by the superior mechanical properties of composites and the ability to customize attachments. The production of prosthetic attachments that are durable, comfortable, and adaptable will be facilitated by future advancements in composite materials and manufacturing processes [56]. Future trends include smart prosthetics and biocomposites, which could reduce environmental impact while maintaining strength and durability.

References

[1] Bronzino J. Biomedical Engineering: A Historical Perspective. In: Introduction to Biomedical Engineering, Eds: Enderle JD, Blanchard SM, and Bronzino JD. Elsevier, Amsterdam, The Netherlands.
[2] Hussain S, Shams S, Khan SJ. Impact of medical advancement: Prostheses. In: Computer Architecture in Industrial, Biomechanical and Biomedical Engineering, Eds: Wang L and Yu L, IntechOpen, London, UK: 2019; 9–28.
[3] Patel A, Taufik M. A Study on Design and Development of Prosthetics by Fused Deposition Modelling. In: AIP Conference Proceedings, Lovely Profession University, 2023, 2800(1): p. 020144.
[4] Nagarajan YR, et al. Single polymer composites: An innovative solution for lower limb prosthetic sockets. Prosthesis. 2024; 6(3): 457–477.
[5] Sehar B, et al. The impact of laminations on the mechanical strength of carbon–fiber composites for prosthetic foot fabrication. Crystals. 2022; 12(10): 1429.
[6] Khamees ZS, Khalil AS. Fatigue and tensile characteristics for composite materials used in prosthetic socket. Iraqi Journal of Physics. 2023; 21(4): 32–44.
[7] Syaifudin A, Suryandari DA, Agustin HCK. Strength assessment of prosthetic pylon considering the geometric of raw material. Journal of Medicine and Health Technology. 2024; 1(1): 50–58.
[8] Bash AM, Othman TT, Oleiwi JK. Improving mechanical properties of laminated biocomposites for artificial lower limb socket. Tikrit Journal of Engineering Sciences. 2023; 30(3): 9–16.
[9] Hashim HA, Oleiwi JK, Hamad QA. Evaluation the Impact and Flexural Properties for Lower Limbs Prosthetic Socket. In: AIP Conference Proceedings, University of Kufa and University of Babylon, 2023, 2830(1): p. 030012.
[10] Mahmood Bash A, Oleiwi JK, Othman TT. The influence of various fiber reinforcements on the performance of transtibial prosthetic socket. Samarra Journal of Engineering Science and Research. 2024; 2(1): 12–27.
[11] Abdalihkwa HZ, Al-Shammari MA, Hussein EQ. Experimental and Numerical Analysis of a New Low-cost Prosthetic Pylon for Amputees. In: AIP Conference, University of Karbala and Swinburne University, 2023, 2631(1): p. 030008.
[12] Bombek M, et al. Mechanical testing of laminated composite materials for prosthetic sockets. Materials and Technology. 2021; 55(5): 655–661.
[13] Oleiwi JK, Hamad QA, Abdulrahman SA. Flexural, impact and max. shear stress properties of fibers composite for prosthetic socket. Materials Today: Proceedings. 2022; 56: 3121–3128.
[14] Xie J, et al. Study on friction behavior at the interface between prosthetic socket and liner. Acta of Bioengineering and Biomechanics. 2021; 83–93.
[15] Gariboldi F, Pasquarelli D, Cutti AG. Structural testing of lower–limb prosthetic sockets: A systematic review. Medical Engineering & Physics. 2022; 99: 103742.
[16] Nizam MAM. Biomechanical Mechanical and Clinical Analysis on the Improvement of Prosthetic Liner Using Polyurethane on Anterior Distal Part of the Residual Limb, Doctoral dissertation, University of Malaya (Malaysia); 2021.
[17] Vinay CV, et al. Study of castor oil-based auxetic polyurethane foams for cushioning applications. Polymer International. 2021; 70(11): 1631–1639.
[18] Sivakumaran T, Nawawi AB. Investigation on prosthetic leg using Design of Experiments (DOE). Progress in Engineering Application and Technology. 2023; 4(1): 741–748.
[19] Fidelis OP, Arowolo AP. Low-cost body-powered prosthesis for transfemoral amputation. Journal of Medical Engineering and Technology. 2023; 47(2): 147–152.
[20] Overmann A, Forsberg J. The state of the art of osseointegration for limb prosthesis. Biomedical Engineering Letters. 2020; 10(1): 5–16.

[21] Savsani S, Singh S, Mali HS. Additive manufacturing for prostheses development: State of the art. Rapid Prototyping Journal. 2023; 29(4): 741–765.

[22] Stenvall E, et al. Additive manufacturing of prostheses using forest-based composites. Bioengineering. 2020; 7(3): 103.

[23] Blatchfordmobility. Available from: https://www.blatchfordmobility.com/en-us/for-professionals/our-products/.

[24] Wei X, et al. New advances in fiber-reinforced composite honeycomb materials. Science China Technological Sciences. 2020; 63(8): 1348–1370.

[25] Ossur. Available from: https://www.ossur.com/en-in/prosthetics/products.

[26] Kumari N, et al. Natural Fibre Reinforced Composites for Industrial Applications. In: Polymer Composites: Fundamentals and Applications, Eds: Moharana S, Sahu BB, Nayak AK, and Tiwari SK. Singapore: Springer Singapore 2024, 301–327.

[27] Trent L, et al. A narrative review: Current upper limb prosthetic options and design. Disability and Rehabilitation: 15(6): 604–613.

[28] Steepergroup. Available from: https://www.steepergroup.com/prosthetics/.

[29] Karthik K, et al. Evaluation of tensile properties on Glass/Carbon/Kevlar fiber reinforced hybrid composites. Materials Today: Proceedings. 2021; 39: 1655–1660.

[30] Ramadhani GA, et al. Advanced composite-based material selection for prosthetic socket application in developing countries. Cogent Engineering. 2020; 7(1): 1745553.

[31] Phillips SL, Craelius W. Material properties of selected prosthetic laminates. JPO: Journal of Prosthetics and Orthotics. 2005; 17(1): 27–32.

[32] Jambhulkar T, Sahu R. Modelling and optimization of hybrid Kevlar/glass fabric reinforced polymer composites for low-velocity impact resistant applications. Forces in Mechanics. 2024; 15: 100267.

[33] Hoque ME, Riham SAH, Shuvo MAA. A cost-effective prosthetic leg: Design and development. Hybrid Advances. 2023; 2: 100017.

[34] Bates TJ, Fergason JR, Pierrie SN. Technological advances in prosthesis design and rehabilitation following upper extremity limb loss. Current Reviews in Musculoskeletal Medicine. 2020; 13(4): 485–493.

[35] Biddiss E, Beaton D, Chau T. Consumer design priorities for upper limb prosthetics. Disability and Rehabilitation: Assistive Technology. 2007; 2(6): 346–357.

[36] Wang M, et al. Design of lower limb prosthetic sockets: A review. Expert Review of Medical Devices. 2022; 19(1): 63–73.

[37] Li Y, et al. A review on the tooling technologies for composites manufacturing of aerospace structures: Materials, structures and processes. Composites Part A: Applied Science and Manufacturing 2022; 154: 106762.

[38] Lotfi A, et al. Natural fiber-reinforced composites: A review on material, manufacturing, and machinability. Journal of Thermoplastic Composite Materials. 2021; 34(2): 238–284.

[39] Rajak DK, Wagh PH, Linul E. Manufacturing technologies of carbon/glass fiber-reinforced polymer composites and their properties: A review. Polymers. 2021; 13(21): 3721.

[40] Fu Y, Yao X. A review on manufacturing defects and their detection of fiber reinforced resin matrix composites. Composites Part C: Open Access. 2022; 8: 100276.

[41] Haghani R, Yang J. Application of FRP materials for construction of culvert road bridges: Manufacturing and life-cycle cost analysis. Rapport. 2016; 2016: 3.

[42] Castro-Franco AD, et al. A review of natural fiber-reinforced composites for lower-limb prosthetic designs. Polymers. 2024; 16(9): 1293.

[43] Muralidhara B, Babu SK, Suresha B. Utilizing vacuum bagging process to prepare carbon fiber/epoxy composites with improved mechanical properties. Materials Today: Proceedings. 2020; 27: 2022–2028.

[44] Bharath D, et al. Tensile and erosion behaviour of medium calcined alumina microparticles on GFRP composites fabricated with vacuum bagging process. Materials Today: Proceedings. 2021; 46: 307–310.
[45] Kim H-S, et al. Optimization for the prepreg compression molding of notebook computer cover using design of experiment and finite element method. SN Applied Sciences. 2020; 2(9): 1619.
[46] Al–Furjan M, et al. A review on fabrication techniques and tensile properties of glass, carbon, and Kevlar fiber reinforced polymer composites. Journal of Materials Research and Technology. 2022; 19: 2930–2959.
[47] Babazadeh J, et al. Effect of glass, carbon, and kevlar fibers on mechanical properties for polymeric composite tubes produced by a unidirectional winding method. Materials Research Express. 2021; 8(4): 045301.
[48] Wang Y, et al. Overview of 3D additive manufacturing (AM) and corresponding AM composites. Composites Part A: Applied Science and Manufacturing. 2020; 139: 106114.
[49] Mobarak MH, et al. Recent advances of additive manufacturing in implant fabrication–A review. Applied Surface Science Advances. 2023; 18: 100462.
[50] Ottobock. Available from: https://www.ottobock.com/en-us/resources.
[51] Symmetry electronics. Available from: https://www.symmetryelectronics.com/blog/defining-additive-manufacturing/.
[52] Ding S, et al. Hybrid layout and additive manufacturing of continuous carbon/glass fibers reinforced composites, and its effect on mechanical properties. Composite Structures. 2023; 319: 117133.
[53] Rout S, et al. Development of improved flexural and impact performance of kevlar/carbon/glass fibers reinforced polymer hybrid composites. Journal of Composites Science. 2022; 6(9): 245.
[54] Dubey D, Singh SP, Behera BK. additive manufacturing of fiber–reinforced composites. Journal of Materials Science. 2024; 59(27): 12219–12256.
[55] Gavette H, et al. Advances in prosthetic technology: A perspective on ethical considerations for development and clinical translation. Frontiers in Rehabilitation Sciences. 2024; 4: 1335966.
[56] Barrios-Muriel J, et al. Advances in orthotic and prosthetic manufacturing: A technology review. Materials. 2020; 13(2): 295.

Index

https://doi.org/10.1515/9783111386027-017

Series: Advanced composites

Advanced Composites
Edited by J. Paulo Davim
ISSN 2192-8983

Published titles in this series:
Vol. 21: Abrasive Water Jet Machining of Composites (2024) Ed. by Sachin Salunkhe, Vikas Sisodia, J. Paulo Davim
Vol. 20: Sustainable Bio-Based Composites (2024) Ed. by Arbind Prasad, Gourhari Chakraborty, J. Paulo Davim
Vol. 19: Green Composites Manufacturing (2024) Ed. by Gurminder Singh, Ravinder Pal Singh, Neeraj Sharma, J. Paulo Davim
Vol. 18: Polylactic Acid Composites (2024) Ed. by Pawan Kumar Rakesh, J. Paulo Davim
Vol. 17: Nanocomposite and Nanohybrid Materials (2023) Ed. by Rajesh Kumar Verma, Devendra Kumar Singh, J. Paulo Davim
Vol. 16: Waste Residue Composites (2023) Ed. by Murahari Kolli, J. Paulo Davim
Vol. 15: Cellulose Composites (2023) Ed. by P. K. Rakesh, J. Paulo Davim
Vol. 14: Hybrid Composites (2022) Ed. by K. Kumar, B. S. Babu
Vol. 13: Plant and Animal Based Composites (2021) Ed. by K. Kumar, J. Paulo Davim
Vol. 12: Glass Fibre-Reinforced Polymer Composites (2020) Ed. by J. Babu, J. Paulo Davim
Vol. 11: Polymers and Composites Manufacturing (2020) Ed. by K. Kumar, J. Paulo Davim
Vol. 10: Biodegradable Composites (2019) Ed. by Ed. by K. Kumar, J. P. Davim
Vol. 9: Wear of Composite Materials (2018) Ed. by J. P. Davim
Vol. 8: Hierarchical Composite Materials (2018) Ed. by K. Kumar, J. P. Davim
Vol. 7: Green Composites (2017) Ed. by J. P. Davim
Vol. 6: Wood Composites (2017) Ed. by A. Alfredo, J. P. Davim
Vol. 5: Ceramic Matrix Composites (2016) Ed. by J. P. Davim
Vol. 4: Machinability of Fibre-Reinforced Plastics (2015) Ed. by J. P. Davim
Vol. 3: Metal Matrix Composites (2014) Ed. by J. P. Davim
Vol. 2: Biomedical Composites (2013) Ed. by J. P. Davim
Vol. 1: Nanocomposites (2013) Ed. by J. P. Davim, C. A. Charitidis